T0234019

A PHILOSOPHER'S GUIDE TO NATURAL CAPITALISM

This book posits that a sustainable future is possible without abandoning Capitalism. In its current form as Consumer Capitalism, the organization of the global economy is clearly unsustainable. But Capitalism is a malleable concept that has assumed a variety of forms since the 17th century, and it can be altered as needed.

In Part I of this book, the author sets out an economic model for a sustainable form of Capitalism, referred to in the literature as Natural Capitalism. In Part II, he abandons exposition in favour of rigorous philosophical analysis and critiques the older but still dominant narrative that underlies Classical Liberalism. The narrative will be reconstructed with great care and analysed to understand why it has been so powerful and enduring, and, of course, why it is no longer appropriate for our current circumstances. In Part III, he investigates from a normative perspective Classical Liberalism and globalized Capitalism and the economic system it licenses. Finally, in the conclusion, the author draws the threads of the discussion together in a way that emphasizes the differences between the two narratives, Classical Liberalism on the one hand and the contemporary version of Progressive Liberalism that nurtures and supports Natural Capitalism on the other.

This book will be of interest to a broad range of scholars and curious laypersons interested in a clear and interdisciplinary presentation of the issues arising out of climate change, including corporate governance, social and environmental policy, declining social capital and the capacity of democratic institutions to deal effectively with sustainability. It will be particularly relevant for students and instructors of philosophy, history, economics, political science, social policy and environmental sociology.

Wayne I. Henry is Associate Professor of Philosophy at the University of the Fraser Valley in Abbotsford, Canada, where since 2009 he has taught courses in logic, the history of analytic philosophy and a course in business, globalization and sustainability. The ideas that inspired the development of this last course also inspired this book. Also, he is a lecturer in philosophy at Langara College in Vancouver, Canada, where he has taught business ethics for many years. Wayne has published in the areas of consumerism, globalized Capitalism and about establishing reasoned discourse with the "unreasonable," among others. He has for a long time been active in the animal welfare movement and activities in Vancouver area.

A PHILOSOPHER'S GUIDE TO NATURAL CAPITALISM

A Sustainable Future Within Reach

Wayne I. Henry

LONDON AND NEW YORK

from Routledge

Cover image: Getty images

First published 2024
by Routledge
4 Park Square, Milton Park, Abingdon, Oxon OX14 4RN

and by Routledge
605 Third Avenue, New York, NY 10158

Routledge is an imprint of the Taylor & Francis Group, an informa business

© 2024 Wayne I. Henry

The right of Wayne I. Henry to be identified as author of this work has been asserted in accordance with sections 77 and 78 of the Copyright, Designs and Patents Act 1988.

British Library Cataloguing-in-Publication Data
A catalogue record for this book is available from the British Library

ISBN: 978-1-032-47148-8 (hbk)
ISBN: 978-1-032-48346-7 (pbk)
ISBN: 978-1-003-38855-5 (ebk)

DOI: 10.4324/9781003388555

Typeset in Bembo
by Apex CoVantage, LLC

For Sherrey, and in memory of my mother, Jean.

CONTENTS

ACKNOWLEDGEMENTS

Since the earliest planning stages, the book has been an ongoing project for several years now, and there are many people to thank. If I have forgotten anyone, please accept my sincerest apologies.

I thank several professors who nurtured my interest in philosophy and the critical skills that have made this project possible. At the University of Victoria, where I did my BA Honours, there is Dr John M. Michelsen and Dr Eike-Henner Kluge. I have tried my best to live up to the model you two set for passionate engagement in the classroom. At the University of British Columbia, where I did my MA, there is Dr Edwin Levy, Dr Steven Savitt and my advisor Dr Richard Sikora. Thanks for your guidance at a critical stage of my academic career. And at the University of Western Ontario, where I did my PhD, there is my advisor Dr Ausonio Marras, the late Dr William Demopoulos and Dr Kathleen Okruhlik, one of the finest human beings I have ever met.

I have had many exceptional students pass through my hands and I thank them all. Particularly noteworthy in connection with this book, though, is a handful of students I have had in the course that was the original inspiration for this book, *PHIL412: Corporations, Globalization and Ethics*. Thanks to Serena Gearey, John Luzia, Cole Parsons, Amanda Penner (who was also my teaching assistant for two years), Derren Roberts and Taylor Wilson. You have sharpened and clarified my thoughts on several topics in this book, and I thank you all.

Thanks to the University of the Fraser Valley for the sabbatical leave during 2019/2020 that got the writing well under way. And thanks to Dr Sylvie Murray, Dean of the College of Arts, for your support for a partial teaching leave in the Fall term of 2021 to finish the manuscript.

To my colleagues in the philosophy department at the University of the Fraser Valley, Anastasia Anderson, Dr Glen Baier, Dr Joseph Carew, Dr Anna Cook, Moira

Kloster, Dr Jeffrey Morgan and Dr Mark Thomson, I can't thank you enough for your support over the years. But thanks especially to Jeff for beers and discussions that have filled in the gaps for me I was previously unaware of.

At Langara College, there is Dr Rana Ahmad, Dr Alexander Boston, Dr Katherine Brown, Dr Liam Dempsey, Dr Kelin Emmett, Dr Richard Johns, Dr Kurt Preinsperg (whom I have known since our days as graduate students at UBC), Dr Kent Schmor, Dr Christopher Yorke and Dr Roger Semmens of the English department. You're a great bunch to have as colleagues. Thank you! I extend my special thanks, though, to Dr John Russell and Dale Beyerstein for lunches and conversations over the years and especially to John for his paper on striving that did so much to clarify and enrich my thoughts on Kant in Chapter 10. Thanks, John!

I have had the great fortune to benefit from conversations over the years on these topics with Gary Hack, Steven Lloyd, Robert Shortland and Dr Emily Collier. Thank for your input and your friendship over the years.

Thanks to several anonymous referees who provided feedback and encouragement and suggested I change the title and the organization of the text to put the part about Natural Capitalism at the front of the book.

Finally, to the two people most deserving of thanks, there is my dear friend and former colleague, Dr Susan Gardner (Capilano University). Sue guided and nurtured me through my first years as an academic and continues to do so to this day. She read and provided extensive feedback on all but the chapters of Part I. Thank you Sue for everything! And to my dearest partner, Sherrey Collier, who read and provided feedback on several chapters, did extensive research into sources for me (for years!) and helped me better understand several of these issues from a legal and policy perspective. Thanks too, Sherrey, for having the patience to put up with me for all these years.

PREFACE

Why this book? And why now?

As I write this, in November of 2022, the world is approaching the end of a two-year period that has been momentous in terms of disastrous climate events. There have been record-breaking floods in Pakistan, summer wildfires in the Mediterranean area and a heat wave in Europe that left many dead and strained infrastructure. And where I live, in Abbotsford, British Columbia, there were floods last November that introduced a new word into the lexicon of many, myself included. An event known as an "atmospheric river" left behind flood damage on a scale not seen in my lifetime and which will take years to fully repair. And this followed in the wake of a summer wildfire season that was similarly unprecedented and which completely decimated the small town of Lytton in B.C.'s Fraser Valley. And this Fall, the quaint Fraser Valley, prime agricultural area and, until just a few years ago, flanked by a coastal rainforest, is officially in drought.

Not surprisingly, given the urgency of recent events, there has been a proliferation of books on topics related, directly or indirectly, to climate change in the last three years or so. It might well be wondered whether another book is necessary. And, if so, what does this book contribute to the dialogue? I believe this book constitutes a vital missing piece to framing the debate, only hinted at in the many books I have encountered. This vital piece is the new narrative required to contextualize how we frame the problems around sustainability and provide a coherent framework for moving forward constructively, together with the arguments that motivate and support it. I will take a rather indirect route to filling out this claim in more detail.

In the year 2000, I discovered a book which, over the years, has had a profound impact on my teaching and, more generally, on my view of the world. This book was *Natural Capitalism: Creating the Next Industrial Revolution,* by Paul Hawken,

Amory Lovins and L. Hunter Lovins.[1] The economic theory they set out in this book is the foundation of the optimistic and hopeful story I will be sharing in Chapters 3 and 4. This story is about the principles that underlie a reimagining of capitalist systems, a new economic model, that has the potential to remake Capitalism in sustainable form.

At the time I discovered this book, I was teaching environmental ethics and business ethics at Capilano College (since then rebranded as Capilano University) and at a few other post-secondary institutions in the Greater Vancouver Regional District, including Kwantlen Polytechnic University. A standard approach to these courses is to focus on a range of topical issues. We begin by getting clear on a number of normative perspectives from the literature and then engage in a debate applying those normative perspectives to a range of issues, including climate change and sustainability.

There is a lot of value in this process, not least in getting the students actively engaged in thinking about the connection between business practice and what we care about, and in understanding the variety of perspectives typically brought to bear on these issues and how we can begin to navigate these differences in finding solutions. I was left increasingly uneasy, though, by the nagging feeling that something essential was missing. At the very least, there seemed little to be said about concrete *practical* solutions to the problems raised: loss of species and biodiversity, the fact that the miseries of climate change are visited disproportionately on the poorest nations, the rising inequality of wealth and opportunity that has accompanied the globalization of a model of Capitalism that has been the biggest single driver of climate change, and more.

Natural Capitalism hinted at the shape of the missing piece. There is a positive story to be extracted from this book, and other sources to be discussed at length later, about how to restructure our economic and social systems in ways to achieve genuinely sustainable societies without abandoning Capitalism. Moreover, as we shall see, this restructuring of economic and social systems will have implications for citizenship, for personal liberty, the common good and what we value and why.

From the outset, I was struck by the hopeful and inspiring optimism of the book, but there is a downside to this presentation. The message conveyed is that we already have the technological means to implement the first iteration of changes that could move us to sustainability in a single generation and that it is inevitable – the changes were already underway and gaining traction and momentum everywhere. What was required of us as citizens was to be enablers, by our consumer choices, informed political activism and, for some of us, equipping the next generation with the knowledge and critical skills that would be needed to guide civilization through its greatest challenge. It turns out, though, that this optimism was premature and rather naive.

In Chapter 4 we will explore this point further in the story of Auden Schendler, a consultant with the Rocky Mountain Institute,[2] and his journey from the front lines of the climate wars to his subsequent disillusionment. I can state categorically, though, it is not inevitable that we will prevail in the battle with climate change. The situation is serious, timelines are tight and even with the relevant knowledge and the technologies that could help us avoid the worst outcomes of climate change, there is no guarantee that we will be successful. The changes required are mostly about our consciousness, our world view: changing how we see the world and our relationship to it. Or, better, the changes must *start* with the alteration of our world view. This is what must happen first and which will carry the economic and social changes forward to achieve the best possible outcomes. *And that is what this book is about and that is the vital piece it is meant to contribute: the new world view (or new narrative)*[3] *that must accompany our sustainability initiatives to provide a coherent framework that rationalizes and supports these initiatives together with the arguments that motivate and support this narrative.*

On the supposed inevitability of it all, I recall that in 2005, in the wake of hurricane Katrina, the environment and the reality of climate change caught the attention of the public in a very dramatic way with the images of devastation coming out of New Orleans. Suddenly the environment and climate change became a focal concern of voters in much of the developed world, and it seemed the pivotal moment had arrived for voters to push governments to undertake precipitous change.

I was in New Orleans that summer, and we had to end our vacation early because news of hurricane Dennis making its way towards the Louisiana coast suggested flights out of the region might become difficult. This was in July 2005, and Katrina would hit the region in August, just a few weeks later. I recall the drive back to our B & B on the night of a violent windstorm and heavy rain that presaged the arrival of Dennis in the next couple of days. The scene through the windows of our taxi seemed to me to resemble a disaster movie, with one of the city's famous street cars lying on its side across two lanes, branches large and small strewn everywhere and rivers of water washing across the streets. The next day I was standing in the yard talking to the owner of the B & B while he picked up debris to tell him we were leaving early. I said something like, "You must be pretty much used to this here." He paused for a moment, then responded by telling me that he had lived in New Orleans his entire life and he was certainly used to severe windstorms and heavy rain and even hurricanes, "but you know," he said, "they never used to arrive this early. It's getting earlier every year. And I think they're getting worse too."

As events unfolded over the next few months, I felt increasingly confident that real and positive action was really going to start happening. Then came the financial crash of 2008, the subsequent election of reactionary ultra-conservative governments in many jurisdictions and the rise of nationalist populism. The election of Trump and the Brexit vote seemed to signal a dramatic retrenchment of public support for initiatives to deal with climate change and the decision by Trump, shortly after taking office, to withdraw the United States from the Paris Climate Agreement while simultaneously undertaking to gut the Environmental Protection Agency and reinvigorate the coal industry seemed emblematic of the change in public opinion. Since then, it has seemed that progress has stalled if not reversed, and the whole project seems fragile, subject to the whims of volatile public opinion and changes in administration.

What happened? What explains the intransigence of climate change deniers? Or even middle-of-the-road voters who accept the reality of climate change and that it is anthropogenic in origin but think that economic concerns take precedence? And how could any programme for positive change ever gain traction in such an environment of extreme political divisiveness and economic inequality?

It was while thinking about these questions in the years immediately following the financial crash of 2008 that I began to focus on the role of narratives in shaping our beliefs, our values and our behaviour. How do we explain the intransigence of climate change deniers in the face of such evidence? How, for that matter, do we explain the distrust of science and politics, made even more evident during the past two years and more of the COVID-19 pandemic? If we expect answers to such questions, and an opportunity to move forward, the individuals raising these concerns cannot be casually dismissed as a "basket of deplorables." And to insinuate that they are merely ignorant and their views aren't worth taking seriously is to essentially slam the door shut on democracy and to give up on the idea that we can settle matters of public policy by reasoned debate. I have a view on these matters that is different from many in my community, but these are my fellow citizens and meaningful action on climate change will require their willing participation. To achieve this, we will need to find ways to engage one another sincerely and respectfully, and for this to happen it's helpful to understand that we all see the world through the lens of a narrative, and sometimes disagreements arise because we rely on different narratives.[4]

Together with some of my colleagues in the Philosophy for Children movement, we worried about how to do this: how to engage others in respectful debate even in cases where the other parties seem, on the face of it, to be unreasonable. It's important to understand that I am not accusing someone of being unreasonable just because they disagree with me. Rather, it is the evident unwillingness to engage in reasoned debate, and worse, to engage in wilful violence to have their way. How does one even start a genuine and mutually respectful debate with someone who refuses to even hear what you have to say? Part of our answer is to acknowledge that we must begin by understanding *their* position, which requires us to explore the underlying narrative. It's a kind of conceptual archaeology, digging up the assumptions that are foundational to the narrative supporting their view and making these explicit. Only when we have truly heard each other can we begin to explore where we really differ on these foundational assumptions and what might be said to achieve whatever rapprochement might be possible.[5] In a democracy, success is never guaranteed, but if we are to expect any progress at all, it is my view that we must start here.

And so, this book begins with the hopeful story of what our world might look like as genuinely sustainable. I will usually refer to this new narrative as Biosphere Consciousness, a somewhat awkward phrase perhaps, but I'll say more in Chapter 4.[6] Keep in mind that Natural Capitalism is the term used to refer to the new economic model at the heart of this sustainable organization of capitalist economies. The narrative, though, is constructed from understanding the assumptions underlying this economic model, together with the consequences that flow from the details of the infrastructure required to implement it. I will leave it to context to make clear whether I am referring to the economic model or the narrative.

This hopeful story is the vital first step that must happen – the alteration of our world view – and it is the focus of Part I (Chapters 2–4). In Part II (Chapters 5–8) we undertake the project of conceptual archaeology noted earlier. We reconstruct the world view that holds us back from making the changes required to achieve this sustainable society. This is the older and still dominant narrative (I will refer to it as Classical Liberalism) which, as I will argue, constitutes the biggest obstacle to effective change by virtue of how it frames the problems we confront and limits what shall count as an acceptable answer. The goal here will be to present it clearly and fairly. To really understand the view, to really hear the other, is to understand its strengths and durability. But we will also take it to task with the object of demonstrating that it has outrun its useful lifetime. It is no longer suited to our current circumstances and is putting us on a collision course with environmental limits in ways that threaten our very survival. There is a better way forward more attuned to today's needs and that brings us full circle to the new narrative.

Thus, we conclude by sizing up the alternatives. In this regard, it seems to me that we are very much in the position articulated by Thomas Kuhn in his analysis of scientific revolutions. When we look at the actual history of revolutions in scientific thought, they manifest themselves as a clash of world views.[7] For example, consider the revolution in physics that ushered in Einsteinian physics, and not much later Quantum theory, in place of the older Newtonian view. On Kuhn's analysis, the older Newtonian view increasingly encountered limitations, difficulties not resolvable from within its framework, that were reframed and resolved from within the new frameworks. The limitations of the older Newtonian view manifested themselves as tensions, in some cases outright contradictions, in trying to get the assumptions of the model into agreement with observations. Resistance to the new model persisted for irrational reasons having to do with sunk costs, vested interests, emotional allegiance to the older model and the fear of unknown consequences that always accompany change. All of these favour a kind of momentum in support of business as usual that persists until the limitations of the older model, and the success of the newer model, reached a kind of critical threshold. If I am right, we are very much at this point now in the debate about climate change, sustainability and how best to proceed. This debate, though, is accompanied by a real existential threat and the question is whether we will reach a critical threshold that tips the balance in favour of the new world view that this book articulates before it is too late.

Which brings me to the issue of where, if at all, optimism and hope are to fit into this discussion.[8] When teaching the topics of climate change and sustainability and in discussions with colleagues about the manuscript for this book, the question I am asked most frequently is, "Are you sincere in your optimism?" The answer is, yes, I am. But do not be deceived. The task before us will not be easy. Even on the best-case scenario with the widespread adoption of the new narrative and a framework for international agreement on strategies for meeting targets to decarbonize the economy by the mid-century, the future will be challenging. There will be the challenges that always accompany adoption and build-out of a new infrastructure and the economic and physical dislocation that inevitably result. Additionally, though, it is clear that the impacts of climate change have already advanced beyond what most scientists were projecting for mid-century or beyond, and positive feedback loops may now be accelerating the progression in ways that will resist our attempts to ameliorate the damage.

Thus, I would be remiss if I did not confess that my views have altered since I began to write this book during my sabbatical year of 2019/2020. We confront a world of our own making (the Anthropocene) that is undergoing a sixth mass extinction event[9] while re-wilding and there will be momentous decisions around retrenchment. As the vulnerable infrastructure of the Second Industrial Revolution crumbles before the onslaught of extreme weather events happening more frequently and with greater ferocity, we will have many hard decisions to make about what to give up and where to relocate displaced communities.

But the new communities we build will be integrated into their natural environment and they will be socially integrated as well. People will live where they work, go to school, play and socialize. Will it be hard? Yes, but it will also make us more resilient, in so many ways. Approached from within the new narrative, the communities we build will enjoy enhanced trust, vigorous participatory democracies, and they will be genuinely sustainable, with diverse and prosperous local communities overlapping in semi-autonomous regions to the edges of contiguous land masses.

My sabbatical and this project were interrupted by COVID-19 in April of 2020 and I was able to take it up again only in the Spring of 2022. It was during this period that I came to realize, in a very visceral way, that the future is fragile and uncertain and success is not guaranteed. This realization prompted a significant reorganization of the text. Initially, I had planned to begin with the project of conceptual archaeology, the critique of Classical Liberalism, and leave the hopeful story of Natural Capitalism and its attendant narrative for the end of the book. The plan was to leave the reader on a note of optimism as it were. I came to realize it would be more impactful and inspiring if I began with the hopeful story. Also, though, reorganizing the book in this way facilitated putting the need for a change of narrative front and centre. And so, I reorganized the text into its present form, but Part I was the last to be written.

It was while researching this part of the book that I rediscovered Jeremy Rifkin, whom I had not read since my days as an undergraduate,[10] while on a visit to a local bookstore in Ottawa. The store had, for some reason, decided to put a stack of *The Third Industrial Revolution* (then already 11 years old), on prominent display.

I immediately made the connection between the economic model of Natural Capitalism and the infrastructural requirements Rifkin writes about in that book, and Part I of this book took its present shape. Searching out more Rifkin, I found *The Green New Deal* from 2019. It was inspiring for me to see the views of this visionary aligned closely with my own on the need for a new narrative, and its overall shape, if we are to achieve genuine sustainability. As he says there,

> At this critical juncture in history, the Green New Deal story lines need to be put together in a coherent economic and philosophic narrative that can create a sense of our collective identity as a species and bring humanity into a new world-view, giving us a glocal heartbeat. Absent the story, all the ideas get lost in a jumble of items, none of which connect to the others. Every idea becomes a fought-over non sequitur, sapping us of the strength for the imaginative leap needed to take us into the next era of history.[11]

This is what I have tried to do in this book. It is my pleasure and privilege to contribute to this, our greatest challenge.

Notes

1　Published in 1999 by Little, Brown, and Company.
2　A non-profit sustainability think tank founded by Amory B. and L. Hunter Lovins, co-authors with Paul Hawken of *Natural Capitalism.*
3　I will use the terms "world view" and "narrative" as stylistic variants.
4　For some examples of recent work by social psychology and cognitive science into this phenomenon, see Jonathan Haidt (2012) and George Lakoff (2009).
5　"Reasoning (or not) with the Unreasonable," in *Analytic Teaching and Philosophical Praxis,* 19(2), 2019, pp.1–10. (Written in collaboration with Dr Susan T. Gardner and Anastasia Anderson.)
6　The term comes, so far as I know, from Jeremy Rifkin, though there are historical antecedents we will discuss at length in Chapter 4.
7　Kuhn introduced the term "paradigm" as a descriptor to refer to what I am using the terms "world view" and "narrative" to refer to in this book.
8　For more on this, see Homer Dixon (2020).
9　Kolbert (2014).
10　Rifkin (1980) *Entropy: A New World View.*
11　Rifkin (2019, p.211).

1
OVERVIEW AND PLAN OF THE BOOK

The thesis of this book is that we can achieve genuine sustainability without abandoning Capitalism as such. But it will require a complete reimagining of the economic model that underlies Capitalism. The sustainable form of Capitalism to be explored in this book (Natural Capitalism) amounts to rebuilding Capitalism on the foundation of a new economic model and infrastructure. The consequences of these changes for our social, economic and political institutions are significant, and success is not guaranteed.

Fortunately, Capitalism is a plastic institution. It is important to remember that Capitalism is a human artefact, invented by us to meet our needs. If it is no longer meeting our needs, or is not doing so optimally, it can be altered to bring it into alignment with our needs. We tend to forget this and in the routines of our daily life we succumb to a kind of momentum built on the unstated assumption that the ways things are presently arranged is just a brute fact of life, a kind of given. This momentum can be exacerbated by an anxiety when we contemplate significant change. I will be arguing that, in fact, the changes required to achieve sustainable societies will also reinvigorate our democracies and bring the social and political dimensions of our lives into harmony with our values. But how is this to happen?

Arguably, Capitalism in its current form (which I will variously refer to as globalized Capitalism or Consumer Capitalism, depending on context) is not meeting our needs very well and is putting us on a collision course with environmental limits. The view defended here is that success will depend on the prior step of embracing a new world view, a new narrative and the simultaneous rejection of the current dominant world view that supports Consumer Capitalism (Classical Liberalism). This new narrative reframes our problems, opens alternative solutions not even conceivable from within Classical Liberalism, and provides a coherent framework

DOI: 10.4324/9781003388555-1

for our sustainability initiatives, a framework that contextualizes, rationalizes and supports these initiatives. It is the burden of this book to convince you that this is the crucial first step and to set out what this new narrative will look like, at least in broad outlines.

The power of narratives

It is not an exaggeration to say that narratives shape who we are and how we see ourselves, what we value and how we see the world. I have learned this first hand from my experience as an educator. Experience has taught me that if I am to expect effective learning outcomes from my students, I must first capture and hold their attention, and to do this I must engage their imagination. And for doing so, I must find a way to deliver the content of my lectures in a story. Then we can have a fulsome discussion about the story I have told, and in the process of discussion and reflection, engage each other in an activity that continually reshapes the narratives we have about the world, and our place in it, and our relations to one another, for them and for me. (At least, that is how things work when the magic happens.)

Now by a story, I don't mean a piece of purely imaginative fiction, though it might be. I mean a story in the sense of information organized by a narrative. A narrative has structure and thematic elements that connect the bits of content together into a seamless and unified whole. If done well, it has a direction that proceeds from the statement of a problem or question to a conclusion, and when you get there, you can look back and see how the conclusion connects to the beginning and brings a sense of closure to the narrative by, perhaps, offering a potential solution to the problem or an answer to the question asked (who killed the butler, anyway?). This is how we organize information in our minds, and it is how we are able to remember this information and use it to make sense of the world. Subsequently, this organization of information into a narrative structure in our minds shapes how we perceive events. The narratives in our head determine salience, underwrite our ability to see patterns and make inferences about future events and influence how we value events in our lives. Daniel Dennett has argued persuasively that our identities are constructed as narratives, continually shaping how, as individuals, we evaluate our lives and rationalize our behaviour.[1]

It is my view, evolved over the last decade or so, that we live immersed in a larger and dominant cultural narrative that we inherit as an outcome of our upbringing and enculturation. In fact, this is what it means to speak of enculturation. It is the absorption and integration of a cultural narrative, including its values. To be more precise, I believe there to be multiple narratives, overlapping one another in various ways, or lying contiguous to one another in some cases, and interacting with each other and the events experienced in one's life in complex ways. For example, assuming you are a business executive, you are immersed in one narrative at work (the corporate culture) that is contained within the larger cultural narrative that characterizes, say, the capitalist democracy where you live and work. But in the evenings, at your gig with the Death Metal band for whom you play drums, you are immersed in a very different culture with very different values and where the narrative might lie contiguous to the larger cultural narrative, perhaps defining itself in oppositional terms relative to the larger cultural narrative. So, in some cases, a localized narrative may be completely contained by the dominant cultural narrative; narratives within narratives, or sub-narratives, if you will. But it is the dominant cultural narrative that is the most general and is largely determinative of our cultural identity and which shapes how we see the world and our place in the world and determines what we value. After all, even the Death Metal narrative *defines* itself in terms of how it opposes the values of the dominant cultural narrative.

I will sometimes refer to this larger cultural narrative as a "world view." I'm not suggesting we all share the same world view simply by virtue of living "here." Because of the complex ways these narratives interact with our experiences, as noted earlier, there is room to acknowledge that each person will see the world from the perspective of a narrative that is personal, but there will be significant overlap despite our individual differences for all persons sharing a cultural heritage. And this larger cultural narrative is, like the individual narrative of one's identity as a person, continually in flux to some degree, though I have come to recognize that they can "settle" into a dominant set of themes that can last for decades, perhaps longer.

These large and complex cultural narratives underpin our social, political and economic systems. Insofar as these social, political and economic systems determine our relations to one another, both within cultures and between different cultures, and our relations to the natural world, they are largely determinative of outcomes. If one's cultural narrative is oriented to view the natural world as something "other," to be brought under human control and exploited for our benefit, then predictably one can expect outcomes that could be characterized as exploitative. If one's cultural narrative emphasizes the unity of humankind with the natural world, and the intricate interrelations of all living things in relations of mutual dependence, then one can expect a very different set of outcomes.

It is the purpose of this book to engage you, the reader, in a process of imaginative reflection by telling you a story about climate change, politics, business and how things might be different. I will be arguing that Classical Liberalism, the dominant narrative characteristic of capitalist democracies, is largely responsible for our present predicament in regard to climate change and the very real possibility of environmental collapse, by virtue of how it shapes our beliefs and our values about the natural world. This is not a causal claim. I am not claiming that Classical Liberalism is the cause of climate change. When I say it is largely responsible for our present predicament, I mean this in the sense expressed here whereby, in shaping our views, attitudes, beliefs and values, it is determinative of outcomes.

Moreover, this dominant narrative is the single biggest obstacle to making the sorts of effective changes to our political and economic systems that might give us the room to avoid the worst outcomes of this environmental calamity. And this is so because, to the extent that this narrative is also built into our identities as individuals, anything perceived as a threat to our world view can seem like an existential threat. If I am right about this, before we can make meaningful changes to these economic and political systems to refashion them in sustainable form, we must first change the underlying narratives that underpin and sustain these systems. We must learn to see the world in different ways. And that is really what this book is about, and that is the unique contribution it is intended to make to the debate about climate change and sustainability.

Looked at one way, the changes required to make our economic and political systems sustainable are relatively straightforward (though not simple), and we have the technological means to accomplish what must be done. On this point, the book *Natural Capitalism* had it right.[2] The biggest obstacle to effective change is the set of beliefs and values that would have the changes required clash with vested interests and make them seem scarier than the potential for climate disaster we seek to avoid. Additionally, there is the problem that the dominant narrative constrains how we frame our problems and sets limits to what will be considered a viable solution. Thus, changing the dominant narrative could be the crucial first step required to save us from the worst ravages of climate change and, as I will go on to argue, might simultaneously be the best bet for reinvigorating and strengthening the principles at the heart of the democracies that underpin our personal freedoms. Both outcomes could significantly reduce human misery (and non-human misery as well, for that matter). As Jeremy Caradonna puts it, quoting John Ehrenfeld, "Sustainability can emerge only when modern humans adopt a new story that will change their behavior such that flourishing rather than unsustainability shows up in action."[3]

My strategy is as follows. I will begin with the hopeful story – an example of the kind of alternative narrative that would, in my view, be the requisite first step to achieving genuine sustainability. This is a story about a bold reimagining of the capitalist systems at the heart of the economic organization of our lives into sustainable form, premised on a very different set of narratives about our relation to the natural world, and to each other, and about how and what we value. As noted, I will be sharing with you a set of economic ideas that go under the label Natural Capitalism, but my focus will be to drill down to the underlying narrative, Biosphere Consciousness. It is not my goal here to tell you what to think. I am not presenting this as the single "right" answer to our climate woes. Indeed, in the conclusion I will be at pains to argue that sustainability cannot and will not look the same everywhere. This is very much a story for the established capitalist democracies of the world, and I present it as a stimulus to our imaginations, the reflective consideration of how things might be different. In any event, the principles of Natural Capitalism, general as they are, are consistent with a wide range of implementations adapted to local circumstances.

Having done this, I will invite you to travel back in time to understand why our present economic and political systems have failed us so egregiously. To do this we will engage in a careful analysis of Classical Liberalism, the dominant narrative that underlies our existing economic and political systems, by tracing it back to its historical origins. The goal will be to understand not only how we have arrived at where we are today with respect to how we view the world and our place in it but also what the narrative implies for our social and political relations. We will then be able to assess this narrative along several dimensions to better understand how and why it is no longer suitable for our present circumstances.

Finally, we will come back full circle to the bold, imaginative and sustainable vision to compare the two narratives for a better sense of the changes that will be required of us. My goal will be to convince you that the changes required are possible, and that it is the power of our imaginations that holds the key. Quoting my friend and colleague Dr Susan T. Gardner,

> We know that change is always frightening! However, if we become acutely aware of the fact that we have always been swimming in a sea of changes that has been created by the human imagination, we will be more confident in taking the wheel of our political/economic ship away from the tyranny of the status quo and head it toward a more sustainable future.[4]

Central themes

This is a book about narratives that is itself a narrative, and there are three overarching themes which together constitute the narrative threads or connective tissues that link the individual bits together. Crucial to these themes is the recognition that Capitalism, as a mode of production and organization of the economy, is an artefact of human invention and, thus, a plastic institution capable of being adapted to a wide variety of circumstances. It is also important to keep in mind that it was designed by us to meet our needs and enhance our lives. We can be assured that it can be altered further if it is no longer meeting our needs or is no longer enhancing our lives in its present form or at least is not doing so optimally. Now in its predominant form, contemporary Capitalism, which, following typical usage, I will hereafter call Consumer Capitalism or, sometimes, globalized Capitalism, is arguably no longer meeting our needs and may be putting us on a collision course with environmental collapse.

So the *first central theme* of this book is a call to arms of sorts – to embrace the idea that the way things are at present is not how they must be. This is an appeal to engage in imaginative reflection, to rethink Capitalism, to imagine what it might look like in a sustainable form and what this might imply for our democracies, our societies and our lives as individuals. It is also an appeal to reflect critically on the narrative that underpins the ways things are to understand how and why it is no longer suited to our present circumstances.

For the moment, I am assuming that Capitalism can be reworked into a sustainable form, and we will consider the arguments later, but there are some who deny this. More generally, we can say that much of the resistance to introducing the sorts of changes that might be required to avoid the worst impacts of climate change and environmental degradation in general turn on the fear that it may require us to abandon Capitalism altogether for some other sort of economic arrangement. As noted, these fears are heightened by a lot of talk on all sides that can leave many feeling like the debate itself is a kind of existential threat; a threat to our livelihoods, our way of life, our very identities. For example, in some quarters, there is again a serious consideration of the possibility of introducing some kind of (post-Marxist) centrally organized economy to deal with the malaise of Consumer Capitalism. In many cases, the assumption underlying these discussions is that Capitalism, as such, is fundamentally not sustainable. That is, there is no workable version of Capitalism possible that can be sustainable in any genuine sense, presumably because the profit motive, at the heart of Capitalism, is itself incompatible with genuine sustainability.[5] And in other cases, some are postulating that we are headed for an inevitable collapse of the environment that will effectively end civilized society as we know it. Those individuals that manage to survive will do so by returning to a kind of pastoral, agrarian lifestyle, grouped together in relatively isolated clusters here and there.[6]

This raises a lot of uncertainty about what our lives might look like in the near future and, thus, a lot of fear. A natural response to uncertainty and fear is denial and perhaps anger and resentment. Personally, I have no desire to live in a society with a centrally organized economy and, in any event, the historical record adequately demonstrates that most such economies have had a worse environmental record than capitalist economies. And I have even less desire to live in a small, isolated society with an agrarian/Feudal economy. But these are worst-case scenarios and there is, in my view, abundant reason to be hopeful. While it is no longer possible to avoid some of the bad consequences of climate change and general environmental degradation – indeed we are experiencing them now[7] – there is, in the view of many persons well informed about the facts, still time to avoid the worst outcomes. But this will require us to find the political will to move forward with a bold plan that brings people together and this, in turn, requires that we *have* a bold plan that will inspire and motivate people. Quoting Caradonna again, "Transforming society from industrialized unsustainability to a social and economic system that is sustainable requires a consensus-building and viable blueprint for the future."[8]

Thus, the *second central theme* of the book is to recognize this need for a bold new plan and to understand why. It is needed to provide the context for framing our problems and groping our way towards sustainable solutions; it is needed to rationalize and support our initiatives; it is needed to give them a focus and coherence they would otherwise lack and it is needed, as Caradonna notes, to coalesce public opinion on what must be done. I will provide an example of what such a viable blueprint might look like, and this is the bold new vision I referred to earlier as Natural Capitalism, together with its supporting narrative (Biosphere Consciousness) and the arguments that motivate it. This will be the central focus of Part I of the book and I will say a bit more about this shortly, but it is intended as an illustrative example and a stimulus to the imagination.

You will have noticed that I am being quite circumspect in how I frame these themes, because we are all well aware that many of the issues concerning the impact of economic activity on the environment remain under dispute. But in fact, I think the evidence is beyond dispute that the situation is grave and that it is the organization of our economies that is largely to blame. We will consider this evidence in due course, but in addition to its environmental impacts, there is another important dimension to the dominant version of Capitalism that undermines the quality of our lives as political beings sharing our lives with others in civilized societies. As we will see, for many people the world over, their lives are in the service of Capitalism (and the people who profit from it) rather than the other way around. In its most virulent form, Consumer Capitalism, and the model of globalization that supports it, can be oppressive and corrosive of communities. It acts like a solvent dissolving and weakening the institutions and principles of democracy wherever its power grows unchecked.

The ways in which Consumer Capitalism is corrosive of our democratic institutions are various but include the following. First, the emergence, in recent decades, of privately funded "think tanks" that purport to be providing objective research into issues directly related to public policy and regulation but which in fact are, in many cases, dedicated to the production of biased materials that seek to direct public discourse and, thus, to shape public opinion. In other words, these think tanks seek to sustain and strengthen the dominant narratives that oppose making effective changes to our political and economic systems.[9]

There is also the capture of politicians directly by means of campaign financing and of politics itself by means of well-funded lobbyists. Thus, in the worst cases, government is run in the interests of the capitalist plutocracy, the 1 percent as they have been anointed recently, and to the obvious detriment of the overwhelming majority of our citizens. Joseph Stiglitz cites data that establishes that the United States has gone from being one of the best societies to guarantee equality of opportunity and social mobility to being the worst among the G7 nations, and this change has happened in just a little over four decades.[10] Moreover, there are significant resources devoted to determining the outcomes of elections through a variety of means.[11]

Thus, the *third central theme* of this book is that reworking Capitalism to make it sustainable is also what is required to strengthen and reinvigorate our democracies. It will have the effect of bringing the economic, social and political dimensions of our lives into alignment with our values and this is the focus of Chapters 7 and 8 of Part II and of Part III of the book.

Some terminology clarified

Before going further, it is important that we clarify some of the terminologies used so far and anticipate some that we will need in the following chapters. In Part II we will discuss theories that emerged in the 17th and 18th centuries and this will require us to have at least a rudimentary understanding of the historical circumstances of the period that includes the end of Feudalism, and the subsequent transition to Mercantilism and, ultimately, to Capitalism. Feudalism begins to come undone in the late Medieval era, at the end of the 13th century, and initially this is centred in the city states of the Mediterranean area. I take Feudalism to mean an organization of the economy in which all land is held and owned by the crown but is distributed to various nobility in exchange for military service. All ownership, though, is ultimately in the hands of the crown and all distributions are at the discretion of the sovereign. And everyone else outside of sovereignty and nobility is utterly without possessions or even personhood in the modern sense of being possessed of certain inalienable rights. Thus, all serfs work the land of the nobility in exchange for sufficient of the excess produce as is required to sustain themselves and their families. Moreover, should it become necessary for the crown to mobilize military forces, it is these serfs who will be organized and equipped by their lord to undertake military service directly. As this system begins to unravel under the influence of wealthy entrepreneurs enriched by trade, a Mercantile economy begins to emerge.

Mercantilism is the name of an economic theory and mode of economic organization which holds that power is directly proportional to the balance of wealth held by a nation. So one nation is more powerful than another, all else being equal, if its aggregate wealth is greater. Thus, as a mode of economic organization, Mercantilism imposes the imperative on nations to accumulate wealth at all costs and at the expense of their neighbors. Spain in the 16th century is an illustrative example. In the years immediately following Columbus's discovery of the new world, Spain abandoned whatever was left of its Feudal economy in favour of the extraction of wealth, often in the form of gold, silver and sugar from the new world, typically in ways involving bloody genocide and decimation of indigenous cultures. Of course, Mercantilism also focused on trade with other nations outside the colonies – sugar was a valuable tradable commodity, for instance – but this trade was always focused on generating a favourable balance in ways often involving protectionist policies. Mercantilism can be a zero-sum game in this sense. The real significance of Mercantilism for our story, though, is the role it played in the emergence of a powerful and influential class of private capitalists (the entrepreneurs of the Mercantile economy) who pressed for open access to markets everywhere, thus leading indirectly to the emergence of Capitalism.

Capitalism in the broadest sense is a form of economic and political organization that is based on private property relations and free market exchange. So Consumer Capitalism, as a specific form of Capitalism, is designed to meet its growth objectives by the direct stimulation of consumption by "end users." The implicit idea is that, at some point, Industrial Capitalism can no longer meet its growth objectives by extracting further efficiencies from the production process itself. There is a diminishing return on production efficiencies as production platforms reach maturity. Make no mistake, there are further efficiencies to be achieved, as evidenced by such recent examples as robotics and artificial intelligence, but at some point, they are no longer sufficient to reach growth targets deemed to be minimally necessary to adequately sustain requisite levels of profits. At this point, Consumer Capitalism emerges as the solution to the problem of growth.

As indicated earlier, Natural Capitalism is the name of the economic model for a sustainable form of Capitalism. Also, I will speak of the Third Industrial Revolution, a term coined by Jeremy Rifkin to designate the revolution of our economic, social and political systems that is anticipated as a result of the deployment of the infrastructure required to implement the economic model of Natural Capitalism.[12] Finally, I will follow Rifkin in using the term Biosphere Consciousness to designate the narrative that is extracted from reflection on the principles of Natural Capitalism and the foundational pillars of the Third Industrial Revolution. This is the new narrative that will contextualize and support our sustainability initiatives. Though this is a somewhat awkward term, I don't have a better one at hand. As will be explained in Chapter 4, the term Biocentrism, while it might slip more easily from the tongue, already has an established meaning that is narrower in scope than what is needed for our purposes.

Another term that will get a lot of use is "markets." In its broadest sense, a market is a venue for sellers and buyers to meet and negotiate an exchange, often in the form of a good or service being exchanged for money. But the term "market" can be used in a variety of ways, and some of these ways can be quite abstract. By that, I mean to say that markets, as such, can be quite concrete, such as the Farmer's Market you visit on the weekends to get your fresh produce, or more abstractly, we can speak of the Canadian market when we wish to make generalizations, or draw inferences, from studies of Canadian market activity as a whole. In addition, we can speak of various markets, such as the labour market, the real estate market, the auto market and others when speaking of various sectoral markets, as they are known. And sometimes, we will use "market" to refer to markets considered as abstract, theoretical entities, such as when we are discussing the features presumed to define what it is to be a market. Unless there is need to clarify, I will leave it to the context to determine how the term is being used in any particular instance.

Lastly, a word about my choice of labels for the world views to be considered: Classical and Progressive Liberalism. I use the term Classical Liberalism because it traces its origins principally to the two individuals central to the Western Liberal tradition, namely John Locke and Adam Smith. This view has been resurgent in recent decades and will be contrasted with the view dominant in the decades immediately following the Second World War, which I will call Progressive Liberalism. It must be born in mind that I use this comparison and contrast to extract a narrative supportive of a new version of Liberalism adapted to our present circumstances. In any event, I use these labels reluctantly insofar as labels can mislead as much as help, but some convenient way of referring to them will be handy. Moreover, I have been at pains to avoid using the term Neoliberalism to denote the view I am calling Classical Liberalism for several reasons. First, and as noted earlier, it seems to me that this resurgent view I am calling Classical Liberalism looks directly to the classical

texts of the Liberal tradition, those of Locke and Smith, for inspiration and supporting arguments. Additionally, though, "Neoliberalism" seems to me to have become
something of a loaded term. Assuming we could come to some agreement on what
is meant by Neoliberalism, I might even agree that the view overlaps with what I am
calling Classical Liberalism to some degree. But wishing to avoid tainting the argument, as much as possible, I prefer to skirt this debate.[13]

The plan of the book

Immediately following this Introduction, the remainder of the book is divided into
three parts. Part I, titled Natural Capitalism and Sustainability, consists of three
chapters (Chapters 2–4). In Chapter 2, we begin with a discussion of the notion
of sustainability itself. Before we can set meaningful goals and undertake to make
significant changes to our economic and political systems, we must have a very firm
grasp of what we expect as outcomes, and for this we will need to be as clear as we
can be on the operational meaning of "sustainable" and what it will mean to say of
an economy that it is sustainable.

In Chapter 3, we delve into Natural Capitalism in some detail. I will be relying
on the book by Hawken, Lovins and Lovins for my exposition, together with a few
other sources. In this chapter my goal is straightforward exposition. Particularly
for readers not already familiar with this model, it will be helpful for the discussion that follows to have a good grasp of the essentials. For those already familiar
with the model, the value added here will, I hope, lie in the examples chosen to
illustrate the basic principles of the model. I have tried throughout to find contemporary examples that indicate the synergistic nature of the principles when fully
implemented. For me, it is these examples that continue to inspire hopefulness. We
pause, at the conclusion of Chapter 3, to compare what is, to the best of my knowledge, the fullest realization of the principles and the spirit of Natural Capitalism,[14]
with our definition of sustainability from Chapter 2 to convince ourselves that
Natural Capitalism, as an economic model, is genuinely sustainable. As a successful
firm, the example constitutes an existential proof of the claim that we can achieve
sustainability without abandoning Capitalism as such.

Keeping in mind my remarks about sharing a story that can engage our imaginations and inspire us, my ultimate goal is to drill down to the underlying narrative,
Biosphere Consciousness; to lay bare the assumptions at the heart of this view so
as to be able to compare it later to the assumptions underlying Classical Liberalism.
This is the task of Chapter 4 as we set the stage for an exploration of what this view
implies for our capitalist economies, for our democracies, indeed for our identities as
individuals. As we will see, there is an intimate connection between Natural Capitalism, as an economic model, and the pillars of the Third Industrial Revolution as set
out by Jeremy Rifkin.[15] The Third Industrial Revolution is about the infrastructural
requirements for a sustainable economy that corresponds to the principles of Natural
Capitalism, what the infrastructure must be like to align with and implement our

economic model. As such, the principles that underlie Natural Capitalism and the Third Industrial Revolution, taken together, imply a narrative that rationalizes and explains Natural Capitalism and justifies the policy prescriptions that flow from it.

In Part II: Classical Liberalism: the conceptual foundations of Consumer Capitalism (Chapter 5–8), we explore the philosophical theories that constitute the conceptual foundations of Western capitalist democracies. And from these conceptual foundations is formed the dominant narrative of Consumer Capitalism, the ideology I am calling Classical Liberalism. Classical Liberalism, then, stands in the same relation to Consumer Capitalism as Biosphere Consciousness stands in relation to Natural Capitalism. When I speak of this as an ideology, I do not mean to imply any judgement or negative connotations. Rather, my intention is merely to indicate that it is constituted as a system of ideas, a set of theories and their accompanying narratives (the sub-narratives alluded to earlier) that provide support for the prevailing economic and political systems that characterize Western capitalist democracies. Indeed, Biosphere Consciousness is an ideology in the same sense. As such, Classical Liberalism is a very powerful set of ideas that are both prescriptive and justificatory. This system is prescriptive, for example, of social and economic policy – lower taxes, deregulation, minimization of government involvement in the economy – and of corporate policy – globalization of the workforce, liberalization of financial markets, maximization of productive efficiency, and so on. This same system of ideas also provides the resources for robust philosophical justifications of these policy prescriptions and the actions that flow from them.

Classical Liberalism, as we will see, is founded on two philosophical theories that we will consider independently, though we shall also spend some time exploring how they work together to constitute this very robust ideology. These constituent theories are John Locke's theory of private property and Adam Smith's theory of the laissez-faire market.[16] Locke's theory of private property is the focus of Chapter 5. We begin by setting Locke in historical context to understand his philosophical motivations. When we understand what Locke took himself to be doing and why, we will be in a better position to evaluate the merits of his theory, with respect to both his own goals and the suitability of his theory to our own times. We will carefully reconstruct Locke's supporting argument and explore the implications of his assumptions.

I will argue that Locke's assumptions about the value added by labour as defining property relations have largely shaped how European cultures have viewed the relative value of nature. The view easily encouraged by what has been, for the most part, the dominant interpretation of Locke's theory is that nature is a set of resources to be extracted at our pleasure and turned into the valuable commodities created by human labour and ingenuity.[17] Coincidentally, and rarely even mentioned, nature is held to be a "sink" for the wastes that inevitably accompany human production processes. Moreover, the same labour theory of value implies that wherever nature lies fallow and/or unoccupied, it has not yet fallen within private property relations, which is to say it is still in the *commons*.[18] This implication of Locke's theory has been used rather extensively to justify colonialism and expropriation of aboriginal lands for several centuries and arguably still does so. Lastly, and intimately connected to this last point, Locke's assumptions imply that any society organized by such private property relations is a meritocracy. That is, in a society defined by such private property relations, the wide disparities of wealth are, in principle, justifiable, at least where there is no coercion or fraudulent activity of any kind. This view continues to shape our narratives about indigenous citizens and the most disadvantaged members of our societies to this day.

In Chapter 6, we turn our attention to the free market theory of Adam Smith. Following the methodology employed with Locke, we begin by first getting clear on Smith's historical context and the main motivations driving the construction of his theory. Smith, who is writing nearly a century after Locke, has a very different set of concerns. Capitalist market economies are now well established in Britain and the Netherlands, and indeed in much of Europe, but monopolies have proliferated that diminish the efficiency of these markets. The monopolies act as an impediment to fair competition, effectively serving to concentrate ever more capital into fewer hands. The net result is an inefficient distribution of resources and capital that has the effect of limiting economic growth well short of what it might otherwise have been. Thus, Smith is focused on a theory of the organization of the market that can be expected to maximize market efficiency and explain why monopolies are bad for everyone. It is a theory of the conditions that can be expected to stimulate fair competition between market participants, thereby maximizing the efficient distribution of capital and resources. Under such conditions, we will all be better off by virtue of the enhanced economic growth that will inevitably result.

As with Locke's theory, we will also explore some of the consequences of Smith's theory. It is my view that Smith has been widely misunderstood; so this will require us to carefully distinguish between what Smith has actually argued for and what his interpreters have attributed to him. One prominent example is the confusion, or better conflation, of "self-interest" with "selfishness." Smith emphasized that a necessary condition of market efficiency is that participants must be free to pursue all market transactions exclusively from the motive of self-interest. Classical Liberals of the late 20th century have had a tendency to read Smith as advocating for the view that market participants must be left free to act with maximal selfishness. Thus, the creed 'greed is good.'[19] I will argue that Smith is best understood as arguing in favour of leaving individuals free to act on their self-interest, which he characterizes as sound practical judgement.

Having said that, though, it is undeniable that one can extract policy recommendations from Smith's theory. Among those that have been particularly prominent since the re-emergence of Classical Liberalism as the prevailing view in Western capitalist nations are an emphasis on privatization of the public sector as much as possible, deregulation of the private sector and the minimization of taxes and the size of government. The ultimate goal is to minimize the involvement of the government in the economy as much as possible. As always when discussing policy, the devil is in the details and we shall spend some time getting clear on the range of views that are possible on these policy recommendations to set the stage for the discussion in Chapter 7 of what these Classical Liberal policy recommendations can look like when taken to their logical conclusion.

And so, in Chapters 7 and 8, we turn our attention to how these implications of Locke and Smith have played out in the resurgence of Classical Liberalism since the late 20th century. In Chapter 7, Contrasting visions: Classical versus Progressive Liberalism and the ideal state, we compare the Classical Liberalism of Robert Nozick with the Progressive Liberalism of John Rawls. It turns out that, although they share many assumptions, they come to very different conclusions about how the Liberal state should be organized. I will argue that the contrast sets a choice before us that Joseph Stiglitz has characterized as the ideological battle of our times. It is the choice between continued allegiance to the minimization of the role of government in favour of the liberalization of markets versus a return to something resembling the mixed economies that characterized Western democracies in the 30 years or so between the end of the Second World War and the mid-1970s.

I say "something resembling" because it's not a matter of being merely revisionist. Insofar as I am claiming that there is a need for a bold new vision of a sustainable future that can inspire us, I would be remiss were I to fall back on appeals to return to the "good old days." My point is merely that a sustainable future will require a coordinated global response, and this, in turn, will require the participation of nation state governments and their citizens everywhere. There is no avoiding this, in my view, and it implies increasing government involvement. The increasing role of government being argued for is, as we will see, the appeal for careful and sensible regulation of the markets to avoid market failures and fiscal policies that have the effect of directing resources to new and sustainable practices. But this is imagined taking place within the context of a frank and honest appraisal of our present historical situation.

In Chapter 8, Corporate governance and the limits of corporate responsibility, the various threads of Part II come together in an extended discussion of corporate governance and corporate social responsibility, contrasting the Shareholder Theory of Milton Friedman, with the Stakeholder Theory of R. Edward Freeman. In addition to giving us an opportunity to review and consolidate the discussion of Classical Liberalism thus far, the objective here is to convince the reader that a range of views is possible concerning our ethical obligations to one another as individuals and concerning the ethical duties of corporate entities to the civilized societies that host them. Moreover, these alternative views have consequences that impact us as citizens; consequences that demand our involvement. Specifically, the objective is to understand that responses from within these frameworks raise important questions of value (in the ethical sense) that demand our attention as critical, informed and engaged citizens. We have set the stage for an extended discussion, in Part III, of what is likely the most consequential issue of all to arise out of these considerations: the relationship between Classical Liberalism and our values, in the fullest sense of what matters to us and why.

And so, in Part III: Classical Liberalism through a normative lens (Chapters 9 and 10), we will examine Classical Liberalism and its various sub-narratives from a variety of normative perspectives. The goal is to get a sense of how the theories of Locke and Smith work together very effectively for the two purposes noted earlier, namely for making important policy prescriptions and providing powerful justificatory arguments to support those prescriptions. As we will see, though, these theories stand in need of the support of a normative theory to be fully successful in these crucial roles. Every ideology must appeal to some normative theory to connect its prescriptions to our values, but, as it turns out, there is a variety of normative traditions that compete for our allegiance. In the cultures where Classical Liberalism prevails, these traditions consist predominantly of Utilitarianism and Kantian Duty Theory.

In Chapter 9, the focus is on the Consequentialist family of theories known as Utilitarianism. We begin with the most influential version of this theory, known as Rule Utilitarianism, attributed to John Stuart Mill. We will trace its development into a contemporary version known as Preference Utilitarianism. This latter version is most likely to judge Classical Liberalism favourably, though even here the agreement is far from perfect. Preference Utilitarianism, arising out of responses to difficulties raised for Mill's version, is probably the normative tradition best suited to the purposes of giving a moral dimension to the arguments supportive of the policy prescriptions of Classical Liberalism, but it is, in one important respect, at odds with the conceptual foundations of the ideology. Utilitarianism denies that there are such things as inalienable rights which Locke's theory presumes as the very foundation of property rights and the freedoms required for genuine agency.

In Chapter 10, we turn to Kantian Duty Theory, a very influential theory that is probably closer to the moral intuitions of most people and which shares with Locke the view that people are endowed with rights to autonomy and should be free to choose for themselves. As it turns out, though, for Kant this implies that persons have an intrinsic dignity that must never be sacrificed for any goals we might set for ourselves. Thus, the supposed benefits of globalization come under intense scrutiny. The benefits of globalized Capitalism can never be justified in cases where there is abuse of individuals and, in particular, cases where these individuals become the *means* to benefit the wealthy few.

Finally, in the conclusion (Chapter 11), we draw the threads of our discussion together in a way that emphasizes the differences between the two narratives, Classical Liberalism on the one hand and the contemporary version of Progressive Liberalism that nurtures and supports Natural Capitalism on the other. It must be understood that this is not the Progressive Liberalism of the Roosevelt New Deal; it is a thoroughly contemporary ideology adapted to our present needs.[20] What it shares with the older Progressive Liberalism is the idea that markets cannot be left unregulated to manage themselves. Government will be required to set the terms of engagement by means of sensible and transparent regulation that is framed as broadly as possible to facilitate innovative approaches to compliance and fiscal policies that stimulate investment in sustainable practices while discouraging unsustainable practices.

Additionally, though, it is pointed out that this new and hopeful model is applicable to only the established capitalist democracies of the developed world. It is not suggested that sustainability will or should look the same everywhere. Indeed, it would be naïve to expect so. Rather, sustainability, and the initiatives required to achieve it, will have to be adapted to local circumstances – the historical, political, cultural and economic circumstances of each jurisdiction. The developed world will have a prominent role to play, though, in taking leadership on forging international consensus and being seen to be undertaking to do the most they can in their own jurisdictions, not the least. And they will have a role to play in lending assistance to developing economies to ensure development there does not resemble what it has been for the developed world. In other words, assistance will be required to decouple development and poverty reduction from the hydrocarbon economy. This they can do by sharing best existing technologies and practices. And on both points, forging international consensus and sharing best existing technologies and practices, there is a moral argument that these overtures are their due for centuries of exploitation. It is both an opportunity and an imperative to abandon the neocolonial practices of the post-Cold War period given further urgency by the evident fact that time is quickly running out to make significant changes to avoid climate disaster. It is argued that these conclusions will seem natural from within the perspective of the new narrative. If these conclusions strike the reader as naïve, it is requested that you patiently wait for the arguments.

And so, we proceed.

Notes

1 Dennett (1991). Dennett goes on to make some contentious claims about the nature of consciousness. I am not presuming to enter into that debate here.
2 As discussed in the Preface [Hawken et al. (1999)].
3 Caradonna (2014, p.236). The original passage is from Ehrenfeld (2008).
4 In personal communication.
5 For a recent example, see Wolfgang Streeck (2016).
6 For example, John Michael Greer (2016).
7 The most recent data as analysed by NASA reveals that, despite COVID-19 severely restricting much economic activity, 2020 tied 2016 as the warmest years on record. And 2019/2020 saw the worst forest fire season on record in several jurisdictions, including California and Southern Australia: https://climate.nasa.gov/news/3061/2020-tied-for-warmest-year-on-record-nasa-analysis-shows/ (retrieved October 13, 2022).
8 Caradonna (2014, p.235).
9 Many sources are available on this topic, for example, Donald Gutstein (2018).
10 Joseph Stiglitz (2012). The views of Joseph Stiglitz will be discussed in greater detail in Chapter 7.
11 For some insight into this, and how far it has progressed in the United States, see Jane Mayer (2016).
12 Rifkin (2011).
13 I have recently come across a book published in 1986 titled, *The Evolution of Rights in Liberal Theory*, by Ian Shapiro and published by Cambridge University Press. It is an excellent book and highly recommended, in which the author also discusses the views of Locke and Smith (which he refers to as Classical Liberalism) and the views of Robert Nozick and

John Rawls, to which we turn our attention in Chapter 7. While the overlap in content reflects to some degree our shared concerns, Shapiro's main focus is quite different. In any event, these chapters were already written and I have not attempted to address Shapiro's insights or incorporate them here.

14 I am not implying that it is the only example.

15 Rifkin (2011).

16 I am not for a moment denying that many streams of influence run into Locke and Smith, and I am not wanting to minimize the importance of these influences (e.g., the French Physiocrats, who coined the term *laissez-faire*). But, with respect to subsequent influence on the re-emergent Classical Liberalism, it is my view that our energies are best spent on Locke and Smith.

17 At this early stage, I deliberately avoid discussion of the role of God in all of this. Locke certainly claimed that God had given to us (i.e., humans) the bounty of nature in common, with the intention that we rely on our labour and ingenuity to enhance our lives. Secular theories might dispense with the role of God in providing humanity with the bounty of nature: we simply find ourselves here, as it were, and surrounded by the raw materials of nature. But the assumption that nature, as such, is merely a set of resources to be used at our pleasure does not change, claims about stewardship notwithstanding.

18 "Commons" being the technical term used to designate land or expanses of the natural world that are not yet privately owned. In his theory, Locke speaks of a State of Nature, which he hypothetically supposes to be a state of humanity before the advent of an organized division of labour. One might suppose this to be a state where humanity exists as hunter-gatherers and there is not yet any land actively cultivated. In such a state, all of nature exists in commons, that is, not yet privately owned by anyone. Today, commons are relatively rare but, as an example, international waters and Antarctica are held in common. Additionally, as an anonymous reviewer brought to my attention, many indigenous groups have kept common areas as part of post-colonial negotiations, though I am not sure how extensive these might be.

19 This is the view excoriated implicitly in the movie, *Wall Street*, starring Michael Douglas and directed by Oliver Stone (1987). This phrase is uttered by Douglas (as Gordon Gekko) as he walks through the crowd at a lavish dinner hosted for Wall Street executives. See the clip at: https://www.youtube.com/watch?v=VVxYOQS6ggk (retrieved October 14, 2022).

20 Though some have referred to it as the Green New Deal, for example, see Jeremy Rifkin (2019).

PART I
Sustainability and Natural Capitalism

2

GENUINE SUSTAINABILITY AND WHAT IT WILL REQUIRE OF US

Before we can begin to talk about what a sustainable culture might look like, we need to have a precise understanding of what we mean by "sustainable." As we shall see, one definition commonly used by nation state governments and the United Nations (UN) is seriously flawed insofar as it carries little in the way of implications for how we might set goals or change our behaviour. Moreover, it does not provide any guidance or metrics for how we might measure our progress. We go on to introduce a definition that, it is argued, will meet these desiderata and ultimately the model of Natural Capitalism to be introduced in the next chapter will be assessed with reference to how it performs against this definition.

How is "sustainable" to be defined?

We begin, then, with the definition of the term "sustainable development" from the Brundtland Commission. The Brundtland Commission, named after its chairperson, Gro Harlem Brundtland, was convened in 1983 under the auspices of the UN General Assembly to articulate a vision of sustainable development that could coalesce international agreement around a plan to significantly reduce environmental degradation while simultaneously lifting the developing world out of poverty. The goal of the commission was to construct a road map of the basic principles for economic development that could achieve these two goals. The commission completed its work with the release of its report, Our Common Future, in 1987. The underlying assumptions of this report are that meaningful action on the environment cannot proceed without addressing the issue of poverty and that this, in turn, cannot be addressed without continual growth of capitalist markets worldwide. It is, after all, capitalist markets that best deliver prosperity and the efficient distribution of resources. Moreover, under the terms of such a new "sustainable" road map

DOI: 10.4324/9781003388555-3

for economic development, the competition that can be found only in capitalist markets would drive the innovation required to develop the technological resources that could achieve this growth sustainably. The Brundtland definition is as follows:

> **Sustainable development** is development that meets the needs of the present without compromising the ability of future generations to meet their own needs.[1]

The goals to lift the developing world out of poverty, while simultaneously ending or even reversing environmental degradation, are laudable, but the resulting framework has arguably met with limited success. This is, in my view, owing largely to the definition of sustainable development at the heart of this model. Tying sustainability to the notion of development and the globalization of capitalist markets in particular has had the effect of licensing the accelerated extraction of resources together with the concomitant production of excess goods and burdensome waste that has significantly worsened and accelerated environmental degradation. Indeed, the model of sustainable development at the heart of this definition has had the effect of rationalizing the globalization of capitalist markets and a business-as-usual approach to the conduct of Western capitalist economies. It does this by rationalizing a particular model of globalization that favours liberalization and deregulation of markets, and "business as usual" has come to mean continual expansion of the production of consumer goods with little regard for environmental limits. Moreover, the enormous increase in offshore production and globalized supply chains has greatly increased the burden of shipping, packaging and all the additional waste associated with getting these goods to markets, now often very distant from the production site.

Looked at from our present perspective, one might reasonably ask why it was ever thought that such a model could be sustainable. To begin to understand this, look again at the definition of sustainable development given earlier. To say that sustainable development is "development that meets the needs of the present without compromising the ability of future generations to meet their own needs" is to say nothing at all. Or, at the very least, it says nothing that carries any operational significance. This is so for several reasons. First, we have no idea what the needs of future generations will be, beyond vague notions about necessities such as adequate food, water and so on. But second, and intersecting with this first reason, is the assumption noted earlier that the innovation of capitalist markets will continue to find means of meeting our needs. If we run out of fresh potable water, innovation will turn to discovering technologies that desalinize ocean water at affordable costs. If fish stocks deplete, the development of new technologies around fish farming and genetically modified organisms will address the problem. If we run out of arable soil, hydroponics and other technologies, perhaps not even envisioned yet, will be developed to fill in the gaps in food production. The idea is that technology can always find substitutes for the resources we use, even for non-renewable resources. Thus, it is believed that, in principle, the potential for economic growth is unlimited.

There are significant problems with this comfortable assumption. Not least is the complete disregard for the unforeseen consequences of our technological innovations, and there are always unforeseen consequences. Reflection and debate on these have been common at least since the publication of *Silent Spring* by Rachel Carson in 1962 and the revelations of the ecological and human harms arising from the widespread use of the pesticide DDT beginning in the 1940s.[2] Despite the ongoing dialogue, though, there has been little change with respect to our attitudes. Instead of seeking ways to diminish our use of pesticides and herbicides to favour alternative methods of agriculture, by, for example, careful use of crop rotation, intercropping, cover crops and use of natural pesticides and herbicides, our first inclination is to seek a substitute product, in this case, an alternative pesticide or herbicide, such as Roundup. Not surprisingly, the substitute products themselves often have unforeseen consequences that show up in the fullness of time and the process reiterates. It's not my purpose here to wade into the debates raised by specific examples of controversial technologies, such as fish farming and genetically modified organisms. I am well aware there are arguments supportive of these technological developments. I will be satisfied with remarking that there are always unforeseen consequences of our technologies and that they are at least sometimes bad, indeed sometimes very bad.

Additionally, though, there are unforeseen consequences that flow from upsetting the balance of natural ecosystems, a kind of collateral damage as it were. Given the complexity of the interconnections between species and their habitat that together constitute an ecosystem, it is likely that significant alterations can quickly ramify, cascading into one another and ultimately causing disruptions that we could never foresee with any degree of clarity and that may be irreversible, including the loss of species.[3] It is surely better, for example, to rely for our fresh potable water on the resource nature provides. We can, perhaps, meet our water needs by desalinizing ocean water, but what happens to other species that rely on these freshwater resources we've left in a spoiled state? And if they disappear, what happens to the species that depend on them as a food source?

And connected with this, another problematic aspect of sustainable development as defined by the Brundtland Commission is to give insufficient attention to the essential services provided by nature. By this, I mean such things as the production of oxygen, the sequestration of carbon dioxide, regulation of weather cycles, production and retention of topsoil and the detoxification of waterways, air spaces and soils, to name just a few. These services, essential to life and the integrity of the biosphere, are provided by heathy functioning natural ecosystems (forests, oceans, coral reefs, wetlands and such), and there are no known substitutes. As noted by Lovins et al., the respected journal *Nature*, "conservatively estimate the value of all the earth's ecosystem services to be at least $33 trillion a year."[4]

In relation to these natural services, the problem with development under the terms of the globalization model is twofold. First, there is the loss of natural habitat and natural resources – deforestation for urban development, logging and mining operations; the loss of arable soil to desertification as a result of our agricultural practices; the loss of wetlands due to phosphate run-off from agriculture; the loss of coral reefs to bleaching due to rising carbon dioxide levels and subsequent acidification of the oceans and the significant extinction of species on a scale not seen since the Permian-Triassic extinction some 2.5 million years ago. Because of this shrinkage of natural ecosystems and loss of species, there is less wild space in nature to provide these vital services. Moreover, the accumulation of ever more waste from the increase of globalized production activity and supply chains, as noted here, overwhelms those natural systems that remain, further impacting negatively on their ability to carry out these services. In other words, the increased levels of pollution have the effect of significantly reducing the *functional capacity* of natural ecosystems to provide the vital services all living things rely upon, in ways additional to the direct loss of these natural spaces themselves. But simultaneously with this loss of natural ecosystems and decrease in their functional capacity, there is the simultaneous increase on the demand side for both. So we are caught between the double pincers of increasing demand for natural resources and services on the one hand and the diminished capacity of nature to provide on the other. One can only ask how this might be thought to be compatible with not compromising the ability of future generations to meet their needs.

The real problem at the heart of this definition of sustainability is the assumption, rarely challenged or even acknowledged within the world view that we are calling Classical Liberalism, that the natural world is a resource here for us to use at our pleasure to better our condition but is itself conceived to exist as something "other."[5] Or, more accurately, we conceive of ourselves to be something other than nature. We command nature, by means of our technologies, to do our bidding, and we imagine that we have it within our power to shield ourselves from the consequences of our technologies. And, again, if the consequences of our exploitation of nature includes running out of some basic resources, our technology will give us the means to find substitutes. This illusion that we are somehow separate from, and in control of, nature and, thus, in control of our destiny, is a comforting illusion reinforced by our urban lifestyles. But make no mistake: it is an illusion and one that obscures, or even makes impossible, an accurate understanding of our relation to nature.

For a particularly graphic representation of the impacts of this conceit, let me introduce Global Footprint Network, an international non-profit organization founded in 2003.[6] This group publishes, every year, what has become known as Earth Overshoot Day, the day of the year that humanity has effectively used up the available budget of renewable resources and services provided by natural systems for the year. The idea is that we can calculate the total production of natural goods and services every year based on the best available data of the extent of natural resources and their functional capacity, and this annual production figure constitutes the total budget of natural and renewable capital available for that year. And remember, this includes such things as oxygen, purification of water, fish, lumber – everything that our lives and economies depend upon – and this budget must meet the needs of *all* living things. And, of course, as any wise household manager would affirm, we must ensure that usage in a year comes in well below the total budget available if we are to safeguard the health and productivity of the system as a whole. But we are profligate; we use all the available natural capital every year and, in fact, we overconsume egregiously, so that we are now in overshoot, or deficit spending, for much of the year and have done so since about 1970 or so when Earth Overshoot Day was calculated to have arrived on December 30. By 2020, it was calculated to be August 22, and this was a significant improvement over the previous year when it was calculated to be July 29, temporarily reversing a trend of earlier arrivals of the date since the calculations began to be made, probably due to the interruption of economic activity by COVID-19 pandemic. For 2021, it is calculated that the overshoot day arrived on July 29, returning to the trend before COVID-19 pandemic.

Even if one questions the assumptions that underlie these calculations, it is unlikely that they are so far off the mark as to move the date by more than a few weeks, or perhaps months. Take the most generous assumptions you like and it is clear we are still in overshoot and that is not sustainable. Clearly another definition of "sustainable" is needed if we are to adequately frame the problems and set meaningful targets for ourselves.

Consider, then, the following definition: we will say of a *system* that it is *sustainable* if:

a) it minimizes the extraction of non-renewable resources and the production of waste as much as possible within existing systems (including zero tolerance for toxic waste[7]);
b) it moves continuously in the direction of further minimizing both the extraction of non-renewables by seeking to replace them with renewable alternatives and the production of waste by repurposing such products in useful ways as our systems evolve;
c) it draws down renewable resources at a rate below their replenishment by natural processes;
d) it produces non-toxic waste at a rate below the level at which it can be absorbed into the environment, and, with respect to such toxic waste as may be presently unavoidable, we seek the means to detoxify it before releasing it into the environment.

To begin, note that this is a definition of *sustainable systems*. It is our economic systems, production systems and our societies that must be retooled to make them sustainable. In my view, it makes no sense to speak of sustainability independently of systems. Further, we have decoupled the notion of sustainability from the notion of development. I will expand on the implications of this definition by considering each clause in turn.

With respect to clause a), any requirement to minimize the extraction of non-renewable resources and the production of waste is exactly contrary to standard practice under the terms of globalized Capitalism. Indeed, the narrative underlying

Consumer Capitalism connects the production of wealth directly with resource extraction. Resources must be extracted to be turned into value-added products by human labour. Thus, the growth of the economy demanded by "sustainable development" requires that the extraction of non-renewables expand continuously as the foundation of this economic growth that will supposedly lift the world out of poverty. Any call to reduce extraction of resources is seen as a threat to the entire programme of sustainable development: no resources extracted, no products; no products, no consumption; no consumption, no jobs; no jobs, no prosperity.[8] And little regard is given to the inevitable generation of waste that accompanies this continually expanding extraction and production unless specifically required to do so by law, because the maximization of profits requires that we extract these resources as cheaply as possible.

This emphasis on the maximization of profits is a guarantee that the process will be extremely wasteful including, in many cases, the production of toxic waste products. A particularly revealing example is Mountaintop Removal Mining, which, as the name implies, is the extraction of a resource (usually coal) that leaves the landscape devastated by literally removing the tops of mountains. The process begins with huge machines that pull trees from the ground and collects them into huge piles to be set ablaze, releasing millions of tonnes of carbon dioxide and ash into the atmosphere. Huge quantities of explosives are then planted into the exposed mountaintop and ignited, reducing the mountaintop to rubble. This rubble is fed into huge machines, called draglines, that push the debris out of the way, destroying streams and local waterways and leaving a landscape incapable of supporting forest cover for decades, if not centuries. But this leaves exposed the coal seams that would otherwise be buried deep beneath the surface, which are now extracted very cheaply.[9]

So this first clause carries real implications for demanding sweeping changes in our practices. Moreover, it gives us guidance for measuring our progress: progress will be measured by developing appropriate metrics that will allow us to compare systems with respect to resource extraction and waste production per unit of productivity. So, with our metrics, we will say that one system is more sustainable than another, all else being equal, if it extracts less and/or produces less waste per unit of end product than other systems producing a similar product or service. A refined system of metrics along these lines could produce sustainability scores for comparable systems that would facilitate easy comparison. For example, we might compare energy production systems in the ways indicated by looking at the entire process, from first inputs to ultimate outputs, for each system compared. We might take as our comparison set coal-fired energy production, hydropower, nuclear power, wind turbines and solar farms and compare total consumption of resources as inputs (including construction of the infrastructure amortized over its expected productive lifetime) together with waste products generated, in relation to some fixed measure of power generated, gigajoules per hour perhaps. So the energy system with highest sustainability score will be the system that performs best on our metrics of resource extraction and waste production per gigajoule.

None of these are intended to deny that other considerations will be relevant to any final decision about which energy system, or mix of energy systems, a jurisdiction should use, such as ability to meet expected peak demand, reliability and so forth. But our system of metrics would allow us to produce sustainability scores for any mix of systems we wish to consider. We could, for example, compare a mixed energy system with base power generated by wind power and hydropower as a backup system during peak demand or when the wind is calm, with another system employing wind power as base and nuclear power as backup.

The design of appropriate metrics will depend a lot on the specific systems being compared, and other relevant considerations will include how essential the goods and services are and the risks involved. Dependable power is essential for any economy, and risks associated with even the best coal-fired power generation exceed those associated with wind power, though wind power is not without risks of its own, especially for birds. Additionally, we must bear in mind that any efforts exerted to decrease the risks associated with a technology increase its costs. As a further precaution, we could require that any system that produces toxic waste, and where there is no means presently available to detoxify or repurpose those products, be ruled out. Or, alternatively, if such a system was significantly better than a competitor system on the measures of extraction and the production of non-toxic waste and came in significantly cheaper than other comparables, we might impose rigid timelines to find ways to eliminate the toxic waste by either detoxifying or finding substitutions for the inputs to the process that eliminated the production of toxic waste without impacting negatively on the original sustainability score. We will see examples of this in Chapter 3.

This raises an important point about policy decisions in regard to production systems and energy systems in particular. The argument has been made that our economies are committed to the ongoing use of hydrocarbons until at least the middle of this century. Thus, regardless of any concerns we might have about climate change, any attempt to eliminate or even significantly reduce hydrocarbon consumption at this point would be to invite economic stagnation, if not the utter ruin of our economies. And so, it is said, we must build new pipelines to carry hydrocarbon products to market more efficiently and to reduce risks associated with transporting these products by rail. But this is a facetious and self-serving argument. It is worth keeping in mind the issue of sunk costs, that is, costs associated with the development of infrastructure that must be abandoned before the useful lifetime of that infrastructure is reached. The same point applies to the costs associated with discovery and development of non-renewable resources that will require these infrastructures. So, if we commit the resources to building these pipelines to carry hydrocarbon products to market, we are committing to the continued use of that pipeline and, thus, the continued extraction of hydrocarbons, for decades to avoid the pain of sunk costs of abandoned pipelines and hydrocarbons left in the ground.

I would argue that the question of whether our economies require the continued reliance on hydrocarbons, and the investments in infrastructure this reliance will require, is a matter of policy, not a foreordained feature of the economies in and of themselves. If we build the pipelines, then "yes" we will be committed to the continued extraction of hydrocarbons for the foreseeable future, but it is and should be a matter of public debate to decide this. It's not an inevitable fact about the world. I might add, when comparing energy systems on costs, the petrochemical industry benefits from subsidies, direct and indirect, that make these systems appear cheaper than they are. According to one source, these subsidies may be as much as $5.3 trillion (American) in post-tax subsidies per year, a staggering sum to contemplate.[10] Additionally, if the costs of environmental damage (an example of what I am calling an indirect subsidy) were internalized into these calculations,[11] there is little chance that they could compete favourably against energy systems based on renewables. And so, it is my conclusion that these metrics of sustainability, and the possibility of direct comparison of sustainability scores for comparable systems, significantly expands the policy options available to us.

Regarding clause b), there are several strands to take note of. First, there is the imperative to search continually for ways to replace the use of non-renewable resources in our production systems with renewable alternatives. Examples range from the elimination of chemically produced dyes in favour of soy-based inks to the reduction of steel and concrete in building construction to favour wood and other natural fibre products. In cases where elimination of non-renewables in not possible, the goal is reduction of their use as much as possible. A familiar example is reusing printer paper by means of laser pulses from the printer to remove the toner from an earlier print job. A particularly intriguing example concerns concrete, the second most consumed product in the world after water. A technique presently under development injects carbon dioxide into concrete during preparation of the mix. The carbon dioxide mineralizes by chemically bonding with the calcium ions of the cement.[12] The resulting structures use significantly less cement, are stronger, and there is the additional benefit of sequestering this carbon dioxide in the structures. It is estimated that, once this process is matured and widely used, it could sequester as much as 500 mega tonnes annually.[13]

A second strand to take note of is the imperative to move continuously in the direction of eliminating the waste products of our production systems by repurposing them. Essentially, they become useful products that are no longer treated as waste. There are different ways this might be done, from selling, trading, or giving these waste products to other manufacturers who might have a use for them in their production stream, to finding ways to implement them at later stages of the same production process that generated them. This may require significant re-engineering and re-imagining of the process but can significantly reduce, and in principle eliminate, waste as a by-product of production.

A third strand to emerge from this last point is that we are incentivized to take a larger and long-term perspective on our production systems with an eye to designing production sites that gather businesses that could, by their proximity to one another, more easily share with one another those by-products formerly treated as waste. Think of this as a designed production park, if you will, that would essentially constitute a system of smaller systems (i.e., the independent businesses) that is designed to function as an integrated whole with the explicit goal of maximizing sustainability scores for these production systems. We feature an example of such a system in Chapter 3.

Additionally, though, these designed production parks, and the individual systems within them, would be subject to the goal of ongoing iterative improvements over the long term. This points to an underlying feature of clause b) which is the overarching goal of continually drawing down the extraction and use of non-renewables and the simultaneous reduction of waste products, with the ideal being the complete elimination of both. This is not merely an aspirational goal, though, insofar as it works in tandem with the development of metrics, noted in our discussion of clause a), to continually refine our best practices. And it signals a significant shift in perspective from the short-term perspective inherent in the maximization of profits to a long-term perspective that focuses on the sustainability scores of systems. More to the point, then, this clause underscores the fact that our project is never completed; it is ongoing. At every iteration, we act on the best practices and technologies available to us, observe the performance of the systems carefully as measured against the metrics we have developed and seek to implement improvements to further refine the processes and the metrics at the next iteration. We can also expect continual improvements of the techniques and instruments we use to measure system performance in relation to our metrics. Working together, these iterative improvements can move our systems as a whole to exponential growth of our sustainability scores.

Now insofar as we are searching continually for ways to substitute renewables for non-renewables in our production systems, we must be mindful to not drawdown renewables at a rate that exceeds the ability of ecosystems to replenish them, and this is the point of clause c). A useful way to think about this, hinted at earlier in our discussion of Earth Overshoot Day, is to think of the total stock of renewable resources as capital, specifically natural capital as distinct from financial or technological or human capital.[14] Like any other kind of capital, natural capital can grow over time if invested well and properly managed. And proper management requires, at a minimum, that we do not drawdown these resources at a rate that exceeds the functional capacity of ecosystems for renewal and growth. More prudently, we will want to draw down this natural capital at a rate that is significantly below the ability of ecosystems to replenish them; restrict ourselves, as much as possible, to drawing down from the dividends returned to us as interest on the original stock of natural capital, as it were. As noted, our current practices have us reaching overshoot not much past the midpoint of the year. We are spending all the interest and burning up our original stock of capital. Clause c) is an imperative to reverse this trend.

Clause d) focuses our attention on such waste as already exists or which has not yet been eliminated as by-products of our systems. Residues of toxic waste that threaten natural ecosystems are widespread and concerning, such as tailings ponds associated with the tar sands and residues of mining operations that are often highly acidic and contain huge concentrations of arsenic, mercury and other toxic substances. Hird reports that the former Giant Mine, in the region of Yellow Knife, Canada, produced more than seven 7million ounces of gold during its operational lifetime from 1948 to 2004 but has left behind 237,000 tonnes of arsenic trioxide.[15] Remediation efforts are urgently needed for such sites, and they are numerous; but equally urgent is the need to eliminate the production of toxic waste as a by-product of our systems.

More generally, the point of clause d) is to reinforce the imperatives to deal with waste products by the best means available depending on the kind of waste involved. In the case of non-toxic waste that cannot be eliminated or otherwise repurposed as per clause b), we must ensure that this waste is not released into the environment faster than natural systems can absorb and neutralize it. Essentially, when natural systems absorb such waste products, they are being repurposed by nature, perhaps by being used as food by some species, or they become the inputs to the growth of natural systems, or perhaps by being sequestered by natural processes. But enormous care is required to ensure that we are not also encouraging the growth of species in ways that disturb the balance of natural ecosystems, for example, phosphate run-off that feeds out-of-control algal growth. A better strategy in this case would be to regard phosphate run-off as toxic waste and adopt a policy of zero tolerance requiring significant changes to our agricultural practices.[16] This is already possible with existing technologies and, therefore, such waste is avoidable.

As a general principle, such damaging expulsion of waste products into the environment as *can* be avoided *should* be so avoided. A point that emerges here is that what is considered toxic waste depends very much on a sophisticated understanding of the natural ecosystems involved and the interconnections between the various species. And, as noted earlier, we must acknowledge that, in many cases, it will be impossible to foresee the long-term consequences of our effluents into an ecosystem and clause d) urges us to rely always on the Precautionary Principle.

In the case of toxic waste that cannot be eliminated at this stage from processes producing essential goods, we must work resolutely to find mitigation strategies that detoxify this waste before releasing it into the environment. It must be remarked that constructing huge tailings ponds of toxic sludge is not a mitigation strategy. It is a case of forestalling detoxification, usually because of the costs involved, and leaving the problem for future generations to deal with. Allowing companies to deal with waste in these ways is a subsidization of bad practices insofar as they are permitted to externalize these costs from their balance sheets. Clause d) would require us to end such illicit and often unrecognized subsidization of wasteful practices.

It is worthwhile to take a moment to reflect on the 3Rs slogan, "Reduce–Reuse–Recycle," in light of our definition, and doing so reveals much to underscore the role of narratives in the public domain. Intuitively, the slogan is meant to imply a prioritizing of behaviours related to reducing waste by prioritizing reduction of our consumptive habits in the first instance and leaving recycling to deal with what waste remains after reduction and reuse have eliminated as much waste as possible.

In practice, though, this priority has been reversed to emphasize recycling which has come to function as a kind of salve on the public conscience. Moreover, as Hird points out, recycling has been bound up with a public narrative that places the responsibility for our waste problems squarely on the shoulders of households: we are running out of landfills, and the solutions lie with us. The narrative has been constructed to imply that reducing one's waste designated for landfill by redirecting as much as possible to recycling has become the mark of good citizenship. She notes that the growth of waste management as an industry since the late 1960s has gone hand in glove with municipal governments outsourcing their waste management to such companies, and these same companies often own and operate the landfill sites and the recycling facilities. So, in this case, what counts as good citizenship aligns perfectly with their corporate interests. This public narrative is not only misleading but also functions to misdirect public attention from the real problem, which is our consumption habits and the industries that feed these habits.

Relying on the best available data, she notes that, "we may reasonably estimate that MSW [Municipal Solid Waste – WIH] accounts for less than 1 percent of Canada's solid waste production."[17] The remaining waste, roughly 99 percent, is the result of extraction and production processes, together with a significant contribution from military operations. Moreover, much of the MSW is downloaded onto households by the industries that market and sell us the goods. The increase in packaging required by models of online commerce supports enormous profits. And for present purposes I am ignoring the significant increase in the generation of greenhouse gases (GHG) resulting from delivery of these goods directly to our homes. Moreover, the narrative is designed to present waste management as a technical problem that requires a technocratic solution, which in turn rationalizes municipal governments outsourcing waste management to waste management companies. The subsequent increase in Gross Domestic Product (GDP) resulting from the growth of the waste management industry means that the good citizenship involved in recycling is at the same time supporting jobs in the community. Yet, in all this, we are never coming to grips with the real problem of waste flows. Our attention is redirected from a critical attention to our extraction and production systems and our consumption habits in ways that virtually guarantee we will never make meaningful progress towards establishing sustainable systems.

By contrast, our definition of sustainable systems is a demanding imperative to change our practices by placing the emphasis on reduction in the first instance: it is the imperative to alter our policies and practices to significantly reduce the drawdown of renewable resources and the production of waste while simultaneously striving to eliminate altogether the further extraction of non-renewables and the production of toxic waste. Any concerted attempt to meet these goals will result in a set of production systems and governance systems that will scarcely resemble the ones with which we are familiar today. Over time this would have the effect of radically changing our societies and ways of life, and this point raises a further aspect to this discussion of sustainability to which we must now turn.

Political sustainability

As a discipline of study and as a discourse, sustainability is currently undergoing rapid development. A cursory search will reveal a range of approaches, from three principles or factors to as many as six presumed to constitute the conceptual foundations of sustainability. There is broad agreement, though, that economic, societal and environmental issues must be included in any meaningful discussion of the topic, sometimes referred to as the triple bottom line.

Social sustainability focuses on issues related to achieving social equity, such as fair and equal access to quality education and healthcare and the protection of individual autonomy and respect for rights, as these relate to achieving social cohesion and unity of purpose in pursuing our sustainability goals. Economic sustainability focuses on issues arising out of the organization of the economy in ways that will deliver a living wage and safe, respectable work conditions for everyone while simultaneously respecting environmental limits. We see these concerns at the heart of the definition of sustainability coming out of the Brundtland Commission as noted earlier. The answer implicitly given there to the issue of social sustainability is that development that can guarantee lifting the world out of poverty will address these concerns. And they rightly note that we can never address environmental sustainability if we do not simultaneously address poverty. And their implicit answer to the issue of economic sustainability is to promote a model of globalization that will unleash the powers of Capitalism to both create wealth and inspire the innovation that will be required to do so sustainably, which we have found wanting.

I argue here that the actions required to meet the goals that come out of our definition of sustainability will require what I shall call *political sustainability*, and that this incorporates and addresses within its scope our concerns about economic and social sustainability.

I will first address the priority I am giving to the notion of political sustainability. I begin with the claim that action on climate change will require a globally coordinated effort and this will require the participation and cooperation of nation state governments. There are some who deny this. They claim that action is happening at the grassroots level everywhere – in relatively isolated pockets of local activism, or sometimes in political units smaller than nation states such as the state of California, or the city of Vancouver – and this will continue regardless of what nation state governments decide to do. And it is sometimes suggested that this local action is sufficient on its own, or will be, to get the job done.[18] In many cases, this attitude arises out of disillusionment with politics and this is understandable. Anyone familiar with the history of international climate agreements reached through the auspices of the UN since the first such environmental conference was held in Stockholm in 1972 will know that the obstacles are many and progress grindingly slow, possibly too slow to do what must be done within the relevant time frames. The story of the United States' participation in the Paris Climate Agreement is illustrative. The conference concluded with a legally binding agreement that was adopted by 196 parties on December 12, 2015 (not all of these parties were nation state governments). The United States, then under the Obama administration, was one of the signatories to this agreement. When Trump took office in January 2017, he immediately began signalling his intention to withdraw the United States from the agreement and this was made official on June 1 of that year. Subsequent to taking office in January 2021, President Biden re-engaged the United States by signing the relevant documents in the first month of his presidency. The conclusion many draw is that if we wait for politics to do what must be done, all hope is lost.

However, it is my view that a different lesson is to be drawn from this. Sustainability is a global problem and the effects of climate change do not respect boundaries of any kind. As the world's second largest contributor of GHGs and still the world's largest economy, it is clear the targets agreed to can never be met without US involvement and leadership. This underscores the need for what I am calling political sustainability. It will not be sufficient on its own to ensure success, but it will be essential to that success. And this because it is only nation state governments that have the powers to set meaningful targets and the tools to design and enforce the policies that will be required to meet these targets and the fiscal resources to encourage investment in sustainable infrastructures and to force corporations to internalize costs associated with unsustainable practices.

Moreover, nation state governments are required to address the principles at the heart of the social and economic pillars noted earlier. It is only nation state governments that can, for example, guarantee fair and equal access to quality education and healthcare or put laws in place to guarantee a fair and liveable wage and safe working conditions for workers not to mention the protection of basic rights and entitlements. Most importantly perhaps, it is only nation state governments that can, by virtue of the scope of their powers and abilities to communicate with the public, bring their citizens together in the shared enterprise of retooling our economies and our societies. My point, then, is that political sustainability, as I characterize it in the following, is the *sine qua non* of social and economic sustainability.

But what do I mean by political sustainability? I do not mean merely political legitimacy, though that is certainly part of it. A government will enjoy political legitimacy if the governed accept its governance on the basis of being assured that it genuinely represents their collective interests. It is not likely to ever be the case that everyone will agree unanimously with any particular policy or law enacted, but there will be popular acceptance of the regime if it is generally regarded as fair in attempting to balance the interests of citizens consistently with its broader mandate of governance and international relations. So political legitimacy will be essential to social cohesion. But what might be required in addition to political legitimacy to achieve what I am calling political sustainability? At the very least, it will require a nation state to embrace a vision of well-being for its people that includes the principles of social and economic sustainability and that will persist across changes of administration. We need, in other words, a unity of purpose and commitment to the goals of sustainability that can transcend, or survive, temporal changes of government. However much one administration might differ from its immediate predecessor, there must be ongoing commitment, at the very least, to the foundational principles of economic, social and environmental sustainability and to working cooperatively with international partners to advance these goals.

As a first statement, then, we will say of a political system that it is sustainable if it enjoys legitimacy with the governed and with the international community (its peer nations) and it can guarantee a commitment to the principles of social, economic and environmental sustainability that will persist across changes in administration and ensure ongoing cooperation with its international partners on advancement of sustainability goals.

On this last point, if we think of governments at various levels (city, state/province, nation state, larger federations, etc.) as systems within systems, then the point must be made that we will also require political sustainability at the international level to facilitate the kind of cooperation needed to make meaningful progress on our goals. And it is the vision of well-being at the heart of, and which sustains, the principles of social, economic and environmental sustainability, which is the key to political sustainability. This should not be taken to imply that application of the principles will everywhere be the same. The principles are deliberately framed at a very general level to permit each jurisdiction the discretion to implement the general principles in ways adapted to their particular circumstances and level of economic development (a point to which I will return in the final chapter).

How is this to be achieved? Keeping in mind the point about the vision of well-being at the heart of political sustainability, I believe Caradonna addresses this issue implicitly when he says,

> If the sustainability movement hopes to achieve real change, then it must galvanize public support without becoming associated with an established political ideology or party. Sustainists have been largely successful at projecting political neutrality by conveying the idea that living sustainably is good for *everyone*. The broad interests of humanity and the well-being of the planet transcend the narrow and self-interested ethos of political parties. Thus the challenge is to have a politically active movement without becoming politicized.[19]

While it is true that Caradonna is here speaking of sustainability movements, and not nation state governments, I think the view is consistent with the one I am pressing here. In particular, I take his point seriously that it is movements that will have to inspire political action. In the first instance, the responsibility falls to these movements (including NGOs, local neighbourhood activists, student movements, the Occupy movement and others) to press for meaningful political action on the part of their representatives at all levels, from civic to international.[20] Given how deeply entrenched the model of globalization at the heart of the Brundtland report is in politics throughout much of the Western world, it is unlikely that political activism will be initiated there without a persistent and urgent demand from a concerned public.

Also, I think Caradonna is right to suggest that this demand will have to be accomplished without these movements themselves becoming politicized. Or, better perhaps, they will have to do so by transcending party politics because they will have to create this popular demand by bringing together the interests of the overwhelming majority of citizens, those dubbed "the ninety-nine percent" by Occupy activists, many of whom will identify their interests with allegiance to a particular party or political ideology. Looked at this way, the question is transformed: how do we achieve sustainability of a politically active movement that can unite the entire globe in the project of sustainability and transcend political divisions and the interests of individual nation states without itself becoming politicized? In other words, how can we get sustainability movements to bring about political sustainability? The answer leads us back to narratives.

At the present time, depoliticizing the language (discourse) of sustainability is not possible under the prevailing narrative of Classical Liberalism. It makes this impossible in principle insofar as the discourse is politicized by the very way the issue is framed within the narrative.[21] Worse, not only it is politicized but it is done so in a way that maximizes division by virtue of the way this framing triggers defensive responses. Thus, it will never be possible under the prevailing narrative of Classical Liberalism to achieve the social cohesion and unity of purpose required to do the hard work that must be done to achieve sustainable societies. And so the role of sustainability movements must be to change the narrative.

So how to change the narrative? That is the principal issue at the heart of this book, but we must first have in hand a new narrative that will inspire our imaginations and trigger our feelings of empathy and solidarity. As Caradonna puts it, "Transforming society from industrialized unsustainability to a social and economic system that is sustainable requires a consensus-building and viable blueprint for the future."[22] And this is the blueprint I present in the next two chapters. The purpose of Chapter 3 is to articulate the general principles of the economics at the heart of this sustainability model, and in Chapter 4 we will articulate the underlying and supportive narrative.

Notes

1 https://sustainabledevelopment.un.org/content/documents/5987our-common-future. pdf [retrieved October 15, 2022].
2 Carson (1962).
3 For some insight into this, see Leopold (1966). In particular, the chapter titled, "The Land Ethic," found in *Part IV: The Upshot*. I take up the topic of the interconnectedness and interdependence of ecosystems in more detail in Chapter 4.
4 Lovins et al. (2007, p.174).
5 We explore the details of Classical Liberalism and the model of globalization it supports in Part II.
6 https://www.overshootday.org/ (retrieved October 15, 2022).
7 At present, zero tolerance for toxic waste must be treated as an aspirational goal. Hence, the qualification in clause d).

8 Durning (1992).
9 https://earthjustice.org/features/campaigns/what-is-mountaintop-removal-mining (retrieved October 15, 2022).
10 Rifkin (2019, p.109).
11 This is exactly the point of a carbon tax: to internalize at least some of the costs associated with hydrocarbon-based energy systems.
12 "Cement" refers to the bonding agent of calcium compounds, usually consisting mostly of limestone, whereas "concrete" refers to the mix of cement and the various aggregates that give it structural strength, often consisting largely of sand and/or gravel.
13 Two articles are provided in the following links: https://www.plant.ca/features/a-cure-for-carbon-putting-co2-to-work-in-concrete-manufacturing/ and, https://www.forconstructionpros.com/concrete/equipment-products/concrete-materials/article/21205418/strengthening-concrete-by-injecting-carbon-dioxide-c02 (retrieved October 15, 2022).
14 See Hawken et al. (1999) Introduction, for a useful discussion.
15 Hird (2021, pp.5–6).
16 A point that will be underscored in Chapter 3.
17 Op. cit., pp.15–16. On this point, compare Hawken et al. (1999, p.48).
18 I don't have a specific source for this claim, but it is one I hear rather often in conversation with students and others. A good source for a description of some of this localized activity, though, is by journalist Chris Turner (2008). This is an inspiring book and it must be noted that Turner does not suggest local activism will be sufficient to address all of the issues arising out of environmental sustainability. See also, Suzuki David and Holly Dressel (2002).
19 Caradonna, p.251.
20 A few interesting stories about youth activism: https://www.nea.org/advocating-for-change/new-from-nea/greta-effect-student-activism-and-climate-change and https://www.nationalgeographic.com/magazine/article/greta-thunberg-wasnt-the-first-to-demand-climate-action-meet-more-young-activists-feature (retrieved October 15, 2022).
21 This claim will be explored further in Part II.
22 Caradonna, p.235.

3

NATURAL CAPITALISM

In this chapter my goal is to explain the economic principles that underlie a genuinely sustainable capitalist economy, the model known as Natural Capitalism, and illustrate how these principles work in practice with some well-chosen examples. The proof that this model is sustainable will follow from the consideration of some real-world examples of successful implementations of these principles as measured against our definition of sustainability from the previous chapter, which we get to in the final section of this chapter. It should be noted that this will be simultaneously a proof that Capitalism itself *can be* sustainable. Ultimately, though, my goal is to use this exposition to extract a narrative – the narrative of Biosphere Consciousness – which I will be doing in Chapter 4.

I will be relying on several sources for the present exposition. First, and most pertinent, is the book noted in the preface, *Natural Capitalism: Creating the Next Industrial Revolution*, and an article by the same authors titled, "A Road Map for Natural Capitalism," together with an additional source from Paul Hawken.[1] Also, I will refer to the work of Jeremy Rifkin, a truly visionary figure on the sustainability scene since at least 1980 with the publication of *Entropy: A New World View*.[2] Rifkin, more than anyone else I am familiar with, has focused on mapping out the infrastructure that will be required for a genuinely sustainable capitalist economy and which results in what he calls the Third Industrial Revolution. His status as a visionary, though, is manifest most obviously when he is working out the consequences, short- and long-term, of this infrastructural change for our industries, our communities, our politics and, most importantly, our world view. It is on this last point, the consequences of moving to the infrastructure of the Third Industrial Revolution for our world view, that our concerns overlap in anticipating what the narrative will look like that will support and sustain genuine sustainability, and so he will play a bigger role in Chapter 4.

DOI: 10.4324/9781003388555-4

A new focus for productivity

The central insight that informs Natural Capitalism is that we must abandon our singular focus on labour productivity to focus instead on resource productivity.[3] Like any mature platform, Industrial Capitalism, and its recent variant Consumer Capitalism in particular, is now confronting significant limitations that arise from the fact that the assumptions of the model are in tension, if not outright contradiction, with our current circumstances.

As discussed in Chapter 2, Capitalism since its inception has been committed to continual growth, and Consumer Capitalism seeks to sustain this growth by the direct stimulation of consumer demand. There are at least two features of our present circumstance that threaten this growth.

First, and one would have thought this would be obvious, we live on a finite planet. This manifests itself with increasing urgency as a shortage of commodities and ever-rising costs associated with dealing with waste by-products. On the face of it, this conclusion might seem to be contradicted by the evidence insofar as lower commodity prices on the market indicate that at least some of these commodities are more readily available than ever before. But stepping back to take a larger perspective, this is deceptive. In addition to some commodities that are in short supply in absolute terms, such as fresh potable water and rare earth metals (used in increasing quantities to make batteries for the entire range of electrified goods, including cars and computers), consider oil which is still the principal source of energy for our systems. Arguably the wildly fluctuating price of oil on the markets is not an accurate reflection of its real cost nor of its relative scarcity. Even setting aside the influence of political factors exogenous to the markets themselves, we can see that increasingly our supply of oil is gotten from expensive technologies such as tar sands extraction, fracking and offshore oil platforms that significantly reduce the return on investment (ROI).[4] Additionally, the extraction of resources in these ways is exacting an ever-greater toll on the environment and imposing increasing risks, as made starkly evident by the explosion of the Deepwater Horizon oil platform in 2010 and the subsequent oil spill, the worst accidental oil spill ever recorded at four million barrels according to some estimates. Never was it said more truly, "the low hanging fruit is gone."

The conclusion is that the shrinking availability of commodities and increasing costs of extraction place downward pressures on the growth expectations of capitalist markets everywhere. And although many of the costs associated with extraction and waste flows continue in many cases to be externalized, the biosphere that is our shared environment is, for practical purposes, a closed system and those wastes and their consequences are still with us. Think for a moment of the devastation resulting from more frequent and more violent storms, lengthier and more destructive forest fire seasons and the increasing food risk associated with climate change. It is evident the bill on externalized costs is now coming due and arguably exceeds the benefits, at least for most of us. And the very real possibility of environmental collapse and widespread species extinction in our lifetimes means no one will escape the consequences much longer if swift action is not taken.

The second feature of our present circumstance that threatens growth arises directly out of the myopic focus on labour productivity. Since the dawn of the (first) Industrial Revolution,[5] attention has focused almost exclusively on the productivity of labour for the understandable reason that labour was the biggest single cost on the production side. Thus, any increases in labour productivity would enhance profits. Historically, increases to productivity have resulted from the application of technologies to the production process.[6] Throughout the early decades of the post-war era, the jobs displaced by technological advancement produced other jobs higher up the skills inventory. There were dislocations, to be sure, but with a robust and growing education system and government support for education and skills training,[7] opportunities to retrain for higher skilled and better paying employment were plentiful.

This is, in principle, how the system is supposed to work and is what explains the faith of the Brundtland Commission in the globalization of development economics to lift the world out of poverty. The same principles applied globally encourage the export of lower skilled production jobs to other places where the labour is cheaper. This has the virtue of driving down labour costs, making cheaper goods available in abundance for markets in developed countries, and it provides workers with regular employment and the income to stimulate demand in the newly burgeoning markets of developing countries. Meanwhile, the workers displaced here are retrained and bumped up the skills inventory, earning higher incomes to buy the goods now produced more cheaply and in ever-increasing abundance elsewhere. Demand is stimulated everywhere and there is a steady progression of labour up the skills inventory.

While this might be how it is supposed to work in principle, there are at least two problems that have emerged simultaneously with the onset of climate change as a noticeable phenomenon impacting on our daily lives and the relative scarcity of commodities that undermine this simple picture of endless growth and prosperity for everyone. First, governments everywhere now find themselves constrained in their spending and no longer support educational initiatives as they once did, and many students find the rising costs of education and burdensome student loan debts to be a disincentive to upgrading their skills. And second, even for those that do invest the money in their education, opportunities are drying up. The application of technology today is focusing increasingly on artificial intelligence (AI) systems and robotics that displace not only many low-skilled workers in production systems but also those higher up the skills inventory. As Rifkin notes, productivity gains from the application of technologies are now having impacts far beyond the usual manufacturing sector to include white-collar service industries, retail and even medical diagnostics, healthcare delivery and tax and legal services.[8] The prediction is that tens of millions of jobs across every sector of the economy are at risk of being displaced by AI technologies by the middle of this century. And so,

The conundrum is that if productivity advances brought on by the application of intelligent technologies, robotics, and automation continue to push more and more workers to marginal employment or unemployment around the world, the diminishing purchasing power is likely to stifle further economic growth. In other words, if smart tech replaces more and more workers, leaving people without income, who is going to buy all of the products being produced and services being offered?[9]

The consideration of these two features that place downward pressures on our expectations of continual growth underscores the late maturity of Industrial Capitalism. As Hawken et al. (1999) point out, the situation is different now than it was in the immediate post-war years, and greater returns can be expected from shifting our attention to enhancing the productivity of the resources we use. As the authors make the point,

> With nearly ten thousand new people arriving on earth every hour, a new and unfamiliar pattern of scarcity is now emerging. At the beginning of the industrial revolution, labor was overworked and relatively scarce (the population was about one-tenth of current totals), while global stocks of natural capital were abundant and unexploited. But today the situation has been reversed: After two centuries of rises in labor productivity, the liquidation of natural resources at their extraction rather than their replacement value, and the exploitation of living systems as if they were free, infinite, and in perpetual renewal, it is people who have become an abundant resource, while *nature* is becoming disturbingly scarce.[10]

In speaking of extraction cost versus replacement cost here, the authors are underscoring the point made in the previous chapter that the present market price of many commodities is kept artificially lower than their real cost by market mechanisms and market failures that allow the costs associated with resource depletion and despoliation of the environment to be externalized. This masks the fact that we are actually in radical deficit spending on the balance sheets of natural capital. And the only way to resolve the problem is to begin to value natural systems at their replacement value. If Capitalism is to be refashioned in sustainable form, the focus will have to switch to maximizing resource productivity, and this insight is the ultimate foundation of Natural Capitalism.

The four principles of Natural Capitalism

But what do we mean when we speak of resource productivity? As we will see in what follows, it is getting the most we can from every bit of resource utilized. If we are going to take some resource from the bank of natural capital, then we must commit to getting the maximal service that can be obtained from that resource, which implies minimizing waste. We will systematically explore the four principles that collectively define Natural Capitalism and establish that it is a truly sustainable alternative to Consumer Capitalism.

First principle: maximize resource productivity

The first principle mandates that we strive to focus consistently on maximizing resource productivity by seeking always to get the most of the desired output relative to resource inputs. Whether we are speaking of outputs as goods or services, the goal is always to get the most we can from the inputs used, which is the same thing as to say that we seek always to eliminate avoidable waste. Ultimately, our systems must evolve to the point where we regard all waste as avoidable, or perhaps we should say, waste becomes unthinkable. There are two strategies that work together to help us achieve this goal. One is to reimagine production and delivery of services and goods from a system perspective. And the second is to embrace and implement the best-existing technologies and practices.

When we speak of re-imagining production and delivery of goods and services from a system perspective, the idea is to approach the design of these systems with the goal foremost in mind of maximizing resource productivity by analysing the system as a *whole* so as to have that effect. To put the point metaphorically, it amounts to stepping back to look at the system as a whole with the goal of (re-)imagining how we might design the system differently in ways that will produce more from less. Success in accomplishing this perspectival shift requires us to leave behind the old perspective that focused on labour productivity with all its attendant design principles. Until we do this, our imagination will be constrained by limiting assumptions. Hawken et al. underscore this point in reference to Edwin H. Land, who in 1948 invented the Polaroid camera. The Polaroid camera was an astonishing thing. It produced photos that would emerge from the camera and develop as you watched.[11] The versions I used to sell during my days in camera sales (the Polaroid SX 70) produced remarkably clear colour photos in about a minute. The point here is that before Land could produce his remarkable invention, he had to abandon the preconception widely shared at the time that photos must come from film that one takes to a processing facility to have the negatives turned into pictures, which in those days took about a week or so. Land had to first have the boldness of imagination to wonder if it might be possible to produce pictures directly from the camera in your hands. It is reported of him that he would often remark that, "Before you can have a new idea, you first have to stop having an old idea."[12] This is, to my mind, the point about narratives all over again. A new narrative abandons the old and limiting assumptions and resets the frame of reference. Looking at the problem from this perspective can result in surprising, even astonishing, productivity gains and savings.

It will help to have an illustrative example, so consider the case of Interface [from Lovins et al. (2007)], a manufacturer and supplier of flooring services, commercial and residential.[13] Upon construction of a new production facility in China in the 1990s, the company availed itself of the opportunity to rethink the design of the factory with an eye to eliminating waste. The manufacture of their products at that time required delivery of fluid inputs at various stages of the process that can only be delivered from a set of reservoirs by a system of pipes and pumps. Previously, design of the factories had proceeded by first situating the reservoir tanks, usually at the periphery of the facility, permitting organization of the work floor in ways that would maximize labour productivity, presumably some version of Fordism. But this in turn requires a network of piping and pumps to deliver the inputs to the appropriate location when needed. Perforce, the pipes are required to trace a route through the facility that often requires a lot of bends and long distances, which in turn requires massive and powerful pumps (rated at about 95hp). In addition, because of the bends and the distances travelled, the pipes must have thick sidewalls to sustain the incredible pressures required and the wear of the fluids passing through them. All of this requires an infrastructure that is materials-intensive, requires enormous energy to operate and has a significant overhead in terms of maintenance.

When constructing the new factory, it was decided to place the reservoirs near where they would be needed. As a result, short lengths of pipe, mostly straight, could be used to deliver the inputs, and using pipes with a larger diameter further reduced the pumping power required. Thus, much smaller, more efficient pumps could be used. Rated at 7hp, the new pumps represented a 92 percent reduction in power required to deliver the inputs. Additionally, the shorter, straighter pipes could be made with thinner sidewalls and, so, were much less materials-intensive for a given length, and they cost less and were easier to insulate thereby "saving an extra 70 kilowatts of heat loss and repaying the insulation's cost in three months."[14] The net result is a factory that is significantly less materials-intensive in its construction, requires less energy to operate, produces much less heat loss and waste generally and requires less maintenance. And note the importance of relying on the best existing technologies: the most efficient lower horsepower and high efficiency motors; sophisticated monitoring systems, sometimes involving AI systems to conserve energy; the highest-efficiency LED lighting designed to enhance illumination where it is needed and not wasted illuminating spaces where it isn't needed and so on.

Similar potential for savings and waste reduction apply to the design of integrated manufacturing systems which can gather together various businesses that can be integrated to benefit one another. This was anticipated in Chapter 2 when speaking of designed production parks, but the present point is that if we keep in mind the idea that systems can be organized into levels, as "systems within systems" so to speak, then we see that the benefits of the systems perspective can similarly ramify at various levels. We will consider an illustrative example, the Kalundborg Eco-Industrial Park, later in the discussion of the second principle, biomimicry.

On this point about system integration at various levels, consider urban design. The system of suburbs and the network of roadways connecting these to urban centres that was built up in the 1950s and 1960s is an example of extreme waste. A typical suburb features wide paved streets with little natural drainage, massive yards with unnatural lawn cover requiring huge quantities of water and chemicals for weed suppression. Typically, they are built around centralized shopping and services that require residents to drive for almost everything, usually resulting in lengthy commutes to work, trips to shopping centres, school, hockey practice and church services. The result has been the intensive production of waste from vehicles and lawn mowers in the forms of carbon dioxide and noxious fumes, rubber from tyres, noise, heat and extensive systems of roadways needing continual maintenance. The houses themselves were built to inefficient standards reflecting the cheap availability of heating and electricity. They also present a brittle and vulnerable face to the environment in many ways, subject to surface flooding in the extreme rainfall becoming increasingly common because the water can find no natural drainage.

A systems approach to urban design will seek the elimination of as much waste as possible and find ways to integrate natural landscape features into neighbourhoods that are resilient to extreme weather events. In fact, the point is better expressed in reverse: we will integrate neighbourhoods into the natural environment. The focus will be on building communities as neighbourhoods, with parks, bike lanes and pedestrian paths and with restricted access for cars and a mixture of commercial and residential structures that have people live where they work, play, attend school and shop. And looking out to the middle of this century, the homes in these neighbourhoods will be their own power stations relying on a mix of renewable energy sources that include solar photovoltaic, geothermal and small-scale wind power.[15]

Additionally, the buildings themselves will be built according to much higher efficiency standards to the point where, depending on location, many will rely on passive heating and cooling and will maximize the use of natural lighting during daylight hours. Together with the best existing technologies in energy-saving appliances, triple-glazed windows and insulation to reduce overall energy consumption by as much 75 percent or more relative to present high-end standards. Many of these homes will be net energy producers, selling surplus energy back to the grid for use by others, which they will do via an intelligent energy distribution grid that can adjust the flows of energy according to immediate feedback data. And every home will have its own energy storage, often the family vehicle, or perhaps a hydrogen fuel cell, for occasions when demand exceeds what is available.

Ultimately, the point of whole system design is to optimize the components of a system together *as a system* with the goal to reduce waste to an absolute minimum rather than optimizing each component individually and then assembling them into a system. The change in perspective here is to start from a clear vision of the purpose of the system. We must always ask, "What is the system being designed to accomplish?" Optimizing the whole results is a system that can wring every available iota of productivity from the resources used by virtue of the beneficial synergies that result.

Another benefit, as the many examples cited by the authors demonstrate, is that the initial costs of construction often come in lower, and this is additional to the enormous savings going forward as a result of the elimination of waste, enhanced productivity, reduced energy costs and reduced maintenance costs. Quoting the authors on this point,

> At the heart of . . . the entire book, is the thesis that 90 to 95 percent reductions in material and energy are possible in developed nations without diminishing the quantity or quality of the services that people want.[16]

Second principle: biomimicry – closed-loop systems

The second principle, ideally working hand in glove with the implementation of whole system design and of best existing technologies and practices, adds the desideratum to design our systems in ways that mimic biological systems, where the central insight here is that in biological systems there is never any such thing as waste. So in addition to asking "What is the system being designed to accomplish?" we must go on to ask, "How would nature do this?"

The integration and complex interconnections between all the parts of biological systems are such that everything that lives and ultimately dies is reabsorbed into the system by becoming a part of the nutrient system upon decay, often in a series of stages. The story of the ecology of dying and dead trees in mature, old growth forests is a study in such complex relationships. A series of studies of old growth Fir forests in the Pacific North West, and done over a period spanning decades, reveals that decomposing trees can take more than 400 years to be fully reabsorbed into the environment and go through as many as nine distinct stages in doing so.[17] And at every stage a robust and active ecology is supported by the decaying mass, not only serving as food for some but also providing such services as cavity nest sites, nesting platforms, food cache, site of courtship and mating and more. The complex interactions amongst the thousands of species is supportive of rich diversity, and each fallen tree is like a bountiful savings account of nutrients and organic matter. The trees in such a forest, living and dead, "are linked together in the living machinery of a forest."[18] "When a fallen tree decomposes, unique new habitats are created within its body as the outer and inner bark, sapwood, and heartwood decompose at different rates."[19]

For example, at a relatively early stage in the decomposition of a tree, the outer bark may be infested with termites. The burrowing activity of the termites permits the entry of other life forms into the interior of the tree mass leading to the next stage of decay, but the story of the termites and the trees is itself indicative of the rich interrelationships between species. These termites can digest the wood because of a relationship between three species: the termites themselves and cellulose-digesting protozoa and nitrogen-fixing bacteria that live in the gut of termites. The protozoa digest the cellulose with the aid of the nitrogen-fixing bacteria, and the waste products of the cellulose digestion in turn becomes food for the termites. Again, we see systems within systems, within systems, all integrated and interdependent. "The continuum of a fallen tree is composed of, and driven by, an increasingly complex network of simultaneously developing minisystems – all interdependent."[20]

Of course,

> [u]ltimately the entire tree is incorporated into the soil. It has gone full circle, having been formed as a product of the soil and of photosynthetically captured carbon, it now returns to the soil through release of carbon by decomposition.[21]

And therein lies the insight that inspires biomimicry: nature in the form of a mature, old growth forest operates as a *closed loop*. As a system, there is little in the way of nutrients or other inputs to the system (apart from the sunlight, of course), nor is there much output, in the sense that little is lost from the system. Rather, all the essential nutrients, including potassium, nitrogen, carbon, and all the rest, are cycled continuously within the system and all the while this cycling supports a rich living biodiversity. Even the hydrological cycles and weather systems are intrinsic to the system, being shaped by and giving shape to, the local ecosystem. It is like a localized niche of negative entropy; the ecosystem as a whole maintains a high level of complexity and order over a period that, if left undisturbed, can span millennia and in apparent defiance of entropy with little in the way of energy inputs other than the sunlight.[22]

Thus, when we ask how nature would accomplish our design purposes, we must look to how mature ecosystems close the loops and try to model that in our systems. As Benyus notes, when we pay close attention to mature ecosystems, they are consistently characterized by complex web-like food chains that link the various trophic levels into complex interrelationships.[23] There is nothing like a simple linear input of nutrients into the system matched by an output of waste from the system, as characterizes most of our existing infrastructure and production systems. Additionally, though, these mature ecosystems are characterized by a lot of diversity, to be contrasted with our intensive monocrop agriculture. And the life cycles of many of the organisms are long and complex, as indicated in the forest example just considered. A Douglas Fir typically lives for more than 750 years and can spend more than 400 years in decomposition before being returned entirely to the local environment.[24]

And so, systems designed in accordance with this principle are also called closed-loop systems. As noted earlier, the ultimate goal is to make waste unthinkable within the system. Anything presently considered as waste is to be reconceived as a resource, *repurposed* to become the input to some other part of the process or some other process altogether. Only when no other purpose can be conceived for some output will we consider it to have exhausted its productivity, and at this point it is imagined that its extended productivity will have rendered it harmless (i.e., non-toxic) and like the Douglas Fir it will be returned to the environment as a basic nutrient.

Returning to the point noted earlier, namely that this strategy will work hand in glove with the implementation of whole system design, we see the leading-edge innovation of this insight manifested in the design of production parks that collect businesses together in a manner that can be mutually supportive by the active exchange of their outputs with one another in ways that will enhance maximizing productivity and the elimination of waste. The most fully realized version of such an eco-park is perhaps the Kalundborg Eco-Industrial Park, known as the Kalundborg Symbiosis.[25] This park collects together 12 public and private companies, including Kalundborg Utility, all actively sharing water, energy and materials. As stated on the website,

The main principle is, that a residue from one company becomes a resource at another, benefiting both the environment and the economy. Having a local partnership means that we can share and reuse resources, and that way we save money as well as minimize waste.

A similar point would apply to the complete redesign of our agricultural practices along closed-loop parameters and whole system design, a story which I take up in detail later in the section on restoration of natural capital.

A further point about biomimicry, as Benyus notes (p.97), is that life manufactures its products under life-friendly conditions. The examples are by now familiar but remain humbling. The beautiful pearlescent interior shell of the abalone is a substance prized for its beauty and amazing hardness.[26] It is harder than any manufactured ceramic, made from organic polymers and inorganic minerals, in ocean water at life-friendly temperatures, and it self-assembles! Or spider silk, used to make webs, is flexible, stronger than Kevlar and made inside the spider's body from the digested exoskeletons of its prey.

On this point, biomimicry benefits from working symbiotically with the adoption of best existing technologies. For example, the application of enzymatic processes to facilitate reactions that would otherwise require massive inputs of energy and/or toxic chemicals. Enzymes are biocatalysts, proteins essentially, that can be used in many applications to replace inorganic catalysts such as acids, bases, metals and metal oxides, which are almost always toxic when released into the environment. The development of enzymatic processes is presently a very active area of research and involves recombinant gene technology to drive enhancements. A further benefit is that enzymes can often be reclaimed and, thus reused, whereas the inorganic alternatives are usually single use.[27] Presently used in the production of pharmaceuticals, detergents, paper and other pulp products, brewing, food processing and biofuels, they hold out enormous promise for wringing productivity out of our resources at enormous savings of energy while greatly reducing waste.[28]

A very different kind of example, cited by Lovins et al. is the case of Motorola which found itself in a predicament when chlorofluorocarbons were outlawed following the discovery of ozone depletion that led to the Montreal Protocol (more on this later). These highly toxic substances had been used extensively to clean circuit boards from residues left behind by soldering. Searching for an alternative solution, the company for a while embraced terpenes made from orange peels. Since that time, though, the company has completely redesigned the soldering process so it no longer requires cleaning, thereby eliminating the need for cleaning solvents of any kind.

Here we see several themes illustrated. First, there is the implementation of both whole system (re)design to realize a process that is closer to closed-loop production and significantly eliminates a lot of toxic waste. And it underscores the point that the goal is forever receding into the future: we can continually improve at every iteration. We act, observe the outcome to learn and introduce further refinements that get us even closer to our goal: the complete elimination of waste.

Third principle: cradle to cradle – a new business model

The business model of Consumer Capitalism is built around centralized production by huge multinational companies that produce goods for sale to consumers as end users. To be sure, small- and medium-sized businesses play their role, but these large, hierarchical and vertically integrated firms and their supply chains are the backbone of the system. The point about being an "end user" is that responsibility for the product falls to the consumer for disposal at the end of the product's useful lifetime or when it simply isn't wanted anymore. So the process is a linear system that involves extraction of resources, often at enormous environmental costs as we have seen, to be processed and delivered to producers who take the various inputs and produce mass quantities of products which must then be delivered to markets, all of it involving complex supply chains and middle-level firms handling logistics, delivery and sales. And the linear chain ends with the products in landfill. The model favours a highly centralized and vertically organized corporate culture at every level to extract profits from production at scale.

Now, as noted, the point of Consumer Capitalism is to encourage consumption in ever-growing quantities to ensure capitalist markets can meet their growth targets. To this end, two strategies are particularly important, which work in tandem to stimulate this continual growth in consumption. First is the enormous investment in advertising as a direct stimulus to creating what is called "artificial need."[29] And the second strategy is planned obsolescence.

It is estimated that just under $650 billion (Am) will be spent directly on advertising globally in 2022. This is a staggering sum and does not even include the sum spent on research into how to maximize the effectiveness of advertising. We can say with confidence, though, that ads work. Aside from the obvious point that such staggering sums would not be invested absent the assurance that advertising works, we have the evidence of research.[30] The strategy that has been discovered to be most effective is to stimulate a desire that typically works away at the psyche over time in an indirect way, aided by repeated exposure to the relevant ads,[31] until it becomes an irresistible need. It is planting the seed of an idea which, based on repeated exposure to advertisements that evoke positive feelings associated with the product, will over time manifest itself as a persistent desire for the thing now fetishized as a need (as Marx would put it). No one is immune to this.

There is a case study frequently discussed in business texts because it illustrates so powerfully how effective advertising can be: the advertisement for the Apple Macintosh placed in the 1984 Super Bowl.[32] In the late 1970s, as personal computers were just beginning to get a foothold in the market, they appealed mostly to a relatively small market of hobbyists. Manufacturers such as Tandy/Radio Shack, IBM, Pet Commodore, Texas Instruments and others started marketing relatively simple devices that weren't particularly useful, since there wasn't much you could do with them, truthfully. A friend of mine purchased a Pet Commodore unit in 1976 with 16K of RAM that connected to a cassette tape unit for input and data storage and to a black and white TV as a monitor. There was a simple version of a word processor and a spreadsheet on a few cassette tapes and a simple version of the programming language BASIC was built into the ROM. But it was very cool and many of my friends wanted one, and so did I. This burgeoning desire, restricted as it was in those days to hobbyist nerds and techno geeks, grew in that community without much in the way of advertising outside of industry magazines. But as the market grew and saw the entry of Apple and others into the marketplace, this began to change. In retrospect, the Super Bowl ad marked a turning point for Apple and for personal computers generally. In 1984, when the Super Bowl ad ran, the penetration of personal computers in the North American market was 8.2 percent and consisted almost entirely of PCs. By 1997, it was approaching 37 percent, and by 2016, it exceeded 89 percent, and as of 2022, Apple computers enjoy 13.5 percent of global market share.[33]

What about planned obsolescence? Planned obsolescence has indeed come a long way from simply building things to fail. Of course, one strategy still widely used is to slowly shorten the usable lifetime in tandem with carefully probing public attitudes and expectations. Mistakes can be made, trust damaged, brands devalued, but we all know it continues. An example is a scandal that rocked Hewlett-Packard a few years ago involving the toner cartridges in its inkjet printers.[34] A far better strategy is to find a way to get consumers to *want* to buy a new unit because they've grown tired of the old one or to encourage owners to insist on name brand parts. The car industry presents a valuable lesson here.

The first auto show in the United States was held in Madison Square Garden on November 3, 1900.[35] Eventually, it became an annual event, giving auto makers an opportunity to showcase their wares and stimulate curiosity and demand in the public. Particularly after the Second World War, it became a refined tool for stimulating demand for the latest and greatest, by building in new features, often not more than mere cosmetic changes, that made older models seem dated if not obsolete, and all of it surrounded by the hoopla of the Annual Auto Show, now a major event. For some time now, the average period for fleet renewal has been about six years. Of course, a car is a major purchase, but the result has been, and continues to be, a sector that struggles with the consequences of a periodic business cycle. From the seller's perspective, the issue is always a matter of how to shorten the period between business cycles.

As it turns out, a similar strategy has been refined by the computer industry which has managed to shrink the timelines to less than three years as an average. A big part of this, I think, is that periodic software updates precondition consumers to the idea that the product is fundamentally ephemeral. But it's not possible to overstate the role played by successfully engineering product releases as major media events. Think of the media attention focused on the release of the latest iPhone and the spectacle of people lining up, often overnight, to be first in line for the new model. The addition of new features, apps and computing powers that can make older phones perform slowly or become incapable of some tasks results in an environment that continually pushes consumption to ever higher levels. The average time between releases is in some cases less than a year. The iPhone 12 came out on October 24, 2020, the iPhone 13 on September 24, 2021, and the latest (at time of writing) is the iPhone 14 released on September 16, 2022. And in the United States, the average time consumers held onto their phones before switching was, in 2018, just barely more than two years (24.7 months).[36]

Moreover, and insofar as we are the end users of these products, it is up to us to dispose of the older product, often still perfectly serviceable, and this usually means it is "recycled." Regrettably, as we saw in Chapter 2, the narrative surrounding recycling is not only a means of downloading responsibility onto consumers to carry the burden of waste disposal, but it is also in many cases a merely comforting illusion insofar as little of what is redirected to recycling is actually recycled. It often sits in huge piles contaminating soils and underground water aquifers, or in some cases is incinerated releasing dangerous contaminants into the atmosphere. The CBC reports that the UN has released a report for e-waste for 2020. Globally, a record 53.6 million tonnes of e-waste was disposed of that year, redirected to recycling, often involving fees imposed on consumers at point of purchase, allegedly to ensure proper recycling. The report tells us that best estimates are that a mere 17.4 percent was actually recycled.[37]

What is the way out of this endless cycle of excessive consumption and waste production? Clearly, in addition to anything else we might do, a new business model will be needed to realign the incentives of all parties with our sustainability goals. To this end, the business model at the heart of Natural Capitalism is about replacing the sale of goods to end users with the sale of services and putting the burden of ownership and disposal on the producer, where it properly belongs. It is also about putting the production of goods under the direct demand of clients (client "pull") as opposed to mass production, which typically results in surplus production that must then be cleared (batch and queue).[38]

When we speak of aligning incentives with our sustainability goals, a central aspect of this is that it can virtually eliminate the traditional business cycle and replace it with a steady and predictable cash flow. There are several specific implementations of the general principle to replace sale of goods to end users with the sale of services, but in all cases the outcome is the same. It's about developing "a relationship [between client and service provider – WIH] that provides a continuous flow of services to meet the customer's ever-changing needs [and which] automatically aligns the parties' interests, creating mutual advantage."[39]

As an example of one such implementation, I am going to tell you a somewhat idealized story about how such a model might work in relation to photocopying services. I say "idealized" because we can expect the details of specific contracts to differ. As an academic I can tell you that educational institutions have an enormous need for copying services and likewise for law firms, hospitals, healthcare delivery services and many more. When I began my career in the mid-1990s, I worked contractually at a variety of institutions and the situation seemed to be the same everywhere. A bunch of machines spread across campus, with cabinets full of paper and toner cartridges. The machines were under continual heavy use by faculty and staff, who were typically often rushed and under duress, and in most cases, completely ignorant of the proper use and maintenance of such complex machines. Thus, the machines did not last long. I would guess it was typically about three to four years before the machines were performing so poorly, they had to be replaced, at enormous expense and short-term disruption for the institutions. And then there was the day-to-day tedium of using these machines: line-ups at the machines, with frustrated faculty fretting about being late for class while watching their colleagues trying to figure out how to undo a paper jam or install a new toner cartridge. (Off to find the maven in the department who had cultivated the rarefied knowledge about the machine's strange inner workings.) Often, there would be at least one machine with a sign taped to it announcing, "Out of order," and it could take several days, sometimes as long as a week, before a technician would arrive to service the unit. This was enormously costly for the institutions, both in terms of the direct costs of maintaining the machines and the reduced productivity, and it was stressful for faculty and staff.

Since those days many institutions have entered into contracts with various service providers, companies such as 3M and Canon. Instead of buying the machines outright, the institutions buy copying services from a service provider. A request goes out for bids. Consultants visit the institution and analyse their needs. How many machines? How many of these should be colour? How many can be black and white? How will they be distributed around the institution? They determine paper and toner requirements and submit a bid to provide a complete copying service package that includes regular servicing visits (which includes stocking up paper and toner cartridges) and an 800-number prominently posted on each machine in the event of a breakdown, with a guarantee of service within 24 hours. And, at the end of the contract, the provider is responsible for taking the machines away and all of this for a predictable monthly fee.

This is a win-win situation. As an institution, the only interest is ensuring the capacity to make copies as required. They aren't interested in owning finicky machines and the faculty and staff that use the machines certainly don't want to be challenged with machines that don't work as expected. And for the service provider, there is an ongoing relationship with the client which, if properly nurtured, often results in renewal of the contract. And, as noted earlier, it also smooths out the business cycle insofar as the periodic sale of machines is now replaced by a regular ongoing monthly fee for service provided. Additionally, it puts the competition and the incentives where they should be. On this model, service providers are incentivized to offer the best service possible at the best price while remaining responsive to the customer's changing needs. Moreover, the producer is incentivized to produce machines that will last longer and provide reliable service throughout the life of the contract – after all, "we want that contract renewed." All of this underscores the sense in which this model aligns the interests of consumers and producers.

This example concerns a commercial application, but there is no reason this couldn't be applied to a wide range of services for consumers including, but not limited to, home heating/cooling services, refrigeration services and cooking services, or all kitchen appliances for that matter, and certainly automobiles.[40] The same basic model applies. Let's say you want the most up-to-date kitchen appliance services at the best price, and you don't want to be bothered with occasional interruptions of service because the unit malfunctions and you're required to wait on a technician to visit. Under the present system of consumer ownership, as the previous point about planned obsolescence illustrated, expectations have been set so that in many cases consumers don't even bother to pursue servicing the unit after a period of about four to five years. Consumers assume there's no point in putting more money into a machine that has probably reached the end of its usable life, and in any event, a service visit just to determine if the unit can be repaired at all is typically about $70 (and you have to wait, sometimes as long as a week). It's easier to just buy a new one, and for a small additional fee you can have them take the old one away.

Imagine instead you phone a couple or three service providers for a consultation. They come to your home, analyse your needs, based on family size, what you wish to spend, what you regard as essential, what you consider negotiable and so on and submit a bid for a contract to install the best and most up-to-date appliances that will meet your needs, with an 800 number to call in the event of a problem and a guarantee that a service technician will visit within 24 hours, plus regular service visits on some schedule sufficient to guarantee the machines are always working as they should. Even more likely these days, service of the units is done remotely by software updates and continual feedback to the service provider indicates when a home visit is required (the internet of things).[41] And when the contract ends, and assuming you subsequently go with a different service provider (presumably because you get a better deal) the previous service provider is responsible for taking the units out of your home, which they can refurbish and lease to the next client. Again, it's win-win.

A different implementation of the same general principle is illustrated by the example of third-party apps that facilitate getting more usage out of otherwise underutilized goods. Thus, consider Zip Car. Zip Car is a service that puts a contemporary spin on car rental services by putting car owners and potential renters together via an app. A car represents a significant investment yet often goes underutilized. One recent study found that car owners use their vehicles for an average of only nine hours per week. This means the vehicle spends about 159 hours per week sitting idle. If an owner wants to make their car available for rent when not being used, Zip Car can facilitate this. As a service provider, Zip Car is providing the connectivity that puts car owners together with renters in a way that is safe, secure and provides the necessary feedback for keeping such transactions honest and fair. In addition, they establish price schedules and such, so all parties have the relevant information. A single vehicle can, in principle, service the needs of several users. Zip Car's website claims that every Zip Car takes the equivalent of 13 cars off the road.[42]

Other examples of extending the use of underutilized resources are numerous, and Airbnb and Uber spring to mind. Uber represents a very different way of wringing utilization out of a privately owned vehicle while creating employment for the vehicle owner, and Airbnb does a similar thing for recreational properties or even additional rooms in one's main home.

These latter examples raise an important point, though, namely that advocating for greater utilization of underutilized resources is not the same thing as advocating for Uber or Airbnb or any other particular application of the model. It is well known by now, or should be, that Uber, among other app-based services (e.g., Skip the Dishes), has been accused of exploiting their "contract workers" in various ways, principally by refusing to acknowledge them as employees. One scholar has done an extensive study of this exploitation in relation to Uber, and it is sobering indeed.[43] In my view, this is not a challenge to the soundness of the general principle but speaks to the need for sensible regulation of the market. Under such regulation, whatever shape that ultimately takes, the various service providers will be sorted according to how well they are able to align their values with those of their clientele. It is noteworthy that even among car-sharing apps, there is a wide variety of implementations out there and some of them, like Communauto, are non-profit.[44]

A final remark, and in relation to the point about privacy noted earlier, it will take time for legislators, policy wonks and courts to wrestle with these newly emergent business models and their implications, just as they must wrestle with implications of social websites and how they are to be regulated to protect against misinformation, data mining, invasions of privacy and fraudulent activities. Here, too, we can expect the regulatory environment to improve over time as we learn and refine our policies to meet our goals.

Having said, there is much that sensible regulation can do right now to facilitate the switch to a business model that aligns our needs as consumers with our sustainability goals. In particular, there is a need for fiscal incentives to encourage more investment in alternative energy technologies and the supporting infrastructure, as well as the gradual rescinding of subsidies that have the perverse effect of encouraging unsustainable practices, particularly in relation to mining, exploration, energy production and agriculture (more on this later in relation to agriculture in our discussion of the fourth principle).

On the policy side, there is a need for low-burden policies such as "cradle to cradle" and "take back" legislation of the sort already in place in some jurisdictions.[45] The point of the legislation is to make the producer responsible for taking back the product at the end of its serviceable life. Other policies that could enhance this further include laws that require major appliances and automobiles to be up to 95 percent recyclable. So couple these policies together with the "service and flow" business model, and you have effectively incentivized producers to make their products to last as long as possible by cutting the legs out from planned obsolescence while aligning the interests of producers with the interests of their clients and the interests of both with the sustainability goals of society generally.

The opportunities for improvements as we move forward are endless. Modular design can not only facilitate recycling but also result in products easily updated by simply plugging in a new module, for example, updating the "brain" in a photo-copying system or perhaps replacing the heat exchange unit in a smart home heating system for another that is more efficient. As Benyus puts it, the government sets the boundary conditions to place commerce in an environment where *behaving in ways that are sustainable becomes the best way to be competitive*. And to take up directly the point about being low burden, these policies must be transparent, clear, easily implemented and should be framed broadly in ways that are non-prescriptive so as to leave it to businesses to find their own solutions that will maximize competitiveness and profits while respecting the goals of sustainability.[46]

What all these implementations of the basic business model of Natural Capitalism have in common is that they significantly reduce consumption of resources while having a negligible impact on consumption of services.

Fourth principle: restoring natural capital

When significant progress has been made implementing the first three principles, we will have accomplished at least two things. We will have reduced significantly both the extraction of virgin resources from the environment and the production of waste. This will have the effect of reducing the pressure on the environment, allowing it to begin to heal. Evidence from several sources indicates that when pollution pressures on the environment are eased, it can bounce back quite quickly.[47]

This was our experience with the implementation of the Montreal Protocol to deal with the depletion of the ozone layer as a result of Chloroflourocarbons (CFCs) and Hydroflourocarbons (HFCs) released into the atmosphere. The Protocol was agreed to and signed in 1987 and brought into effect on January 1, 1989, and since that time the ozone-depleting compounds in the atmosphere have declined significantly and the ozone layer has begun to regenerate itself. It is by no means fully restored, but lowering levels of the insulting compounds has provided the chance for renewal.[48]

We are unlikely to ever get to a perfect solution in a single step. Rather, we should set our sights on implementing the best possible solution we can, observe the outcomes and improve at the next iteration. I draw the following lessons from this example. First, the international community was able to recognize the gravity of the problem and come together to work cooperatively for a solution. Second, we were able to use the best science to diagnose the problem and devise an effective strategy to be implemented. And third, and directly relevant to my present point, the natural environment demonstrated an amazing resilience once the pressure of pollutants was eased. A similar observation came to our attention in this last two and a half years of COVID-19 pandemic. As a result of reduced commercial activity and travel, pollution levels declined during the lockdown periods, and the result was evident in cleaner rivers, streams, and air quality in our largest cities.[49]

Let me return to the story of hurricane Katrina shared in the preface to reinforce these points. When Katrina struck, the devastation was of a magnitude that left many awestruck and shaken and as many as 1,836 people dead. The storm surges overwhelmed the levees and in the post-disaster analysis the conclusion drawn by many was that it was the failure of the levees that was the problem. But a further twist enters at this point in the story. It turns out that the wetlands that for millennia had provided a natural buffer for the Gulf coastline had been in steep decline for many years as a result of a host of factors that includes clearing of wetlands, diversion of rivers, agricultural practices that drained pollutants into the Mississippi and even the construction of the levees themselves. The huge reduction in wetland expanse and depth meant they were no longer sufficiently robust to adequately buffer the coastline against storm surges. Thus, the levees were left to carry the burden against a force of nature they were never designed handle.[50]

The restoration of natural capital, as per the fourth principle of Natural Capitalism, could accomplish much in the way of protecting the Gulf coastline against future storm surges by actively intervening to restore the wetlands. We could dedicate resources, including generous summer work programmes for students, to assist nature with the process of natural restoration by, for example, cleaning coastlines of waste and debris, establishing protection perimeters around sensitive areas just beginning to re-establish themselves and direct cultivation of the species that make up these wetlands.

Another example that deserves our attention, though, is the dramatic loss of arable topsoil and the increased food risk that results. Soil erosion and the significant loss of pollinator species, especially bees as a result of pesticide use in agriculture, are among the most important factors cited in scientific studies as the causes of increased food risk in recent years. It is within our power to do much to address both of these contributory causes in our efforts to restore natural capital. Here the focus will be on reinvesting in the soil that produces our food, but many of the practices we will be discussing will also have beneficial consequences for pollinator species.

High-quality topsoil is usually dark brown in colour and should crumble easily in the fingers, not like sand but into small loose clumps tied together by the fine filamentary structure of mycorrhizal fungi. It is usually a mixture of sand, clay, silt deposits and decomposed organic matter called humus. The humus is constituted of dead plants, leaf debris and dead insects. The dark colour of the best soils is an indication of a high concentration of nutrients in the humus; generally, the darker the soil, the higher the concentration of organic nutrients in the soil. The mycorrhizal fungi themselves enter into a symbiotic relationship with the root structure of plants which enhances the ability of these plants to draw nutrients and water from the soil in exchange for rich carbon exudates from the roots of plants, which in turn feed the fungi.

The fungi enhance the density of the nutrients that the plants require by stimulating microbial activity in the soil that draws phosphorus and nitrogen from the soil and fixes these nutrients in a form that is bioavailable to the plants growing in the soil. The mycorrhizal fungi are also essential to soil integrity by virtue of the role they play in *soil aggregation*. Soil aggregation is the process that aggregates, or brings together, the primary particles of the soil (sand, silt and clay referred to earlier) and the organic matter (the humus) into the small and loose lumps that you feel between your fingers when you crumble the soil. The structural integrity of these aggregates is a measure of the stability and resilience of soil to such exogenous impacts as wind and rain. The more stable the aggregates, the more resistant the soil will be to erosion.

Another feature of soil that matters is the abundant diversity of life that includes, in addition to the fungi and bacteria already noted, worms, nematodes and protozoa. Worms aerate the soil and, when plentiful, their extensive channelling loosens the soil and greatly assists with drainage, and soil in this aerated condition can hold more water during periods of little rain. Additionally, the worms feed on the plant debris, playing a prominent role in turning it into the rich organic matter that is humus. And then there are multitudinous nematodes, bacteria and protozoa that live in the soil, cycling nutrients and contributing to the organic richness of the humus. As with the example of the forest in our discussion of the first principle, it is the diversity of life in the soil, in all its intricate interdependent relations, that makes soil fertile. It is life itself that creates the rich soil that in turn sustains life.

Farming methods introduced during the Green Revolution put all this at risk because these methods essentially kill many, if not most, of the living organisms in the soil (and many organisms that live above the soil as well, including pollinator species). The Green Revolution, otherwise known as the Third Agricultural Revolution, promised to feed the world's poor with the expectation of dramatic increases in yields. It brought into play the intensive use of phosphate and nitrogen fertilizers, chemical-based pesticides and herbicides, together with monocrop agriculture practised on a hitherto unknown scale and ultimately ushered in the widespread use of genetically modified organisms. Initially, the promise of increased yields seemed to be true but has since that time stagnated or even declined for many of the most important food crops. The fertilizers have raised the acidity of soils and killed off the beneficial microorganisms and mycorrhizal fungi, and regular ploughing, together with the heavy use of pesticides, has decimated worm and nematode populations. The result is depleted soils that lack the structural integrity to resist erosion.

These depleted soils, lacking the natural aeration of earthworms, are easily compacted by the heavy traffic of farm machinery and dry out in times of little rain, forming a crusty surface. Thus, heavy rains are no longer absorbed into the soil. The rain pools on the surface, ultimately washing topsoil downhill to the lowest elevations and leaving higher elevations with thin soil. And poor soil aggregation from the decimation of the mycorrhizal fungi can leave topsoil loose and low in organic content, a situation made worse wherever ploughing is routinely in use. In this condition, it is easily dried out during times of little rain and is carried away by winds. Moreover, monocrops are more susceptible to being decimated by pests adapted for that crop.

The solution has been to rely on increasingly heavy use of fertilizers to do the job that can no longer be done by depleted soils and increasingly heavy use of pesticides and herbicides for pest and weed control to stay one step ahead of the evolutionary adaptations of these organisms, not unlike the arms race under détente. The result is a vicious cycle that further exacerbates the problem as yields decline. Meanwhile, we are losing arable soil on a scale that boggles the imagination. According to some estimates, we may be losing as much as 36 billion tonnes a year because of erosion, deforestation, land use practices and bad agricultural practices that rely on heavy use of these chemical inputs.[51]

The good news is not only that the restoration of topsoil is within our power but also that it can happen much quicker than we previously suspected. Moreover, like Natural Capitalism itself, it is relatively simple in practice. We already have everything we need, including the knowledge and the tools. It embodies an approach to agriculture that brings together some traditional practices with the best technologies and practices coming out of contemporary research.

I am going to share with you a wonderful and inspiring story from a book, *Growing a Revolution: Bringing Our Soil Back to Life*, by David R. Montgomery.[52] The general principles of soil restoration, which following Montgomery I will variously refer to as Restorative Agriculture or, sometimes, Conservation Agriculture, are (1) minimal disturbance of the soil, (2) growing cover crops and leaving crop residue to ensure the soil is always covered and (3) diverse and complex rotations of these cover crops and the food crops.[53] As we will see, these principles are general directives that require adaptation to local circumstances. As Montgomery puts the point, examples of restorative agriculture can be found everywhere and on every continent: people growing different crops in different soils and different climates – some organic, some not – and some incorporating livestock into their operations, others not. But "they all operated according to a common set of principles."[54]

The first of these principles is minimal disturbance of the soil. In practice, this means take ploughing out of the equation. On the face of it, ploughing the soil has much to recommend it. It supresses the weeds by decimating them before planting; it makes planting much easier and facilitates uniform fields of monocrops which, in turn, make harvesting easier. The downside is it also destroys the filamentary structure of the mycorrhizal fungi. This combined with the use of fertilizers and their impacts on soil acidity noted earlier undermine soil integrity making it increasingly susceptible to erosion. And the fertilizers essentially have the effect of turning the crops into junkies, dependent on these chemical fertilizers for their phosphorus and nitrogen that was previously made available to them from the soil by the symbiotic relations between the mycorrhizal fungi and the bacteria that would fix these nutrients into bioavailable form for uptake by crops. The problem with fertilizers, in this connection, is that when the nitrogen and phosphorus are provided in high concentrations by fertilizers, the crops turn off the exudates that feed the mycorrhizal fungi, essentially starving them, leaving crops totally dependent on the fertilizers, driving the cyclical process of ever-increasing amounts of these products needed to maintain yields. And not surprisingly, the disturbance of the soil from ploughing, together with the higher acidity of soils, also decimates the bacteria, protozoa, worms and nematodes that populate the soil. There is no longer any diversity beneath or above the soil, and this is never good.

Eliminating ploughing is, thus, a huge step in the reclamation of agricultural practices that can restore our soils and crop yields. This is made significantly easier with modern technologies that include crop planters that cut a small slit into the soil (through the cover crop residues), deposit a seed and, if desired, can place a small, targeted amount of fertilizer adjacent to each seed planted. These machines then fold the slot closed behind the planting operation leaving the soil undisturbed to all appearance.[55] This is a perfect example of the use of best existing technologies and practices to enhance productivity. In this way, we can avoid disturbing the soil while simultaneously reducing the inputs of chemical fertilizers. (It is the blanket application of these fertilizers that raises the acidity of soils.) Also, it has the consequence of lower costs for farmers, so it has a direct and beneficial impact on their bottom line. And it is worth mentioning that there is less need for heavy traffic over the crop fields, so less impacting of the soil for this reason as well. We know from experience that adopting the principles of Conservation Agriculture sees a rapid increase in soil microbial activity and soil fauna including earthworm and nematode populations.[56]

Turning to the second principle, the use of cover crops, they are essential to weed control and the restoration of organic matter to the soil. Cover crops control weeds, not by killing them, but by taking up the space that would otherwise be available to weeds, taking away the opportunities for weeds to dominate.[57] They are also relevant to pest control by virtue of the repellent properties of some cover crops. All of this is relative, of course, in the sense that while repellent to one pest, a particular cover crop may attract others. Hence the need for careful planning in a systematic way that embraces whole system design when thinking about which cover crops to plant and how to organize rotations. I will return to this point shortly. Cover crops can also play an important role in making soils resistant to erosion. Soil that is covered and tied together by the root structures of cover crops is much less likely to be eroded even by extreme wind and rain events, which are becoming increasingly common as climate change progresses.

It has also been discovered that cover crops moderate the impacts of climate events in other ways. Soil that has cover crops can be cooler by as much as 10°F relative to uncovered soil during periods of drought and extreme heat. The soil will also retain much more water at these times.[58] One might put it by saying that soil is made more resilient by being kept within a narrower range of temperatures and moisture levels that are more conducive to crop health. And, as if that wasn't enough to make a slam-dunk case, if the cover crops are chosen wisely, they contribute to the organic content of the humus by taking up nitrogen and other nutrients from the soil and concentrating it into their matter. If mulched and left to rot on top of the soil before planting, they contribute to the organic content of the humus, adding carbon and nitrogen as they feed the cycle of nutrients involving the worms, nematodes, bacteria and protozoa in the soil.[59]

And so, coming back to the point about crop sequencing in relation to the third principle of Conservation Agriculture, it is really the implementation of whole system design because, as Montgomery notes, "there is a pattern and a rhythm to crop sequencing."[60] There are many factors that need to be considered, and experience plays a vital role. Farmers must choose cover crops with an eye to the cash crops that will follow. Among the considerations are choosing cover crops for their pest repellent properties and the kinds of nutrients they draw up from the soil. Also, keeping in mind the adaptation of pests, thought will need to be given to how cover crops and cash crops are rotated; we want to avoid a sequence that attracts a particular kind of pest and then persists long enough to provide them with an opportunity to proliferate. This means that crop sequencing will often require complex patterns that will rely as much on experience as scientific knowledge. And,

> [T]here are other considerations too. A deep-rooted crop should follow a shallow one. High-biomass-producing crops should follow low-biomass ones. And a nutrient fixer should follow a nutrient scavenger. In other words, there is a pattern and rhythm to crop sequencing.[61]

Before closing this discussion of reinvesting in natural capital, it's worth taking a moment to consider the dividends for carbon abatement and carbon sequestration that flow from the adaptation of Conservation Agriculture on a global scale. On the abatement side, keep in mind that the chemical inputs, including the fertilizers and the pesticides and herbicides, involve the use of huge amounts of hydrocarbons in their production and transportation. So reducing these inputs to almost zero, which is clearly within the realm of possibility with existing technologies, would have a measurable impact on carbon abatement. Moreover, and insofar as these agricultural practices are less dependent on ploughing and the use of heavy machinery, there would be additional benefits to be realized here, not to mention the lowered costs for farmers. How much are we talking? While comprehensive estimates are hard to find, one study conducted by the US Department of Energy's Argonne National Laboratory estimated that there is the potential for reducing GHG emissions by as much as 70 percent in grain production within the next 15 years.[62]

And then there is the upside for sequestration. The restoration of soils and the increase in the organic content stored there could play a huge role in sequestration. Montgomery reports that, even in its present condition, soil holds more carbon than the combined amount found in the atmosphere and all plant and animal life on Earth.[63] And that, "[b]y the close of the twentieth century, a quarter to a third of all carbon added to the atmosphere since the Industrial Revolution came from plowing."[64] The evidence available provides a range of estimates, suggesting that Conservation Agriculture can sequester anywhere from 0.2 tonne to as much as 1.0 tonnes per hectare per year. On a global scale, this would really matter. And this is taking carbon dioxide out of the atmosphere. This potential for lowering carbon dioxide is an opportunity that can't be ignored. These results also speak to the synergies to be realized from the full implementation of the four principles of Natural Capitalism. Like the natural processes it seeks to emulate, under the model of Natural Capitalism everything is interconnected and the beneficial consequences flow forth in ways that magnify outcomes.

The good news is that implementation of these principles, in a diverse array of conditions and crops, have demonstrated that soil restoration can happen much faster than we have been led to expect. In particular, with our direct intervention, it can happen much quicker than natural processes on their own.[65] The biggest obstacle lies in the fact that the immediate financial interests of the corporations committed to industrialized agriculture are not aligned with our shared interest in maintaining and building the health and fertility of our agricultural soils.[66] In my view this reaffirms the need for a new narrative, a task we take up in Chapter 4. Before that, though, we must convince ourselves that Natural Capitalism is a sustainable alternative to Consumer Capitalism.

Concluding remarks

Now I want to convince the reader that the fullest implementation of the principles of Natural Capitalism can be sustainable; more than sustainable, it could greatly assist with the restoration of natural capital and do so without abandoning Capitalism as such. I will do this by comparing an implementation of the model directly with our definition of sustainability from Chapter 2. The specific implementation I have in mind takes us back to the example of Interface from our discussion of the first principle. We have already seen how, in designing their new factory in China, Interface availed themselves of the opportunity to rethink design from a whole system perspective and, together with embracing the latest and best existing technologies, were able to realize significant savings in terms of inputs (on the order of 90–95%) while simultaneously reducing waste outputs to near zero. I want to pick up the story with subsequent developments at Interface. It is a truly remarkable story of corporate commitment to sustainability.

To fully appreciate the significance of this, let's first contextualize the problem by understanding a few basics about floor coverings in a commercial setting. Consider a law firm occupying, say, four floors in a high-rise building. The floor coverings are looking old and worn, which doesn't take long in the high-traffic areas, perhaps four years or less. Now consider what's required to update these. Proceeding one floor at a time, we must first relocate all the employees, perhaps in the same building if another floor is available for short-term lease or maybe elsewhere if not. These days it may be possible to assign at least some of them to work from home. IT people will be needed to coordinate reconnection of the workers wherever they end up, and document storage facilities may be required to handle sensitive and private legal documents that must remain secure yet accessible. Then a cartage company has to come in and disassemble and remove all the furniture, perhaps to reassemble it in a temporary location, or perhaps to store it until the present location is ready to be reassembled. Then the old flooring is torn up and carted off to landfill. The surface is cleaned and prepared, and finally the new flooring is laid down. The workspace is reassembled, reconnected, the workers moved back in, and it's on to the next floor. The disruption and the lost productivity can't be exaggerated.

I was not able to find concrete figures for the amount of commercial floor-covering going to landfill in an average year, but Lovins et al. report that "Billions of pounds of carpets are removed each year and sent to landfills, where they will last up to 20,000 years."[67] That's 20,000 years! And keep in mind that the high traffic areas that are worn constitute at best some small fraction of the total area. Lovins estimates the worn areas to be not more than 20 percent of the total area. This is waste on a mind-boggling scale.

The Interface sustainability story begins in 1994, when the then-president Ray C. Anderson committed the company to Mission Zero with ambitions to eliminate all waste from the company's operations and thereby better serve its customers while simultaneously enhancing its bottom line. He assembled a Dream Team that included external expertise and this team subsequently developed a road map to achieve zero emissions, zero waste sent to landfills, zero use of fossil fuels and zero GHG emissions.

Mission Zero was accomplished within 25 years, and the company is now launched on an equally ambitious project, Climate Take-Back, which is essentially a multifaceted project in natural restoration. It includes the ReEntry project which has, for more than 20 years, been reclaiming used floor covering and recycling it into serviceable product and the Net-Works project, a cross-sector initiative that is partnered with the Zoological Society of London to buy out-of-service fishing nets from some of the poorest fishing communities to recycle the nylon for use in the production of their Aquafil carpet tiles. As it says on the corporate website the result is,

Fewer ghost nets, less virgin materials and a new source of income for the communities. The partnership has created an inclusive business model with positive outcomes for everyone involved. The programme started in the Philippines and was expanded to Cameroon in 2015.[68]

The journey to Mission Zero, and now beyond, has involved consistent application of entrepreneurial ingenuity that has delivered a succession of innovative products and services that include the first commercial carpet collection to use recycled nylon in the face fibre 'and a 100 percent recycled content vinyl backing layer, glue-free installation systems and the development of carbon-negative backings that actually sequester more carbon than is used in their production. Tracing the major steps of this development in response to the problem noted earlier with commercial settings underscores the point that best practice can improve at every iteration with the proper goals and metrics in place.

The company's response initially was to develop an entirely new floor-covering material and a new business model. The approach was, once again, to think the problem through from the whole systems perspective by first asking the right question – the question that abandons the old perspective to allow new answers. In this case, they asked, "How can we produce a superior product that will best meet our customer's needs while simultaneously eliminating waste at all stages, including the end of the product life-cycle?" The result was Solenium which set a new standard for floor-covering in that it was advertised as 100 percent recyclable.[69] Interface could reclaim used Solenium and, with minimal processing, reconstitute it to produce new Solenium. And on the production side, it used 40 percent less material than other standard commercial carpeting at the time. The product itself was free of toxic materials, including chlorine, extremely stain resistant, resistant to mould and mildew and, in most cases, it could be cleaned with water alone.[70] And it was produced in square tiles that permitted easy replacement in high traffic areas without the disruption otherwise required.

As a compliment to the strategy underlying the development of Solenium, the company developed a new leasing policy, the Evergreen Lease, which is essentially the leasing of floor-covering services along the lines sketched earlier for photocopiers. Consultants from the company would confer with the client on their needs, and for a monthly fee the client would enjoy hassle-free and attractive floor coverings. Interface would regularly inspect, cleaning as required, and replacing worn pieces as needed, and all this for a predictable monthly fee. And at cessation of the contract, Interface was entirely responsible for removing the carpeting, which they would recycle in its entirety into new Solenium. It is worth adding that, insofar as the product was manufactured without toxic inputs, it was essentially free of off-gassing as well; an additional health benefit for employees.

As of 2022, the product line has been further improved, featuring a range of carpet tile, area rugs and luxury vinyl tile made from their new product BioX, which is made from biopolymers and bio-based recycled fibres which, as noted earlier, are *net carbon negative* (they sequester carbon in the biopolymers). The latest iteration of their product, CQuestGB, is made with 97 percent recycled or bio-based content. They also manufacture a range of non-toxic adhesives that are glue-free with no liquids; so VOCs (volatile organic compounds) are eliminated and they are advertised as "no odour," and they feature enhanced resistance to mould and mildew. Their flooring systems are carbon neutral throughout the entire life cycle of the product. You might be surprised to discover the product is also extremely attractive.[71]

Interface has for more than 20 years been in the business of selling floor-covering services while reclaiming its used flooring and reducing significantly the waste otherwise destined for landfill. The loop is closed, and it's a wonderful example of a win–win situation. Here are some highlights from the company's most recent report.[72] Energy efficiency at its carpet-manufacturing sites has improved by 37 percent since 1996. At the manufacturing sites the company owns, 75 percent of the energy used is from renewable sources. At least 48 percent of the material in the flooring products they sell is from recycled or bio-based sources. There has been a 96 percent decrease in the intensity of GHG emissions, a 69 percent reduction in the carbon footprint of their carpeting and a reduction in water intake of 88 percent at its manufacturing sites since 1996.

So here we have a company that has fully integrated the principles of Natural Capitalism into its business model to the benefit of its clients and its own bottom line. From whole system design of its factories and its business model, to the fullest implementation of best existing technologies and practices, it is all designed to realize a closed-loop system that continually improves itself at every iteration. Being bold enough to implement these principles across the organization has paid such huge dividends precisely because they are mutually reinforcing. Consideration of this business model, as a paradigm example of Natural Capitalism in action, provides the opportunity to confirm that Natural Capitalism meets the desiderata we set out in Chapter 2 as defining genuine sustainability. For ease of reference, here is the definition:

We will say of a *system* that it is *sustainable* if:

a) it minimizes the extraction of non-renewable resources and the production of waste as much as possible within existing systems (including zero tolerance for toxic waste);

b) it moves continuously in the direction of further minimizing both the extraction of non-renewables by seeking to replace them with renewable alternatives and the production of waste by repurposing such products in useful ways as our systems evolve;

c) it draws down renewable resources at a rate below their replenishment by natural processes;

d) it produces non-toxic waste at a rate below the level at which it can be absorbed into the environment, and, with respect to such toxic waste as may be presently unavoidable, we seek the means to detoxify it before releasing it into the environment.

In less than 30 years, Interface has eliminated all toxic waste and has reduced wastes that accompany production to the point where they can claim to be carbon neutral throughout the life of their products. And on resource extraction, keep in mind that their main product, CQuestGB, is 97 percent constituted of recycled fibres and/or bio-based content. There is virtually no extraction of virgin resources of any kind for the manufacture of their products. The story of the company over the last three decades, subtending the period that includes the earliest developments of Solenium and their Evergreen Lease to the present era of CQuestGB, is the story of a continual progression in the direction of minimizing the extraction and use of non-renewables, to the point of effectively eliminating their use in favour of the renewable and non-toxic biopolymers.

We turn now to the task of articulating the narrative that nurtures and supports Natural Capitalism.

Notes ·

1 Hawken et al. (1999), Amory B. Lovins et al. (2007) and Hawken (1997). This latter paper is available at: https://www.motherjones.com/politics/1997/03/natural-capitalism/ (retrieved October 16, 2022).
2 Rifkin (1980).
3 Hawken et al. (1999, p.ix).
4 According to estimates, conventional gas and oil operate at a combined energy-returned-on-investment ratio of about 18:1. As a comparison, shale gas performs at about 6.5:1 to 7.6:1, and tar sands oil is in the range of 2.9:1 to 5.1. Source: https://insideclimatenews. org/news/19022013/oil-sands-mining-tar-sands-alberta-canada-energy-return-on-in vestment-eroi-natural-gas-in-situ-dilbit-bitumen/ (retrieved October 16, 2022).
5 Standard practice is to distinguish between the First Industrial Revolution (dated as the period 1760–1840) and built on coal as an energy source and steam power as the driver of industry and the Second Industrial Revolution (dated as the period 1870–1920) and built on oil as the principal source of power and electrification and internal combustion as the driver of industry.
6 A fascinating history of labour productivity and technology is Frey (2019).
7 For example, in the United States there was the Servicemen's Readjustment Act of 1944 (known colloquially as the GI Bill) and the huge build-up of universities, community colleges and training centres, together with strategies to maximize accessibility in the form of student loans and grants.
8 Rifkin (2011, pp.260–263).
9 Ibid.
10 Hawken et al. (1999, p.8); italics original.
11 Strictly speaking, this is true only of the models, like the SX70, released in the 1970s. Earlier models required one to wait for a minute or so and then peel the backing away to reveal the developed picture.
12 Op. cit., p.117.
13 Here is the corporate website: https://www.interface.com/US/en-US/homepage (retrieved October 16, 2022).
14 Lovins et al. (2007, p.176).
15 Recent developments in photovoltaic technology include the development of building cladding, roof coverings that resemble shingles, and even window glazing. And wind power generation is now possible on a scale that accommodate neighbourhoods or even single homes. See Rifkin (2011, pp.98–99).
16 Hawken et al. (1999, p.176).
17 Maser et al., (date unknown) from a publication available at the US Forest Service website. And also from the US Forest Service website, "The Unseen World of the Fallen Tree" (authors not credited, date unknown). Both of these documents are viewable at these links: https://www.fs.fed.us/pnw/pubs/164part2.pdf and, https://www.fs.fed.us/ pnw/pubs/229chpt2.pdf (retrieved October 16, 2022).
18 Maser et al., p.25.
19 Op. cit., p.42.
20 Uncredited article, p.19.
21 Op. cit., p.18.
22 Of course, it is not really negative entropy – it merely appears to be so because we isolate the forest system for analysis. In the context of the larger system that includes the sun, which provides energy inputs, entropy is increasing.
23 We characterize the trophic level of an organism as the position it occupies in a food web, and the food web is the succession of organisms that eat other organisms and may, in turn, be eaten themselves. Thus, the trophic level of an organism is given by the number of steps it is from the lowest level and counting the lowest as level one. As mentioned in the previous chapter, some of the earliest work in this area is by Aldo Leopold (1966).
24 Benyus (1997, pp.251–253).

25 I first read it about it in Benyus. See it here: http://www.symbiosis.dk/en/ (retrieved October 16, 2022).

26 I willingly paid a premium for abalone dot inlays on the fretboards of two of my guitars.

27 https://www.intechopen.com/online-first/76122 (retrieved October 16, 2022).

28 Chapman et al. (2018).

29 I have heard the first use of the term attributed to John Kenneth Galbraith, though I have not been able to verify this.

30 This is a good place to start: https://www.tandfonline.com/doi/abs/10.1080/00913367.2016.1185981 (retrieved October 17, 2022).

31 Think of saturation advertising in connection with this point: running the same ad two, or even three times, in a single ad break.

32 https://www.youtube.com/watch?v=2zfqw8nhUwA (retrieved October 17, 2022).

33 Some data on statistics: https://www.statista.com/statistics/214641/household-adoption-rate-of-computer-in-the-us-since-1997/ (retrieved October 17, 2022).

34 https://www.theguardian.com/technology/2016/sep/20/hp-inkjet-printers-unofficial-cartridges-software-update (retrieved October 17, 2022).

35 https://www.wired.com/2010/11/1103first-us-car-show-new-york/ (retrieved October 17, 2022).

36 https://www.cnbc.com/2019/05/17/smartphone-users-are-waiting-longer-before-upgrading-heres-why.html (retrieved October 17, 2022). Surprisingly, the title of this article is "Smartphone Users Are Waiting Longer Before Upgrading."

37 https://www.cbc.ca/news/science/global-ewaste-monitor-2020-1.5634759 (retrieved October 17, 2022).

38 Hawken et al. (1999, p.127).

39 Op. cit., p.134.

40 This is not anything like the model of auto leasing as it is normally implemented, which is typically not intended to align the interests of consumers and dealers.

41 I am deliberately avoiding the issue of privacy here. As noted later in relation to Uber and Airbnb, it will take years for public policy, legislation and the courts to establish clear ground rules around issues arising out of these new technologies. In the meantime, a client can always decline this part of the service package.

42 https://www.zipcar.com/carsharing (retrieved October 17, 2022).

43 Rosenblat (2014).

44 https://communauto.com/?lang=en (retrieved October 17, 2022).

45 Especially in Europe: https://ec.europa.eu/regional_policy/en/newsroom/news/2016/01/28-01-2016-real-economy-cradle-to-cradle-powering-europe-s-circular-economy (retrieved October 17, 2022).

46 Benyus (1997, p.278). See also Hawken et al. (1999), Chapter 8, in particular, pp.159–169, and Rand (2020, pp.86–88).

47 The use of the phrase "bounce back" should not be taken to imply that ecosystems will be restored to their condition before the introduction of pollutants. In many cases, this will be unlikely. Indeed, some of the consequences of climate change will be permanent, for example, species loss. And insofar as species loss will have impacts on the food chains in the ecosystem, other consequences will flow from this.

48 https://www.nationalgeographic.com/environment/article/ozone-depletion (retrieved October 17, 2022).

49 From the National Center for Biological Information (NCBI – a US government website): https://www.ncbi.nlm.nih.gov/pmc/articles/PMC7323667/ (retrieved October 17, 2022).

50 Also from NCBI: https://www.ncbi.nlm.nih.gov/pmc/articles/PMC1332684/ (retrieved October 17, 2022).

51 https://www.sciencedaily.com/releases/2017/12/171215121055.htm (retrieved October 17, 2022).

52 Montgomery is Professor of Geomorphology at the University of Washington. The book was published in 2017 by W.W. Norton & Company.

53 Montgomery, p.68.

54 Op. cit., p.29. Montgomery's book and his conclusions are the fruit of extended research into restorative agricultural practices, done in part by visiting farms all over the world.

55 Op. cit., p.110.

56 Op. cit., pp.68–69.

57 This is similar to the principle that applies to supplementation with probiotics to restore gut health. The idea is that supplementing with the desired bacterial colonies takes up the space that might otherwise be dominated by the harmful bacterial colonies.

58 Op. cit., p.124.

59 See op. cit., Chapter 6 and, in particular, pp.98–101.

60 Op. cit., p.126.

61 Ibid.

62 https://www.anl.gov/article/changes-in-farming-practices-could-reduce-greenhouse-gas-emissions-by-70-by-2036 (retrieved October 17, 2022).

63 This estimate is based on calculations to a depth of ten feet, so presumably includes more than just topsoil (p.224).

64 Ibid.

65 Op. cit., p.26.

66 Op. cit., p.40.

67 Lovins et al. (2007, p.180). The context seems to indicate these figures are for the United States alone!

68 See an informative video here: https://www.youtube.com/watch?v=DX6Uidpg3VM (retrieved October 17, 2022).

69 The story of Solenium and the Evergreen Lease is featured in Lovins et al. (2007).

70 Op. cit., p.179.

71 You can see the line of products here: https://www.interface.com/CA/en-CA/sustainability/carbon-neutral-floors.html (retrieved October 17, 2022). And while you're there you read their mission statement on sustainability.

72 http://interfaceinc.scene7.com/is/content/InterfaceInc/Interface/Americas/Website ContentAssets/Documents/Sustainability%2025yr%20Report/25yr%20Report%20 Booklet%20Interface_MissionZeroCel.pdf (retrieved October 17, 2022).

4

THE NARRATIVE OF NATURAL CAPITALISM

In this chapter my goal is to articulate the narrative underlying the principles of Natural Capitalism, and it is time to bring Jeremy Rifkin into the discussion. I will be relying principally on his book, *The Third Industrial Revolution*, and the slightly updated story in *The Green New Deal*.[1] Rifkin focuses on the infrastructure that will be required to support a sustainable economy built on the economic principles of Natural Capitalism. An important point about infrastructure in the sense intended by Rifkin in this context is that it is not about bricks and mortar.

> Rather, infrastructure is an organic relationship between communication technologies and energy sources [and transportation technologies, as per Rifkin (2019) – WIH] that, together create a living economy. . . . Infrastructure is akin to a living system that brings increasing numbers of people together in more complex economic and social arrangements.[2]

This point goes to the heart of a theory of the historical evolution of industrialization in which it is the communication system in tandem with the energy and transportation systems that determine the character of an industrialized society by virtue of the way these systems structure economic and social arrangements. More specifically, the nature of these systems determines the character of the political and economic institutions of the societies built on these technologies. Understanding these relationships and the nature of these institutions in turn points to the underlying narrative that rationalizes and supports them.

In the First Industrial Revolution, circa the late 18th century and until about 1830, the communication system was telegraphy and the energy system was based on coal. Coal power drove the development of steam technology to power transportation and industrialization. The development of the infrastructure for all of this

DOI: 10.4324/9781003388555-5

technology is capital-intensive. Coal is expensive to extract and process and is often located in places distant from the industrial processes it ultimately powers. Extracting profits requires operating at huge scales, including the capacity to move the coal efficiently to processing sites and, ultimately, to markets. Now insofar as coal powers steam-driven transportation and industrialization, an extensive infrastructure of railways and ports is needed. All of this technology favours highly centralized and hierarchically organized business and economic institutions to maximize and support this economic activity. In this way, there emerges a system of dominant capitalist players. Governments played a crucial role in facilitating this by building out the infrastructure of rail systems and the telegraphy systems that ran parallel to the rail lines and ports for steam-powered shipping. Again, the system favours not only a tight integration between dominant capital interests but also a highly centralized and vertically integrated public sector.

In the Second Industrial Revolution, from around 1870 to roughly 1914, telegraphy is replaced by telephone and radio (and later supplemented by TV) as the backbone of communications, and oil and gas become the dominant players in the energy system. This energy system drives the development of the internal combustion engine as the principal motive force of transportation and industry. Later comes the electrification of industry. Again, all of this is highly capital-intensive, and there is further intensification of the authoritarian, centralized and vertically integrated economic and political institutions. For example, it requires an intensive investment of public funds to build the infrastructure required for telephony. And in the private sector, it takes an enormous amount of capital to build a network of television stations of sufficient scale to make it profitable. Thus, the emergence of a handful of dominant networks in the middle decades of the 20th century was the predictable result.

To avoid confusion, there is an existing terminology that refers to the Third Industrial Revolution (hereafter: TIR) as an event now concluded and claims we are presently in transition to, or perhaps already up to our necks in, a Fourth Industrial Revolution. This usage refers to the TIR as the period from the middle of the 20th century to roughly the end of that century. This was the revolution in digital technology ushered in with the invention in the 1940s of the modern computer architecture by Alan Turing and John von Neumann, and subsequently the invention of the transistor and the miniaturization of electronic technology this made possible. According to this version of the theory, the Fourth Industrial Revolution begins roughly with the new millennium and the development of such technologies as modern AI, advanced robotics, quantum computing, the internet of things, genetic engineering and the sustainable energy technologies now gaining widespread intrusion into the markets. It is said to be characterized by a blurring of the boundaries between physical, digital and biological domains. The claim is made that, "There are three reasons why today's transformations represent not merely a prolongation of the Third Industrial Revolution but rather the arrival of a Fourth and distinct one: velocity, scope, and systems impact."[3]

I think Rifkin would respond that what they are calling the TIR was not a revolution in the sense used by him because it did not yet have the power to alter the institutional organization of our lives and relationships in ways characteristic of a genuine revolution. However impactful this period may have seemed, and not denying the pace of change that characterized it, it lacked the scope and systems impact of a genuine revolution precisely because it was not yet wedded to the distributed and sustainable energy systems and transportation/energy storage systems required to unlock the full potential of these digital technologies. It was, if I may be permitted, the anticipation of a revolution. It was a revolution of our communication technologies, but it did not unseat the consolidation of large-scale and vertically integrated business practice. If anything, it further intensified the accumulation of capital and the authoritarian, centralized and vertically integrated nature of economic and political institutions. Even a cursory list of the new behemoths of the digital age, such as Microsoft, Apple, Google and Amazon underscores this.[4]

Thus, it was the anticipation of a revolution in the sense that it is only when digital communication technologies are wedded to distributed renewable energy technologies and the transportation systems they facilitate that there will emerge the living economy that will alter our relations to one another and change completely the nature of our economic and social institutions. And the scope and magnitude of these changes underscore the need for a new narrative, a new way of understanding our relationships to one another and to the natural world.

Infrastructural requirements of Natural Capitalism and the TIR

And so, the infrastructure (in the sense intended by Rifkin) that will be required to support a sustainable economy and which aligns with the principles of Natural Capitalism would have consequences for our social and economic institutions that would determine anew the organization of our lives. The changes introduced by the deployment at scale of these infrastructural elements would bring about such change in these institutions as to constitute a thorough revolution. And from these infrastructural and institutional requirements we can extract the narrative that will in turn rationalize and support this society.

We turn to Rifkin's *five pillars*, which constitute the foundational requirements of this infrastructure. First is a shift to renewable energy. The energy systems that drive the TIR will be principally solar and wind with geothermal and tidal playing important supporting roles together with many of the hydroelectric facilities already in place. Many of these latter will be dismantled in the fullness of time and the spaces presently consumed by them will be left to re-wild, in a large-scale example of the restoration of natural capital. This dismantling of hydroelectric facilities is unlikely to begin much before the middle of the century, though, as it will require the wide-scale implementation of the other renewables and the energy-saving technologies that reduce demand as noted in Chapter 3.

Second is to transform the building stock on every continent into interconnected communities of micro power plants with most buildings, from homes to factories and everything in between, generating renewable energies on site. A notable feature of these renewable energies is the rapid evolution of the technologies available. For example, it is now possible to acquire solar photovoltaic cladding, roofing and glazing that can very nearly turn the entire exterior of a building into a solar collector and power generator.[5] In the realm of wind power, there are domestic-scale wind turbines that can be used to power smaller buildings, and geothermal can be deployed in many situations. Together with technologies that significantly reduce energy consumption by, for example, using the best insulation, triple-glazed windows, air exchange systems, superior appliances such as induction stoves and more, individual homes and commercial buildings can be generators of excess energy as measured over, say, an average year.[6]

Detractors will be quick to point out, though, that there will be many times in an average year when individual buildings will not be able to produce enough energy to meet their needs. This objection typically underestimates the reductions in consumption that are possible with best existing technologies (as emphasized in Chapter 3), but the objection deserves to be taken seriously. And it is addressed directly by the third and fourth pillars of the TIR. Third, then, is the deployment of hydrogen fuel cell and other storage technologies in every building where feasible and throughout the infrastructure to store intermittent energies. As part of the refurbishment of the building stock, every building will have its own energy storage system. In the long term, hydrogen fuel cells represent the most efficient means of accomplishing this, but the deployment of this technology at scale will probably take a bit longer (this too is unlikely to happen before mid-century, I suspect). In all likelihood, hydrogen fuel cells will initially find their application in commercial settings before finding wide deployment in domestic settings. In the meantime, in many cases, it will be the family electric vehicle that functions in a dual role as transportation and energy storage (more on this later with Pillar 5). Other innovative storage technologies are being researched and developed continuously for application under specific circumstances. For example, one such technology utilizes underwater balloons.[7]

A full response to the problem of renewable energy being intermittent also drags the fourth pillar into view. This is the deployment of internet technology to transform the power grid of every continent into an energy-sharing intergrid that behaves much like the internet. On this model of localized and small-scale energy production, the widespread but intermittent availability of renewable energies requires that we move to a decentralized energy distribution system. This decentralized system will perforce connect individual buildings, now the primary generators of electricity, to the intergrid via smart meters which manage the distribution of energy within an area. Call these areas *nodes*. Within these nodes, loads will be managed by means of interactive smart metering systems that draw energy from those producing beyond their immediate requirements to sell it to those unable to meet their immediate needs. And all of this happens seamlessly under the control of AI systems that learn and improve their performance over time.

The fifth and final pillar is the conversion of the entire road fleet to electricity. From personal vehicles to transport trucks, they will all be converted to electric plug-in and fuel cell vehicles that can also be employed to buy and sell electricity on the smart and interactive power grid. As noted earlier, the electrified fleet will, in all likelihood, serve the dual role of transportation and energy storage until hydrogen fuel cells can be deployed at scale.[8] Therein lies a point about the synergistic multipliers that can be expected from full implementation of the infrastructure of TIR: the transport fleet, when fully integrated into the architecture of the TIR, serves these multiple roles, transportation and energy storage.[9]

Pursuing this point about synergies a little further, let us experiment in visioning a possible future. One scenario has the fleet replaced, initially, with electric vehicles on the present model of personal ownership of a vehicle serving the dual purpose, noted previously, of family vehicle and energy storage system. This process appears to be gaining momentum in 2022 as electric vehicles gain market share. It is no longer merely the fringe market of early adopters. As 5G develops to maturity, though, the fleet will undergo further changes and autonomous driverless vehicles, now under intense research and development, will likely drive the next wave of innovation. These driverless vehicles will be much smaller and less materials-intensive, constructed as a tough exoskeleton of carbon fibre. Simultaneously with this is the deployment at scale of hydrogen fuel cell technology for energy storage in domestic applications. The need for the family vehicle to function as energy storage evaporates. This fact, coupled with the changes to neighbourhoods that have people living near where they work, play and go to school, discourage private ownership of an investment that spends the bulk of its time (roughly 90%) in the garage not being used. Instead, we will rely on driverless transportation services that will be available on demand using an app on our phone.

Fewer cars on the road, and much smaller and safer in design, will ease congestion and make our roads much safer. And changes to market conditions will, in turn, pull the large auto manufacturers into the new business model of selling transportation services instead of vehicles to end users. When I say "pull" in this context, I mean that auto manufacturers will perforce reorganize themselves in these ways to meet the changing conditions of market demand. The demand for transportation services will continue to grow even as the demand for private ownership of the vehicles declines. Under these circumstances, profits are to be realized from managing the flow of transportation services. The auto manufacturers will own the vehicles and manage the provision of transportation services for a broad range of demands that include transportation of goods to markets. In some cases, they will be leasing fleets of vehicles to third-party auto services like Zip Car, Communauto or trucking companies. In other locales, they may operate transportation services directly. But as the owners of the vehicles, they will be responsible for them for the entirety of the vehicle life cycle. They will be built to last and the carbon fibre exoskeleton bodies and other components will be easily recyclable for further use. The reduction in materials consumption is breathtaking to contemplate.

What we see here is the sort of iterative change moving us continually closer to our goal of a circular and steady-state economy. The synergistic multipliers that flow from the full implementation of all the principles of Natural Capitalism and the infrastructure of TIR mean an economy that is continuously more energy efficient and less materials-intensive. Also, we foresee a declining value in personal ownership in favour of access to services.[10]

We have moreover affirmed the sense in which the internet technology that will be the backbone of the TIR communications systems was ahead of the curve. Arriving as it did ahead of renewable energy systems, it hinted at possibilities but has yet to achieve its potential because to this point it has been shackled to an outdated and highly centralized energy system. The real benefits of these synergistic multipliers will be fully realized only when there is widespread implementation of all five pillars of the TIR, which in turn facilitates an economy arranged on the principles of Natural Capitalism.

The character of the post-TIR world

In describing the infrastructure of TIR I have slipped, insensibly, into the simple future tense, speaking repeatedly of what "will" happen. This conveys the impression that implementation of TIR technology and, by extension, of Natural Capitalism, is inevitable. As we shall see, it is not inevitable, but for now let me emphasize how disruptive the changes introduced would be. The full implementation of these systems would, in fact, constitute a complete reimagining of the energy and transportation systems from the ground up that would drive revolutionary changes in our social and economic institutions.

Formerly, energy providers, in the form of large, hierarchically organized utility companies would produce energy at remote locations relying on large-scale, capital-intensive power-generation facilities. These include hydroelectric dams, coal-fired power plants, natural gas plants and nuclear units. Though the differences between these methods of generating electricity are significant, none of them are terribly efficient. By some estimates, as much as 65 percent of the energy is lost as heat to the environment. The energy that is converted to electricity is then conveyed to customers, often over enormous distances, relying on massive transmission architecture in a one-directional sale of power. There is significant power lost in the transmission as well, and again this loss is given off as heat to the local environment.[11] At times, the system overproduces and the surplus is wasted; at other times, it fails to meet demand and we have brownouts. The infrastructure is expensive to install, costly to maintain and is increasingly at risk as climate change drives extreme weather events.[12] And because of the capital-intensive nature of the infrastructure required, the power utilities themselves often have a *de facto* monopoly. None of these are good for consumers or for the environment.

In contradistinction, the widespread implementation of TIR infrastructure pulls the conversion of energy utilities from their traditional role of energy production and distribution to energy management services in much the way automakers will be pulled to deliver transportation services by the scenario considered earlier. So at the heart of these decentralized energy systems will be the energy service companies (ESCOs). This transformation of power utilities to ESCOs is the realization of the new business model at the heart of Natural Capitalism: the sale of services as opposed to the sale of goods to end users, with all the attendant reductions in waste that result. ESCOs install the interactive smart metering hardware and the AI systems that manage the bidirectional flow of electricity between clients in a service area. These service areas are the ones referred to earlier as nodes. Relying on these interactive technologies ESCOs can manage demand curves within nodes to maximize energy efficiency, dramatically reducing waste. For example, with appliances connected to one another and to the smart meter (the internet of things) the energy system could turn on virtually silent washers and driers at night while the occupants sleep, manage the lighting systems to maximize efficiency and so on. These utility companies are reconfigured as the service providers that use their expertise to manage the energy demands of their clientele in what is essentially an information-energy network.[13]

By virtue of the business model, these ESCOs are incentivized to maximize energy efficiency insofar as their profits are now derived from energy saved, not from energy sold. One of the services they offer, for example, is consultation with companies to help them reduce their consumption by offering customized solutions that effectively turn their client's savings into profits for themselves. The greater the energy savings for their clients, the higher the returns to the ESCOs, underscoring how this move to the new business model at the heart of Natural Capitalism aligns the interests of businesses with the clients they serve. And because the intergrid is networked between individual buildings as the primary generators of energy, every household and commercial building is potentially an energy entrepreneur, a perfect example of how the energy infrastructure determines anew the character of economic relations and institutions.

An additional benefit of such smart systems is enhanced security. At least two distinct types of threat come to mind. First, there is a very real and increasingly perilous risk in relation to large and centralized energy systems simply by virtue of their scale and complexity. And, as noted earlier, this risk is greatly increased by the possibility of extreme weather events that exceed the tolerances of the transmission infrastructure. In 2003, over 50 million people were left without power, in some cases for several days, in Ontario and several states in North-eastern and Midwestern United States. Initially the result of tree branches interfering with transmission lines, a software bug caused the problem to cascade through the system.[14] The derecho storm that struck Ontario and Quebec in 2022 (as noted in Note 12) left more than a million people without power, in some cases for several weeks.[15] But when the bulk of our power is generated by individual buildings using onsite renewable technologies, damage to a single building, or even several, does not cascade through the system.

Second, existing systems are subject to cyberattack and because of the disruptive potential of such attacks, they are attractive as targets. Let us reflect for a moment, though, on the areas serviced by ESCOs and arranged into nodes. This is a kind of lateral organization such that the energy-producing buildings that constitute a node are blockchained together in service areas that stretch across contiguous land masses. Here, "blockchained" means that the system is organized and secured by a digitally distributed and decentralized public ledger. The data structures collected by the smart metering system and managed by the AI systems are stored as blocks within each node and each node will have a replica, continuously updated, of the entire database. It is the redundancy of data storage in encrypted blocks that guarantees the resilience of the system as a whole. If someone hacks into a node and tries to edit the data block, it will no longer match the data blocks at other nodes and the edit will be rejected. In extreme cases of cyberattack, a node can be isolated from its fellows, keeping the problem contained until it is dealt with. And the size of the node to be isolated can be determined dynamically on the basis of the situation. There is nothing fixed about the nodes. Indeed, a single building can function as a node; they exist as an overlapping and contiguous network of nodes continuously reorganizing themselves as needed.

This blockchaining of nodes across continental land masses is what Rifkin calls Continentalism. As opposed to the model of globalization characteristic of the First and Second Industrial Revolutions, premised as it is on the assumptions of Classical Liberalism and the primacy of the nation state, the TIR leads naturally to the continentalization of energy networks and economic activity. The infrastructure of the TIR is collaborative across the network of nodes that reach to the edges of continental landmasses. When individual entrepreneurs (whether family homes, small businesses or larger corporate entities) generate their own energy from renewable sources on site and share this energy in a continuous and multidirectional flow to manage distributions collaboratively, there is no incentive to accept limits on such cooperative activity that serve either the political interests of nation states or the economic interests of multinational corporations. Essentially, each individual generator of power is now a node within a smart intergrid of nodes and borders as political boundaries become increasingly irrelevant, at least in relation to the generation and distribution of energy. As this progresses, it sets in place conditions that favour continental federations, political and economic unions of relatively autonomous states or provinces. An obvious model would be the EU.[16]

Implicit in this is a radical democratization of energy production and availability of high-quality energy for everyone everywhere, which in turn results in what Rifkin calls Distributed Capitalism. The distributed nature of renewable energies ineluctably brings forth a networked and collaborative energy market that is incompatible with centralized and hierarchical control mechanisms. Economic activity is localized around sites of energy production and collaboration based on regional conditions, including working knowledge of these local conditions and best practices. Knowledge is framed as a social construct and a shared responsibility, on the model of Wikipedia and shareware. And this distributed and collaborative market system in its turn brings a more equitably distributed sharing of the wealth that results from the economic activity of such markets.

> The collaborative nature of the new economy is fundamentally at odds with classical economic theory, which puts great store on the assumption that individual self-interest in the marketplace is the only effective way to drive economic growth. The Third Industrial model also eschews the kind of centralized command and control associated with traditional Soviet-style central socialist economies. The new model favors lateral ventures, both in social commons and in the market place, on the assumption that mutual interest, pursued jointly, is the best route to sustainable economic development. The new era represents a democratization of entrepreneurship – everyone becomes a producer of their own energy – but also requires a collaborative approach to sharing energy across neighborhoods, regions, and whole continents.[17]

Summarizing, TIR infrastructure scales laterally and is decentralized rather than intensively centralized. This lateral organization is constituted as a distributed network of overlapping nodes that stretch across continental landmasses such that energy management and economic activity are localized and collaborative. Moreover, the generation and distribution of renewable power is democratic by its very nature. Thus, in the fullness of time, as the infrastructure reaches wide-scale implementation, we expect global governance to be increasingly laterally distributed (perhaps based on local councils) collaborative and democratic.[18] And there is a declining value placed on private ownership as access to services comes to dominate.

The narrative of sustainability

The narrative underlying Classical Liberalism and the economic and social institutions of Consumer Capitalism is concerned with rationalizing the centralization of power and economic activity to support economies of scale and the maximization of labour productivity and extraction of profits. These power structures are hierarchically organized and authoritarian, with an emphasis on message control (i.e., they are non-transparent). There is a reflexive commitment to continual growth, and nature is regarded as something that exists externally to the domain of humans. It is something "other" that is to be brought under control for our sustenance, our comfort and our profit. It is a set of resources to be extracted and turned into the valuable products that enhance our lives and also serves conveniently as a sink for our wastes.[19]

By contrast, the supporting narrative of Natural Capitalism is, in the first instance, premised on valuing nature intrinsically as opposed to regarding it as of merely instrumental value. This view has emerged, I think, from the evolution of systems-oriented thinking that one finds in rudimentary form in Aldo Leopold's *Land Ethic* (mentioned in Note 23 in Chapter 3) and also evident in, for example, *Silent Spring* by Rachel Carson. The view is fully explicit and developed into an autonomous branch of interdisciplinary research by the time of the publication of *The Limits to Growth* in 1972.[20] Systems analysis is grounded in the recognition that systems are to be found everywhere, some entirely artifactual, others natural, and some existing as integrations of artifactual and natural components (anyone with implants or prosthetics is an example of such an integrated system). Under analysis, systems are often revealed to consist of interconnected sets of systems within systems that are defined and distinguished from one another by their structure and functional role. They are subtly responsive to context and typically interact synergistically in ways that result in emergent behaviours, that is, behaviours not expected from any prior knowledge of the causal properties of the constituents of the systems.[21]

As explored briefly in Chapters 2 and 3, understanding this interconnectedness and interdependence of the systems that constitute the natural world is essential to any adequate understanding of the capacity of nature to provide the essential services of life. We know that because of the sensitivity of natural systems to context and their interactions with one another, this capacity is fragile and depends essentially for its healthful functioning on the integrity of natural systems. By "integrity" in this context, we mean, at the very least, that the systems are left to function within the narrow range of parameters to which they are adapted and which are suited to the chemical and energy flows through them, and the diversity of the systems is left largely untouched. We see the effects of violating these constraints when we reflect, for example, on the consequences of rising carbon dioxide levels on ocean systems.[22] When absorbed into the oceans, carbon dioxide raises the acidity of sea water to levels that threaten the shell-forming capacities of molluscs and other shell-forming species and, together with rising water temperatures, results in the bleaching and mass die-off of coral reefs. This in turn threatens the life forms that depend for their lives on the coral reefs, and the mass die-offs ramify through ocean systems reducing biodiversity which in turn spins off consequences of its own.

Thus, the lesson of systems theory for present purposes is that when we interact with natural systems, we introduce changes and because of the synergistic interactions among systems, these changes can result in unforeseen consequences that can quickly amplify by means of multiple feedback loops characteristic of systems. In particular, there will be many cases where these consequences impact negatively on the functional capacity of natural systems to provide the services essential to all life. From these insights, we can reason that nature is to be valued intrinsically.

I must make the reasoning for this conclusion explicit, though, at least in outline, because it is possible for one to insist, on the basis of everything said so far, that our reason for protecting the integrity of natural systems remains merely instrumental. It's not that nature is to be valued intrinsically, but rather we recognize that if we mess it up, it will threaten our well-being. While this might seem like a natural conclusion, it is, I think, mistaken and is likely to lead to ineffectual interventions to address issues of environmental degradation and species loss. Rather like the situation discussed earlier in relation to the Brundtland Report and its definition of sustainability (Chapter 2), this instrumental reasoning inclines us to focus our energies on rationalizing our present course and attempting to resort to palliative technological solutions that will be woefully inadequate to deal with the magnitude of the problems we confront. Indeed, it is likely to worsen the situation. Moreover, as the argument is intended to show, to proceed on instrumental premises is to do a fundamental moral wrong.[23]

Valuing nature intrinsically is adopting a moral attitude with respect to our relationship to nature and the biotic communities that make up the natural world. It is, at bottom, the recognition that we have a moral obligation to these biotic communities to extend the boundaries of ethical consideration to include them and their welfare and that we owe them this consideration because their well-being is to be valued for its own sake and not merely because it might serve our instrumental purposes to do so. For this to happen, we must first acknowledge that our actions have consequences for these biotic communities and, thus, what befalls them is our responsibility. We can decide to act in ways that will enhance their well-being, or we can continue to disregard them and act in ways that harm them, but in either case we will be responsible for the outcomes.

Consider that we are presently in the midst of what scientists are calling the sixth mass extinction event.[24] An instrumentalist might suggest that the extinction of the dinosaurs resulted in the rise of mammals and, ultimately, the arrival of human beings on the scene. Looked at instrumentally, there is no reason for regret. The fact that we are now dominant is bad for the dinosaurs but good for us, and from the perspective of the universe as a whole, neither good nor bad. By parity of reasoning, the instrumentalist approach is good for us pragmatically and irrelevant from any larger perspective. The difference this time, though, is that this extinction is the direct result of human activity and the choices we make and, thus, it is avoidable. We might make different decisions and avoid a lot of unnecessary pain and death. Viewed from the perspective of the universe as a whole, it may not matter, but it should matter to us because we are responsible. That's what it means to be moral, after all. And this would remain the case even if we could avoid the worst consequences ourselves.

So, when we adopt the moral attitude to regard nature as intrinsically worthy, we are deciding to bring nature into the domain of ethical consideration, the "Charmed Circle" as it were. We acknowledge our responsibility for the impacts of our actions and decide to take this into ethical consideration when deciding what we shall do. These impacts are no longer considered irrelevant or of merely instrumental concern.

Alright, but what justifies adopting this moral attitude? The argument is a simple one consisting of four premises.[25] First is the premise that humans must recognize themselves to be members of the biotic community of life on the same terms as other non-human members. Although Taylor's wording is somewhat ambiguous on the scope of this premise, I am convinced it must be intended to apply to us as a species and as individuals. We are just another species, naturally evolved like any other. Evolution has not "led" us to us. Indeed, it is not leading anywhere. We are not the end point of the evolutionary process; it has no end point. And as individuals we must recognize that we are members of the larger biotic community on the same terms with the other individual organisms with whom we share the biosphere, all of us struggling to survive. I take up this point again after introducing the rest of the argument.

Second, as already noted in the discussion of systems theory, the earth's biotic systems are acknowledged to exist as a complex web of interconnected and interdependent systems whose functional integrity as a whole depends crucially on the sound functioning of all the constituent systems.

Third, the individual organisms, whether plant, animal, bacterium or protozoan are recognized as teleological centres of life. Essentially, "teleological" means goal-directed, so these organisms are recognized to be goal-directed individuals striving continuously in all their activity to enhance their well-being and avoid harm. It is important to understand that this striving need not involve conscious goal setting to count as goal-directed behaviour. In this sense, they pursue what is for them their "good," that which enhances their life, and in this pursuit of their good, they can be harmed or benefitted.

Fourth, and last, humans enjoy no special moral status that justifies excluding the other members of the biotic community from ethical consideration.[26] On this last premise, Taylor expresses it as follows,

> Whether we are concerned with standards of merit or with the concept of inherent worth, the claim that humans by their very nature are superior to other species is a groundless claim and, in light of elements (1), (2), (3) above, [i.e., our first, second and third premises – WIH] must be rejected as nothing more than an irrational bias in our own behaviour.[27]

Because this premise is pivotal to the argument as a whole, he goes on to devote most of the remainder of the paper to the defence of this claim. I will not do so here, but I encourage readers to acquaint themselves with Taylor's paper and his argument. I will summarize the point by saying that if we continue to regard ourselves as having some special and unique moral status, then we will persist in viewing nature through the instrumental lens of Classical Liberalism, losing sight of the central fact that we too are a part of the natural world, human societies existing as systems within the larger whole and on an equal footing with other organisms and natural systems.

Which returns us now to the ambiguity in Taylor's argument, as I am reading it. Is the premise that humans are to recognize themselves as members of the biotic community of life on the same terms as other non-human members meant to apply to us as a species? As individuals? Both? As I read him, Taylor's conclusion suggests we have moral obligations to species and to individual organisms; so I'm inclined to read the scope broadly as applying to us both as individuals and as species members. This raises several questions. How is the argument for this conclusion to be understood? And what does it imply for our relations to species?

I speculate the answers to be as follows. First, note that by Taylor's own admission, it is individuals who, in the first instance, are intrinsically worthy and thus deserving of respect for, as he points out, this is an argument that parallels Kant's argument for the intrinsic worth of persons[28] (which we will consider at greater length in Chapter 10). It is individuals, after all, who are goal-directed and for this reason are to be seen as teleological centres of life. It is individuals who are *directly* harmed or benefitted by the consequences of our actions, and it is for this reason that we owe them the moral duty of respectful consideration. But, as Kant notes, as members of the human family we are members of what he calls the "Kingdom of Ends." For present purposes, think of this as a community of goal-directed individuals who, in addition to their needs as individuals, also have needs as a species (the good of the species). On this basis, he argues that humanity as such is deserving of respect.[29]

Taylor's reasoning suggests a parallel that applies to non-human species. Individual organisms are members of their species, their unique family. And each of these families is a kingdom of ends, a family of teleological centres of life, each pursuing what is their individual good and collectively advancing the good of the species. And each species will have its own unique good. It follows that a very important way of treating individuals respectfully is to give careful attention to the needs of the species of which they are a member.

Moreover, it is species that are to be thought of as the systems spoken of in the second premise, the systems whose healthful functioning is central to the functional integrity of the whole. And so, careful attention to the integrity of the bio-system as a whole requires us to attend to the species-needs of the bio-communities that make up the whole and to do so because they matter in and of themselves. They are just as integral to the whole as humans. If it is sufficiently diverse, the system as a whole can absorb the extinction of the Splendid Poison Frog, or even the Pinta Giant Tortoise, and re-establish a kind of adapted integrity.[30] But make no mistake: it could just easily survive the extinction of the human species. In every case, it will be degraded to some extent by the loss of diversity and the repercussions that would flow from alterations to the chemical and energy flows, but we should not be deluded into thinking the loss of humans from the scene would be more significant than any other. Indeed, one might speculate that from the perspective of the long-term integrity of the biosphere, our departure from the scene might be a good thing.[31]

If we accept this, we are led to ask, "What does this require of us in relation to species?" "How are we to understand our moral obligations to something abstract like a species?" Well, we harm the interests of a species, for example, when our actions result in habitat loss or spoliation of their environment. We attend to their species needs – the good of the species – when we take this into consideration when forecasting the consequences of our contemplated actions. And we recognize it to be a moral obligation that falls on us to do so, because these species are to be valued for their own sake.

An illustrative example is the North Atlantic Right Whale. It is estimated that there are now only 340 individuals remaining (as of October 2022). Living as they do in one of the world's busiest ocean shipping lanes, they are vulnerable to ship strikes. This is especially true of infants and their mothers who spend much of their time close to the surface. There is evidence that the sound of the motors interferes with their ability to navigate safely. It is the individuals that are harmed (sometimes still broken-backed across the bows of the ships when they put into port), but it is species that go extinct. If these animals are to be saved, we will be forced to address the issue of their needs as a species. Failure to do this will result in their extinction, and we don't have long. I think Taylor would assert that we have a moral obligation to extend respectful consideration to the individuals and to the species. You might, as a ship captain, make the decision to slow down for the individuals detectable on your ship radar. You thereby extend respectful consideration to these individuals. We extend respect to the species when we adopt a policy requiring all ships to slow down in areas where it is known there are individuals or, perhaps, by redirecting shipping lanes to avoid these areas.

Some readers will remain sceptical of the notion of extending respect to something abstract like a species. And some will be sceptical of the idea of intrinsic worth itself. I'll ask you to keep an open mind until Part III, but for present purposes, I will express the ultimate conclusion as a conditional claim: *if* we regard human life as having intrinsic worth making us deserving of respect and ethical consideration, *then* we must so regard the natural world and the organisms that populate it. And the force of the imperative "must" is that, on the basis of Taylor's argument, there is no good reason to not do so. It is, in that sense, rational to adopt the bio-centric outlook.

Thus, these organisms are deserving of respect, which in this context means we owe them ethical consideration and we owe it to them because they have intrinsic worth on the same basis as ourselves: we are all striving to realize our own good. This is not to deny the very real differences between ourselves and other organisms, but these differences are irrelevant in relation to the features of our existence that should determine ethical consideration. In this regard, it is our similarities to other organisms that take precedence,

> We share with other species a common relationship to the Earth. In accepting the biocentric outlook we take the fact of our being an animal species to be a fundamental feature of our existence. We consider it an essential aspect of "the human condition." We do not deny the differences between ourselves and other species, but we keep in the forefront of our consciousness the fact that in relation to our planet's natural ecosystems we are but one species population among many. Thus, we acknowledge our origin in the very same evolutionary process

that gave rise to all other species and we recognize ourselves to be confronted with similar environmental challenges to those that confront them.[32]

Put these points together and we get a thorough rejection of the premise underlying Classical Liberalism that nature is of merely instrumental worth and is to be brought under our control. And from the recognition of its finiteness and the circular nature of energy flows through natural systems, there is also a rejection of the idea of continual growth as an economic ideal and of linear production systems that generate huge quantities of waste that are ultimately discharged into the environment. These ideas are not only rejected as ideals or objectives but are acknowledged to be completely unrealistic. In place of this outmoded view, we set as our regulative ideals the goal of a steady-state economy based on the principle of circular flows of energy and resources; the circular economy as biomimicry.

All of this and more is implied, I think, by the view Rifkin refers to as Biosphere Consciousness.[33] We use the term "biosphere" to designate the thin film that extends from several metres below the Earth's surface and the ocean floor to the upper atmosphere: the thin film containing all living systems that we have referred to as biotic communities and existing itself as a system, the master system that contains all within it. Agreeing to see nature as having intrinsic worth is the starting point of Biosphere Consciousness, but it embraces much more than this. It is fair to speak of Biosphere Consciousness as the fundamental bedrock of the sustainability narrative. Biosphere Consciousness, then, is that state of awareness of the biosphere as the envelope that contains and supports all life on Earth and, in many respects, is itself to be regarded as a living organism. And so, by the previous argument, the biosphere itself is deserving of respectful treatment. As Rifkin puts it,

> The scientific community's recent insights into the workings of the Earth's biosphere amount to nothing less than a rediscovery of the planet we inhabit. From diverse fields – physics, chemistry, biology, ecology, geology, and meteorology – researchers are beginning think of the biosphere as operating like a living organism whose various chemical flows and biological systems are continuously interacting with one another in a myriad of subtle feedback loops that allow life to flourish on this tiny oasis in the universe.[34]

This idea of regarding the biosphere as itself a living system is reminiscent of the Gaia hypothesis. Initially proposed by James Lovelock, a British researcher, in 1972, the hypothesis is that organisms interact with the surrounding inorganic environment of the biosphere in ways that, as we have seen, form a synergetic and self-regulating system that has resulted in the emergence of the climate and biochemical conditions that make life on Earth possible and which sustain that life; life supporting life, as it were.[35] In relation to our earlier argument, if we think of the biosphere as itself a living organism, then it too is an individual that is to be valued for its own sake and is deserving of respectful treatment and not to be valued merely instrumentally.

On this point of the biosphere being deserving of respect we confront a further ambiguity, though quite different from the ambiguity about scope noted earlier in Taylor. In this case, the problem arises from the fact that the idea of regarding the biosphere as itself a living thing is often couched as a metaphor. Typically, it is said that the biosphere is, in many respects, to be regarded "as though" it was a living thing. Claims based on metaphors do not license the conclusion that the biosphere *is* an individual with intrinsic worth. And it is certainly not a species, not even metaphorically. However, I'm confident the spirit of Biosphere Consciousness can be preserved on this point. We can say we have established a disjunction: either we take the recent discoveries of science to have established that the biosphere is a living thing, an emergent life in the same sense as all other living things or we reserve respect for the individual organisms and species that populate the biosphere and regard ourselves as duty-bound to protect the functional integrity of the biosphere as the means of treating these organisms respectfully. As Kant might have put it, we have indirect duties to treat the biosphere respectfully that arise out of our direct duties to the individuals and species that have intrinsic worth.

This awareness that lies at the heart of Biosphere Consciousness fosters humility and empathy. Humility follows from the recognition that we exist as a naturally evolved species on the same level as others, and the idea that we exist separate from nature is revealed to be illusory. Nature is not ours to control and from the perspective of Biosphere Consciousness the idea seems ridiculous and presumptuous. And empathy is stimulated because we recognize all living things as striving merely to maintain their well-being, their integrity, just as we do. They seek merely to pursue what is their good and by our actions there are consequences that are imposed on them. From this perspective, the casual disregard of the impacts of our actions on their welfare is perceived to be callous and arrogant. Humility and empathy are the appropriate attitudes of Biosphere Consciousness, and they counsel caution in our ethical deliberations.

It was pointed out by the philosophers of the Scottish Enlightenment, Adam Smith, David Hume and Bishop Butler among others, that we are social beings by nature and however much we may be motivated by self-interest, this tendency is counterbalanced by empathy. As Smith pointed out, self-interest is not incompatible with a concern for the welfare of others. Under ideal circumstances, we are possessed of a practical common sense, what he referred to as sound practical judgement, that inclines us to judge a situation from the perspective of self-interest, but this orientation to self-interest is not incompatible with our social natures.[36] As emphasized by Hume, it is empathy that ameliorates our self-regarding tendencies. It is not just that we recognize that treating others fairly encourages reciprocity. We are also, under appropriate circumstances, moved to feelings of empathy, that is, genuine feelings of fellowship and concern for others.

Now as Butler points out, the same empathy that inclines us to feelings of fellowship with others can sometimes function to divide us depending on contextual factors. Our feelings of empathy can be leveraged, whether wilfully or as a matter of circumstance, to strengthen tribal instincts that divide us from some of our fellows, and the outcome depends to a large extent on circumstances. In circumstances where people feel marginalized and disrespected, disenfranchised from the benefits of society and the opportunities available to others, resentment and division are the predictable outcomes. Empathy is muted by feelings of resentment and unfairness and extends no further than the narrow confines of those you believe to share your fortunes and your tribal identity. Under circumstances where people feel valued and respected, and included in the political process that determines public policy and opportunities, empathy is strengthened and the boundaries of one's feelings of fellowship are extended.

Biosphere Consciousness, insofar as it follows in the wake of a reorganization of our lives based on the democratization of energy and economic opportunity and the decentralization of authority that emphasizes local control, encourages and strengthens fellowship with others in our communities. Moreover, insofar as Biosphere Consciousness is premised on the idea that we are members of the biotic communities of life on the same terms as all other living things, our feelings of empathy and the boundaries of our communities are enlarged even further. It becomes natural to feel the boundaries of one's community to extend to these non-human organisms that were formerly regarded as "other."[37] As Rifkin puts it,

> The growing self-awareness of the human race is the psychological mechanism that allows empathy to grow and flourish. As we become increasingly aware of our own individuality, we come to realize that our life is unique, unrepeatable and fragile. It is that existential sense of our *one and only life* that allows us to empathize with others' unique journeys and to express our solidarity. We do this by engaging in acts of compassion whose purpose is to aid another in the struggle to optimize his or her life. To empathize is to celebrate another's existence.[38]

In this quote, when Rifkin refers to the self-awareness of the human race and of our increasing awareness of our own individuality, he is, I think, implying that our awareness is enhanced by Biosphere Consciousness on two levels that mirror the aforementioned point about individuals versus species. On one level, there is awareness of the uniqueness of human life itself; awareness of our uniqueness as a species and the singular and unrepeatable events spanning eons on which our existence depends. The smallest change in circumstances eons ago might have altered the trajectory of evolutionary processes in such a way that humans might never have arrived on the scene. On another level, there is awareness of the fragility of life as an individual. I realize the fragility of my own life as I go through each day striving to realize what is for me my good. This is not a crass awareness of one's individuality as a person, a consumer confronting others in the market, the sort of individuality that divides and facilitates erecting false barriers between ourselves and others and encourages us to view others as competitors for limited resources. Rather it is the sort of awareness of one's individuality that leads to us reaffirm how fragile and precious life is. The underlying sentiment here is captured perfectly by the old saw, "There but for the grace of God go I." The humility at the heart of this expression makes it natural to acknowledge that the same applies to those other individuals with whom we share the Earth, the individual organisms we see striving like ourselves to pursue their good. It is true of species and of individuals that life is unique, unrepeatable and fragile. Circumstances might change at any moment – a meteor might strike the Earth – rendering sentient species extinct *en masse*. But from the perspective of Biosphere Consciousness the callous disregard that leads to their extinction by our actions is seen for the great and avoidable tragedy it is.

Commensurate with this enlargement of our feelings of humility and empathy is a shift in our dealings with others from competition to cooperation. Under the terms of Classical Liberalism and its supporting narrative, we meet each other on equal terms in the marketplace, each responsible for themselves and motivated exclusively by self-interest. Competition is the norm, and it's understood that where one ends up in the scheme of things is a mark of your individual effort and merit. But under the terms of Natural Capitalism and the supporting narrative of Biosphere Consciousness, we are responsible for each other as much as for ourselves as individuals. It is recognized that survival and flourishing will require cooperation.

As a consequence of the distributed nature of renewable energy and its low costs, this collaboration is, in the first instance, localized. It spreads across continents, but this continentalization is the lateral organization of nodes that are simultaneously the sites of energy production and the locus of economic opportunity and activity. It is also genuinely participatory insofar as the localization of economic activity is part of the larger story of the decentralization of authority. This localized control leaves it to each community to adapt economic activity to local circumstances, relying on shared knowledge understood as a social construct and shared responsibility, including knowledge of best practices. As noted in Chapter 3, in our discussion of Montgomery on restorative agriculture, this adaptation to local circumstances works synergistically with enhanced diversity and resilience.

Thus, the community members are engaged and empowered while the enlargement of empathy and humility encourage a sense of belonging and shared destiny. In a world where energy is available everywhere and practically free and economic opportunity has easy entry and exit with almost no transaction costs, there is no longer a competition for scarce resources, and the benefits of the economic opportunity are also more widely and fairly shared. As individuals, we are free to be citizens first, consumers second.

Moreover, as noted earlier, the business model that is the heart and soul of Natural Capitalism aligns the interests of clients and service providers. Under these conditions, cooperation is not only encouraged but proceeds on the assumption that collaboration is win-win, as opposed to zero-sum. Under such circumstances, we can also expect social capital to be resurgent and highly prized, after a period of lengthy retreat in much of the capitalist world since the late 1970s or so.[39] As the narrative drifts from prioritizing intensively concentrated financial capital to decentralization of authority and the democratization of economic opportunity, there is a corresponding emphasis on social capital as the glue that makes successful collaboration possible. Trust, grounded in transparent communications and shared responsibility, is the foundation of social capital. Under a narrative that emphasizes collaboration instead of competition, trust finds fertile ground.

And so, we see that our cooperative ventures are premised on universal and democratic access to renewable energy and the essentials of life. As noted earlier, there is a diminution of the importance of private property. Private property will still exist but increasingly its importance will be eclipsed in favour of access to services. In a world where profits are made from the sale of services and the underlying products are ephemeral platforms continuously updated in the networked internet of things, the thing no longer matters much. It is about quality of life and collaborative relationships with others,

> In a distributed and collaborative economy, . . . the right of access to global social networks becomes as important as the right to hold on to private property in national markets. That's because quality of life values become more important, especially in the pursuit of social inclusion with millions of others in global communities in virtual space. Thus, the right to Internet access becomes a powerful new property value in an interconnected world.[40]

Summarizing, the narrative is premised in the first instance on valuing nature intrinsically rather than instrumentally. This happens when we acknowledge our membership among the bio-communities of life on equal terms with all others. This is tantamount to adopting Biosphere Consciousness as a fundamental moral orientation – the recognition that we owe it to other members of the biosphere to extend ethical consideration to them because their lives matter (they are to be valued for themselves and not because of any instrumental value they may have for us). As a result of this reorientation, we embrace the regulative ideals of a steady-state economy based on the principle of circular flows of energy and resources. Biomimicry becomes the norm for all economic activity.

Communities are based on collaboration and the enhanced trust and social capital that flow from the laterally distributed and fully networked generation of energy and localized economic opportunity. Such communities operate as decentralized authority structures on the basis of full democratic participation.[41] The business model of delivery of services aligns the interests of service providers with consumers ensuring that collaboration is win-win. We enjoy the luxury of being citizens first and consumers a distant second.

There is a shift of values to prioritize reduced consumption and the restoration of nature and wild spaces rather than extraction of virgin resources. The relevant perspective shifts to the long term rather than short term and knowledge is now seen as a social construct and a shared responsibility, on the model of Wikipedia and shareware. Economic efficiency is measured with qualitative rather than quantitative metrics (quality of life rather than GDP – service rather than ownership of goods).

Concluding remarks

I hope that you have found this explanation of the economic model of Natural Capitalism and the five pillars of the TIR infrastructure required to implement it both thought-provoking and hopeful. I have been at pains to argue that this economic model, if implemented, would be genuinely sustainable and without requiring us to abandon Capitalism. But, as should now be evident, implementation of these would alter our lives in fundamental ways, and we should not be deluded into thinking that it is inevitable. It is not and nor will it be an easy thing to accomplish. There is significant resistance in some quarters. Not surprisingly this resistance is found prominently among those who have a vested interest in maintaining existing economic and social systems. Additionally, though, there is a lot of political momentum in favour of the status quo and resistance to change among citizens generally, much of it grounded in anxiety, if not outright fear, about what such a future would be like.

There are also distractions that give the illusion that more is being done than is in fact the case. Consider that the five largest solar farms in the United States produce a combined output of nearly 2,000 megawatts, enough to power roughly three quarters of a million homes.[42] Other large-scale projects are in the planning phases or already under construction. On the face of it, this might seem like a great idea and a significant contribution to bringing renewable energy online in what is still the world's largest economy. However, there is good reason to be sceptical of the value of such projects in relation to the goals of genuine sustainability. They are, in fact, perfect examples of business as usual, subject to all the risks that plague existing large-scale energy projects. It continues the intensive concentration of capital investment partnered with centralized production and distribution of energy for delivery to markets, often over substantial distances, with all of the attendant waste and subject to risks for technological disruption, extreme weather events and attacks from foreign powers and terrorist organizations.

The biggest problem with such projects, though, is that they stand in stark opposition to implementation of the business model that is at the heart of Natural Capitalism. As long as energy companies are in the business of energy production and delivery, there will be no incentive to abandon such large-scale projects and the outdated business model based on maximizing consumption as the route to maximizing profits. As explained in Chapter 3, it is the business model of Natural Capitalism that unleashes the true power of the sustainability revolution precisely because it is this business model that aligns the interests of consumers and service providers and aligns the interests of both with genuine sustainability (i.e., reduced consumption and reduced waste production). Without this, the model can never work. Yet such projects convey the impression that our sustainability goals are being met or, at the very least, significant progress is being made. Even worse, they support the dominant narrative of Classical Liberalism because they suggest that sustainability can be achieved without abandoning business as usual, that is, without real change to our existing economic systems. That is, for many, a very comforting illusion.

Another kind of distraction is the proliferation of smaller-scale pilot projects. Many of these are well intentioned, to be sure and, in some cases, they are an opportunity to explore the limits of what is possible, and that is a good thing. But in many cases, they are driven by the wrong motives. They are framed and undertaken within the narrative that sees sustainable development as coincident with the liberation of market mechanisms. This is particularly the case for public–private partnerships (PPP) that offer opportunities for politicians and private sector players to brand themselves on the green file while facilitating private investment in public infrastructure projects. They attract investment without a concrete plan for how it might be scaled up so as to make a real contribution to reducing consumption and waste. Indeed, the point was never to reduce consumption and waste at all. They can make for a great photo op while facilitating the extraction of profits from consumers, but with no strategy for integrating them into a larger plan they make no meaningful contribution to achieving genuine sustainability. Moreover, there have been some high-profile examples of pilot projects that failed egregiously and have had the effect of tarnishing "green" technologies, thereby discouraging investment in renewable energies and sustainability technologies generally.[43]

As I was researching this part of the book, I came across an online article by Auden Schendler, a former consultant with Rocky Mountain Institute, the nonprofit sustainability think tank founded by Amory B. and L. Hunter Lovins, co-authors with Paul Hawken of the book that inspired much of my thinking in the lead up to this book.[44] Schendler is critical not only of the distractions of "green" projects like the ones discussed earlier but more generally of the idea of leaving it to market mechanisms to lead us to a sustainable future. He notes that, despite all the ingenuity and hard work, climate change has "marched on." More than 40 years after the Rocky Mountain Institute began its work, and at least 23 years since the publication of *Natural Capitalism*, the situation is much worse. His shock and disillusionment mirror my own. The reader may recall, from the preface, I recounted how my own initial enthusiasm was replaced by exasperation: from thinking Natural Capitalism and a sustainable future was inevitable I came to see that there was nothing inevitable about it at all.

Schendler tells a compelling story about corporate duplicity, funding green projects while simultaneously using all the means available to them, with their enormous resources, to maintain the status quo. In this story, the cultivation of pilot projects as deliberately curated distractions is central to the plot. They are intended to displace meaningful action. Meanwhile, behind the scenes, there has been an ongoing misinformation campaign to shape public discourse with the intent of shifting responsibility from the fossil fuel industry, where it properly rests, to the shoulders of individuals. This is a generalization of the point explored in Chapter 2 in relation to waste production and management. As we saw there, decades of shaping the message have displaced responsibility for waste onto the shoulders of ordinary citizens, urged to recycle more while being encouraged to shop online with all the additional waste required to package and deliver these goods to individual homes. And this despite the fact that domestic waste accounts for not more than 1 percent of total waste in Canada (presumably not so different from other jurisdictions). The outcome is the same: the illusion of progress and a sense of complacency permit business-as-usual to continue unabated.

Even worse is the extensive amount of resources expended in corporate lobbying for regressive laws and policies and funding campaigns for politicians who have set themselves in stark opposition to the sorts of measures that are required to achieve real results. Opposing carbon taxes and shifting subsidies from the fossil fuel industry to renewable energies are obvious targets, but there is also the support of regressive laws that impact negatively on our democratic institutions and racialized communities. Examples include laws that make it more difficult for citizens in these communities to vote, or the gerrymandering of electoral boundaries that make their votes count for less, in full knowledge that these people are the ones most likely to demand real and systemic change. From this perspective, green projects amount to "playing into the fossil fuel industry's hands."[45] It is an affront to sustainability and to social justice and, as Schendler rightly points out, it is a moral evil, "Sustainable business practices haven't just been a distraction (bad), nor a dodge of hard, controversial work (sinister), nor even intentionally duplicitous (corrupt). The approach has been *evil* because it represents *complicity*."[46]

Schendler's analysis concludes that we never should have expected market mechanisms alone to achieve genuine sustainability, placing our trust in the business case for environmental action (i.e., "what's good for the environment is good for business"). Real change must be systemic and will require a concerted focus on the locus of power, putting pressure there to demand this systemic change. The corporate sector will never willingly undertake to do what is required to achieve genuine sustainability. They will have to be forced, and this will require nation state governments to play the dominant role. And therein lies the rub and the irony of the situation: the involvement of nation state governments will require citizens to recognize the connection between meaningful action on climate change and social justice, and with determination it will be up to us to sever the connection between corporate power and political power; to recapture political power as it were. So it really is up to us after all, though not in the way we have been told.

And so, new economic models and technologies are not enough. Without the humility and empathy supported by Biosphere Consciousness, these models and technologies will be used to further the model of development rationalized by Classical Liberalism. We observed this model of development to be at the heart of the Brundtland Report (Chapter 2), where it is presented quite sincerely as our best chance for achieving the goals of poverty reduction and genuine sustainability. All the while we are sustaining the illusion that we are making real progress and that we can do what needs to be done with business as usual.

Thus, the real value of a new sustainability narrative is revealed: it is needed to give direction, coherence and unity to our sustainability goals and undertakings, and it is essential to the critique and displacement of the narrative that rationalizes and sustains Classical Liberalism; a new narrative that can unify the issues of meaningful action on climate change with social justice. We need to see our place in the world differently with all the new possibilities this introduces.

And so, as Montgomery asserts in relation to the reformation of agriculture, the biggest obstacle to implementation of sustainable economic models is vested interest.[47] Success will depend on aligning corporate incentives with the social and political goals of sustainability. It will also depend crucially on the democratization of energy and economic opportunity. Advancing these goals will require the active involvement of nation state governments to develop the appropriate regulatory frameworks and fiscal incentive structures. These will be regulatory frameworks that force companies to find their profits in becoming service providers, the ESCO model noted earlier in the case of energy companies, and fiscal structures that incentivize property owners to undertake the necessary retrofits. Thus, in relation to the large-scale solar projects noted earlier, arguably much more would have been achieved if the investment had been made to refit the building infrastructure to produce their own renewable energy and to conserve energy and thereby significantly reduce consumption and waste while simultaneously enabling the democratization of energy production.

And so, it is the burden of this book to make the claim that the first and fundamental step required is to alter our perspective – the adoption of a new narrative that shifts how we measure success. This new vision of success will focus on collaborative ventures premised on a business model of the delivery of services instead of the sale of goods to consumers as end users. And these collaborative ventures will be negotiated on the presumption that such collaborations are win-win for all parties. In a service economy, the economy is a means, not an end in itself. It is the means to an enhanced life of better service and less stress.[48] Accepting this vision will require us to have a fairly clear idea of what this new narrative looks like and this is what I have tried to articulate here. I say "fairly clearly" not as a dodge but as a frank recognition that the narrative will evolve and adapt over time, but I hope I have sketched it in enough detail to inspire confidence.

The process begins with adopting, as a fundamental moral attitude, what we have referred to above as Biosphere Consciousness. Only when we come to value nature intrinsically and abandon the outdated and inappropriate model of valuing nature instrumentally will we stand a chance of achieving sustainability. Viewing the world from the perspective of Biosphere Consciousness is to reconcile ourselves to the utter futility of the belief that nature is ours to control. Rather, nature (the biosphere) is to be seen for what it is, the delicately balanced "envelope containing, provisioning, and sustaining the human economy."[49] As such, its healthful functioning sets the non-negotiable limit to our economic activity.

★★★

Having set out this hopeful alternative, we will now take a step back to reflect on how we have arrived at the present situation of environmental degradation and existential threat. For this, we turn our attention to the dominant narrative which supports and rationalizes Consumer Capitalism, the ideology I am referring to as Classical Liberalism. This view has proven itself to be powerful and enduring in its ability to shape economic and social institutions and, by extension, entire societies. We must understand how it has led us to our present predicament. The narrative will be reconstructed with great care and analysed to understand why it has been so powerful and enduring, and, of course, why it is no longer appropriate for our present circumstances.

This task is the focus of Part II. Essentially, we will be engaged in some investigatory work. The goal is to uncover, and make explicit, the *conceptual foundations* of Classical Liberalism. Call it philosophical archaeology.

To a first approximation I define Classical Liberalism as an ideological view that favours the interests of global free market Capitalism. As we shall see, it serves two important objectives. In the first instance, it is *prescriptive* of both social and corporate policy. By that I mean that it implies that there are certain things we *should* be doing, as a matter of both social and corporate policy and governance to enhance the prospects of success for global Capitalism. And, second, it provides the resources to *justify* these policy and governance prescriptions.

The conceptual foundations of Classical Liberalism are constituted of various philosophical and economic theories which have been with us in various formulations for several centuries. As such, Classical Liberalism is a kind of grand overarching theory meant to form the foundations of political, social and economic theory and policy. In order of their historical introduction, we will consider John Locke's theory of private property in Chapter 5, followed by Adam Smith's theory of the laissez-faire market in Chapter 6. These theories are foundational to Classical Liberalism in the sense that they provide the most basic claims that ground everything else that is held to follow from the ideology.

We go on, in Chapters 7 and 8, to investigate this ideology in its modern formulation as it has been developed in recent decades, specifically to contextualize and support contemporary globalization. In Chapter 7 the focus is on the ideology as it pertains to the ideal state and the proper role of government in civil society. For this purpose, we examine carefully the arguments of Robert Nozick from his 1974 classic, *Anarchy, State, Utopia*. For clarifying what is at stake, we will contrast this view with its progressive alternative as articulated by John Rawls, principally in *A Theory of Justice*, originally published in 1971.

And in Chapter 8 we go on to examine the arguments of Classical Liberalism as they pertain to the corporate sector. For this we turn to the views of Milton Friedman on corporate governance and corporate social responsibility (CSR), relying on a variety of sources. Again, we'll clarify what is at stake in these issues by contrasting Friedman's view, referred to as the Shareholder Model of CSR (alternatively, the Stockholder Model), with the views of R. Edward Freeman, and the view referred to as the Stakeholder Model of CSR.

One final point before we get started. I will be at pains to argue that Classical Liberalism has often distorted the theories of Locke and Smith. Whether undertaken by deliberate design or unwitting ignorance, it is a fact that these theorists, Locke and Smith, would be drawn up short, if not outright offended, by much of what Classical Liberalism has attributed to them and their theories. This point is important because it raises several troubling questions. First, and most obvious, if this misinterpretation has been by deliberate design, then whose interest is served by this? But even setting that aside, we can ask whether there is room to reconsider these theories, individually, and as bundled together into Classical Liberalism. Thus, we can ask whether we might approach them with a fresh perspective to consider whether they need to be amended to better suit present needs.

And so, we proceed in Chapter 5 to commence this investigation with a detailed look at the theory of private property of John Locke.

Notes

1 Rifkin (2011, 2019).
2 Rifkin (2011, p.35).
3 Quoted from: https://www.weforum.org/agenda/2016/01/the-fourth-industrial-revolution-what-it-means-and-how-to-respond/ (retrieved October 18, 2022).
4 This is not meant to deny that a full explanation of the dominance of these firms is a complicated one that has much to do with the changing nature of capital itself: the shift from physical to intellectual capital and the spillover effects that flow from this. See Jonathan Haskel and Stian Westlake (2018).
5 https://www.cbc.ca/news/science/bipv-solar-1.6044485 (retrieved October 18, 2022).
6 Two articles to see here, the first from the US Department of Energy: https://www.energy.gov/energysaver/installing-and-maintaining-small-wind-electric-system and the second from the Government of Ontario: https://www.ontario.ca/page/electricity-generation-using-small-wind-turbines-home-or-farm-use (retrieved October 18, 2022).
7 https://www.sauder.ubc.ca/news/using-underwater-balloons-store-energy (retrieved October 18, 2022).

8 Rifkin (2011, pp.60–61).

9 Op. cit., p.88.

10 Cf., Rifkin (2011, pp.220–221).

11 http://insideenergy.org/2015/11/06/lost-in-transmission-how-much-electricity-disap pears-between-a-power-plant-and-your-plug/ (retrieved October 18, 2022).

12 A recent derecho storm that struck in Ontario and Quebec on May 21, 2022 killed 11 people, and although estimates of the damage are not yet complete, it is widely believed to be the most damaging windstorm to strike the area in decades, perhaps ever. Article from the CBC here: https://www.cbc.ca/news/canada/ottawa/power-utilities-grapple-catastrophic-winds-1.6464692 (retrieved October 18, 2022).

13 See Rifkin (2011, p.54).

14 https://en.wikipedia.org/wiki/Northeast_blackout_of_2003#Duration (retrieved October 18, 2022).

15 https://en.wikipedia.org/wiki/May_2022_Canadian_derecho#Impact (retrieved October 18, 2022).

16 Rifkin (2011, pp.161–162).

17 Op. cit., pp.126–127.

18 On this point about global governance, see op. cit., p.65.

19 This brief characterization will suffice for present purposes. We explore it at greater length in Parts II and III.

20 An updated version is Meadows et al. (2004).

21 An interesting overview: https://en.wikipedia.org/wiki/Systems_thinking (retrieved October 19, 2022).

22 For analytic purposes, we can isolate a system or a collection of systems for study (in this case, ocean systems), but of course the effects of carbon dioxide have consequences much broader than this, and ocean systems are continually interacting with atmospheric and terrestrial systems. Ocean systems too are, at another level of analysis, systems within a larger set of systems that ultimately make up the biosphere.

23 I consider ethical theories and the notion of intrinsic worth in more detail in Part III: Classical Liberalism Through a Normative Lens (Chapters 9 and 10). For present purposes, it is sufficient to deal with the argument at an intuitive level.

24 Kolbert (2014).

25 For this argument, I rely heavily on the presentation by Paul W. Taylor in "The Ethics of Respect for Nature." This is the ethical theory known as Biocentrism [Taylor (1981)].

26 Taylor (1981, pp.206–207).

27 Ibid.

28 Op. cit., pp.202–203.

29 It must be noted that there is much more to it than this for Kant, who would insist that humanity is a "kingdom of ends" because we are the law-givers to ourselves; we exist as a self-regulating community of individuals who legislate the very laws we agree to live by. Thus, for Kant, the most important need for humans, considered as a species, (and in his view, unique to us) is to extend to each other the space, the liberty, to act. That is, the freedom to decide for ourselves to follow the dictates of morality. We will take this up in more detail in Chapter 10.

30 An article on some species that have recently gone extinct here: https://www.treehug ger.com/animals-presumed-extinct-in-the-last-decade-4869347 (retrieved October 20, 2022).

31 Cf., Taylor (1981, p.208).

32 Ibid.

33 This term has the virtue of being descriptive and is used to designate the new narrative supportive of Natural Capitalism. Biocentrism, the view defended by Taylor, is a normative theory with an established usage that is different from, and narrower in scope than, the view being defended here.

34 Rifkin (2011, p.187).
35 Note the year, 1972, the same year as the original publication of *Limits to Growth*. In retrospect, we can see a range of ideas, going back at least to Aldo Leopold and probably further, beginning to coalesce and come into focus as a coherent whole.
36 There is much more to be said on this topic in Chapter 6 when discussing the details of Adam Smith and his theory of the laissez-faire market.
37 For a clear and rigorous discussion of what a society might look like organized on these conceptions of community, see Will Kymlicka and Sue Donaldson (2013).
38 Rifkin (2011, pp.239–240).
39 I take up the topic of social capital and the story of its decline in Chapter 7.
40 Rifkin (2011, p.215).
41 I will return to the topic of full democratic participation and what it will mean in this context in the conclusion when addressing Rifkin's notion of Peer Assemblies.
42 https://constructionreviewonline.com/biggest-projects/top-5-biggest-solar-farms-in-the-us/ (retrieved October 20, 2022).
43 See Tom Rand (2020) for an extensive discussion of the many challenges facing "green" investments and the impacts of failed projects.
44 This article is mentioned briefly in the introduction: https://ssir.org/articles/entry/the_complicity_of_corporate_sustainability (retrieved October 20, 2022).
45 Schendler (2021, p.5).
46 Op. cit., p.6.
47 Montgomery (2017, p.40).
48 Hawken et al. (1999, p.18).
49 Herman Daly as quoted in Hawken et al. (1999, p.9).

PART II

Classical Liberalism

The conceptual foundations
of Consumer Capitalism

5

JOHN LOCKE AND THE THEORY OF PRIVATE PROPERTY

Historical context and Locke's motives

Locke's theory, usually referred to as the Labour Theory of Value, is developed at greatest length in the second treatise of his classic, *Two Treatises of Government*. It is, in my view, not possible to overstate the historical significance and influence of this work, and this in addition to his groundbreaking work in epistemology and metaphysics in *An Essay Concerning the Human Understanding*, for which he is equally famous. Locke is doing a great deal in the *Treatises*. In addition to developing his theory of private property, he is also concerned about developing a theory of the origins, objectives and extent of legitimate government.

Before we can consider the theory itself, though, we must get clear on Locke's motives. This approach reflects a firm commitment on my part that will be evident in everything that is done in the remainder of this book. Specifically, I am convinced that it is not possible to fully understand a profound philosophical theory without first having a very good idea of the author's motives. What is the author's purpose? What is the author seeking to accomplish? Moreover, it is not possible to fully understand an author's motives, understood in this way, without some understanding of their historical circumstances. And so, we begin with a bit of informal history.[1]

Two Treatises was published anonymously in 1688. It was published anonymously for the very good reason that Locke felt it would bring persecution on him, if not worse (the threat of personal harm) as a result of its perceived implications. Locke lived in a very turbulent period of English history. Born in 1632, he lived through the revolution that saw parliamentary government established under Oliver Cromwell. He was 16 years old when King Charles I was executed and, apparently, Locke was within earshot of the unsettled crowd.[2] Subsequent to the death of Cromwell

DOI: 10.4324/9781003388555-7

was the Restoration, which saw Charles II ascend the throne. Tensions were very high at this time between Tories, who, generally, were supportive of the sovereign's right to absolute rule and tolerant of a Catholic sovereign, and the Whigs, who, as a general rule, supported the priority of parliament and significant limits on the sovereign's authority and were strongly opposed to the influence of Catholicism.

Through his association with Anthony Ashley Cooper (the first Lord Ashley), Locke was at the heart of the Whig movement. Influence waxed and waned between the two factions during the reigns of both Charles II (1660–1685) and James II (1685–1688) and led ultimately to Ashley's involvement in plots against Charles II. And vis a vis his association with Ashley, Locke could not escape the suspicion that he too was involved. Ashley was sent to the tower for a time, several people were executed (including Sydney Algernon), and Locke sought refuge as a political exile in Holland for about eight years or so.

Meanwhile, and adding to the tensions, was the publication, in 1680 by Sir Robert Filmer, of *Patriarcha* in which Filmer argued for the unlimited right of the sovereign to rule with absolute power. The underlying theory, the Divine Right of Kings, was premised on the idea of primogeniture, tracing itself back to Adam. The patriarchal right of the sovereign to absolute rule is guaranteed by the fact that he, the king, traces his lineage back to the first-born son of Adam. The king rules the citizens of the kingdom as the father rules the family, hence the title. The timing of the publication of this book is critical, coming as it did at a time when Charles II was attempting to position himself to reclaim power ceded to parliament.

In addition to these political events, though, this was also a time of rapid social and economic change. Consider that, in the 140 years that span the time from Columbus's discovery of the New World (1492) to Locke's birth (1632), Western Europe had been completely transformed from the remnants of a Feudal economy that had begun collapsing in the late 13th century, to an early form of Capitalism referred to as Mercantilism and, in the case of Locke's England, was already beginning the process of industrialization.

Under Feudalism, there were no nation states as we know them now. There were kingdoms, relatively isolated and surrounded by sparsely populated wilderness, with little in the way of urban development. Under the pressures of Mercantilism, and the emerging entrepreneurial class, there was rapid development of urban centres, built mostly around ports as the centres of trade and commerce. As Feudalism came undone, leaving people in a vulnerable state, they left the estates to flock to cities in search of employment. The feedback loop of abandonment of estates to seek opportunity accelerated the decline of Feudalism, giving further impetus to the emergence of the Mercantile economy. This is a pattern familiar to us even today in the recent history of China, for example, as people have fled the agrarian communities to move to major urban centres, Beijing, Shanghai and others.

Simultaneously with the emergence of increased trade and prosperity came the gradual emergence of the nation state, driven in part at least by colonialism and the exploitation of resources in the New World. Adjustment on the social and political levels failed to keep up with this rapid economic change. Commensurate with the emergence of cities and industrialization is the emergence of urban blight, pollution and, for many, desperate poverty.

Locke was himself familiar with the urban poverty and inequalities of wealth that arose at this time. As a member of the emerging rentier class, Locke died a wealthy and influential man, his wealth built largely on landownership and investment in various trade and colonial initiatives. But even as a child, he was among the upwardly mobile and viewed the poverty at close, but safe, proximity. Speaking of Somerset, the place where Locke grew up, Maurice Cranston says,

> Industry had come early to Somerset. Through Bristol the country was close to the greatest source of all English wealth, the sea; while inland, the most important industry of England, wool, was also the most important industry of the county. The Mendips had been mined for lead since Roman times, and for coal since the thirteenth century. . . . The economic situation of the county was unsettled, and bred among some an urge to innovation and among others a yearning for the past. On the one hand there was dire poverty: Butter had risen since the beginning of the seventeenth century to sixpence a pound and cheese to threepence, hungry people attacked the granaries, and many hanged themselves from want. On the other hand, there was increased wealth. The rising prices enriched the rack-renting landlords, the up-to-date farmers who could supply the towns with food, the big tradesmen and anyone else who could turn inflation to his own advantage.[3]

Locke was also familiar with the pollution of industrialization and its consequences. As an asthmatic, Locke found the air in London unbearable and spent the last decade of his life living far from the city. His trips to London, to conduct work there as a member of the Commission for Trade were, perforce, kept as brief as possible.

In the context of these historical circumstances, Locke's goal in the *First Treatise* is to refute Filmer's defence of absolute monarchy and the theory that allegedly supported it, the Divine Right of Kings. Locke must set the stage for his own theory by clearly demonstrating the falsity, if not outright absurdity, of this older view. Having done this to his own satisfaction in the *First Treatise*, in the *Second Treatise* he turns his attention to developing his own theory. In this, his goals are, as he puts it, to discover "the original, end, and extent" of civil government. We might frame Locke's motives in terms of finding the answers to the following questions (which I have framed as neutrally as possible, so as to not presume any particular answers):

- What is the origin of civil society and government?
- What does the legitimacy of civil government consist of?
- What should be the objective(s), or end(s), of civil government?

- What is the extent of civil government? That is, how far does the authority of government extend over the lives of its citizens?
- What is the origin of private property? That is, how does private property come into existence, presumably giving the owner the right to exclude others from the use or enjoyment of what is owned?
- And, in light of the inequalities of property noted earlier, what might justify these inequalities (presuming they can be justified at all)?

Our principal concern here will be with the last two of these motives of Locke's we have identified. We will pause to take note of the other issues to the extent that doing so is necessary to understand Locke on property. For example, there is, for Locke, an intimate connection between private property and the proper goals of government.

The theory reconstructed

To reiterate, the motives of interest to us in regard to Locke's theory of private property are two in number. Explored at greater length, first is the issue of whether there can be an adequate philosophical justification for a democratic right of all persons to own private property, and, if so, what might that justification be? After all, under the patriarchal system, only the sovereign has the right to own property. All dispensations of property to various lords, barons and so forth are at the sole discretion of the sovereign. Now even if one presumes that Locke has, in the first treatise, dispensed with the Divine Right of Kings, it does not follow from this fact alone that the right to own property is extended broadly to the population at large. One might put this point by saying that Locke's arguments to destroy the credibility of Filmer's theory leave a void. There is now a pressing need for a philosophically credible theory of property to fill this void, and the presumption is that it must be some theory that can justify the newly emergent, and increasingly powerful, Mercantile class. Second, given the fact that the economic system that typically accompanies a society based on private property seems inevitably to lead to vast inequalities of wealth, can these inequalities be justified? And if not, what might be the consequences for an economic system based on private property? And how will this new theory of property be made to cohere with a plausible theory of legitimate government? What will be the role of government in protecting the rights of its citizens in their property?

I will now proceed to *reconstruct* Locke's theory in reference to these questions. As such, I mean to indicate that I take responsibility for the interpretation implicit in the following presentation of this argument, indeed for all arguments reconstructed in this book. But let me be clear, I intend to also capture what I take to be Locke's intent. My goal will be to present an interpretation of this argument which, if Locke were present, he would say, "Yes, you have understood me perfectly." Along the way, I will show that, in many respects, the recent interpretation foisted on Locke by contemporary Classical Liberalism has misunderstood Locke's intentions in some important ways, resulting in what is, in my view, a distorted presentation of the theory and its consequences. So I will reconstruct Locke's argument premise by premise, carefully explaining each step as we proceed.

Here, in his own words, is Locke's statement of his first two premises (complete with his dated use of capitals and italics):

[P1] God, who hath given the World to Men in Common, hath also given them reason to make use of it to the best advantage of Life and convenience. [P2] The Earth, and all that is therein, is given to Men for the Comfort and Support of their being.[4]

The crucial point here is that humankind initially find themselves existing in the State of Nature, a world without boundaries or private property. The reference to the State of Nature and the use of the subjunctive mood here is an important one. That is, we do not actually find ourselves in the State of Nature now, though it is presumed that we once did so. And the term "State of Nature" is a technical term used commonly as a conceptual device in political writings in Locke's time to discover the origins of civilized society. Its use is prominent, for example, in Hobbes's classic, *Leviathan*. The idea is that readers are invited to imagine humankind as they might have existed before the advent of civilized society, presumably as small bands of hunter-gatherers lacking explicit political organization and then enquire how and why we might have left this condition to enter organized civil societies.

The essential feature of the State of Nature salient to our present enquiry is that in this state, there is no private property; everything is held in common, as Locke puts it. If this stretches the reader's credulity, it is worth keeping in mind that there are still areas of the Earth held in common to this day, namely international waters and Antarctica, as well as areas designated as such by various indigenous peoples. The Moon and Mars are presently Commons, though this may change in our lifetimes. A further feature of the State of Nature relevant to Locke's motives is that, in this state, humankind lives in a state of perfect freedom and perfect equality. There is no private property, but likewise there are no fences. There are no laws or restrictions on what one might do, and no one enjoys any kind of privileged status that elevates them above others, except such status as might be obtained by brute force. Thus, the question that naturally arises here is, why would we willingly leave this state of perfect freedom and equality to enter civilized society?

It is worth pointing out here that Locke has a very different conception of human nature than Hobbes. Hobbes conceives of human nature as being driven by irresistible motives of self-interest, unrestrained from the most virulent behaviours except for the fear of reprisal or the benefits of being perceived to be otherwise. For Hobbes, civil behaviour is a thin veneer maintained only by force of law and state power. So for Hobbes it is clear why people would wish to leave the State of Nature: it is to escape the state where life is "poor, nasty, brutish, and short."[5] But Locke's conception of human nature is not nearly so dark. We are capable of bad behaviour, to be sure, but also good and honourable behaviour, with the actual outcome depending very much on circumstances. In any event, given the perfect equality and freedom of the State of Nature in Locke's conception, an argument is needed for why we would willingly leave such a state. And, as we shall see, this is intimately connected with property.

Another point that deserves comment in this statement of Locke's premises is the reference to God. Locke was, typical of his time, a devout believer and it may be wondered if this compromises his view, detracts from its plausibility for a contemporary secular audience. I don't think it does, for we can interpret it neutrally, without religious or theological implications. We could say, for example, that humankind, as a naturally evolved species, find themselves in a world that provides abundant resources for our sustenance and comfort, and that, initially, these resources are held in common. This will put some strain on our presentation of Locke's argument, though. This is because, when seen in the broader context of Locke's motives in the *Second Treatise* to discover the origin and ends of civil government, Locke's foundational premise, which he calls the Fundamental Law of Nature is explicitly theological. This foundational premise is stated at various places and in different ways by Locke, but here is one instance of it (quoted at length because of its importance). Speaking of the State of Nature, he says:

> But though this be a *State of Liberty,* yet it is *not a State of License,* though Man in that State have an uncontroleable Liberty, to dispose of his Person or Possessions, yet he has not Liberty to destroy himself, or so much as any Creature in his Possession, but where some nobler use, than its bare Preservation calls for it. The *State of Nature* has a Law of Nature to govern it, which obliges everyone: And Reason, which is that Law, teaches all Mankind, who will but consult it, that being all equal, and independent, no one ought to harm another in his Life, Health, Liberty, or Possessions. For Men being all the Workmanship of one Omnipotent, and infinitely wise Maker; All the Servants of one Sovereign Master, sent into the World by his order and about his business, they are his Property, whose Workmanship they are, made to last during his, not one anothers Pleasure. And being furnished with like Faculties, sharing all in one Community of Nature, there cannot be supposed any such *Subordination* among us, that may Authorize us to destroy one another, as if we were made for one anothers uses, as the inferior ranks of Creatures are for ours. Every one as he is *bound to preserve himself,* and not to quit his Station wilfully; *so by the like reason when his own preservation comes not in competition, ought he, as much as he can, to preserve the rest of Mankind,* and may not unless it be to do Justice on an Offender, take away, or impair the Life, or what tends to the Preservation of the Life, the Liberty, Health, Limb or Goods of another.[6]

There is rather a lot going on in this passage, but note first that the State of Nature, although a state of perfect liberty, is not a state of licence. We are not free to do whatever we want, because there is a Law of Nature that governs behaviour in the State of Nature (though by assumption it is not civil law) and human reason "is that law." It would seem that Locke is convinced that we are all endowed with the capacity to reason (this, in his view, is what makes us distinctively human) and used properly reason leads us to the conclusions that would be licensed by the Law of Nature. This, incidentally, is the reason Locke's conception of human nature is not so dark as Hobbes's. Locke does not deny the role of self-interest in motivating human action, but he is convinced we are capable of reason and curbing the excesses of self-interest within appropriate limits.

At any rate, among the most important of these conclusions are that we are all of us equal (in terms of our moral worth), independent (at the very least this means individually personally responsible for our actions) and without any kind of natural subordination amongst us as humans. As such, we are not at liberty to harm one another in our liberty, health or possessions. Moreover, we are not at liberty to dispose of our own lives either, insofar as we are all of us God's property, the result of God's workmanship. And drawing the reader's attention to the underlined part, put positively this is the claim that we are bound by the Law of Nature to preserve ourselves and, by our actions, to preserve the lives of others as much as possible. In other words, the Law of Nature is not merely the negative admonition to not do harm to others. It is the positive command to do as much as lies within our powers to promote and preserve as many of our fellow humans as possible.

As a final comment on this passage, do note that all the other animals are subordinate to us (i.e., the inferior ranks of creatures) and it is implied that, like the iron ore, oil, gold and other material resources, they are for us to use at our pleasure in meeting the demands of the Law of Nature to support and preserve humankind.

As we can see, God goes to the very foundation of Locke's theory and, thus, any attempt to state his theory in secular terms will do some violence to his intentions. Additionally, as we shall see in Part III, this will present difficulty for understanding how the prescriptions of Classical Liberalism connect with our values. That is, if we are not following the commands of God, then what will be the moral justifications for our actions under the terms of this ideology?

With that cautionary caveat in mind, I rephrase these first two premises as follows:

P1': Before the advent of civilized society, humankind exists in a state of perfect freedom and perfect equality (the State of Nature).

P2': In this state, the resources of the entire world are held in common, without private ownership of any kind.

The questions raised by these premises are put as follows: How does private property ever come into existence? After all, as a hunter–gatherer in the State of Nature surrounded by a world rich in resources, how can I ever lay claim to any of it and presume to exclude others from that to which I do lay claim? More generally, can the very notion of private ownership be given an adequate philosophical justification? Locke's third premise is meant to address these questions.

P3: Though the Earth, and all inferior Creatures be common to all Men, yet every Man has a *Property* in his own *Person*. This no Body has any Right to but himself. The *Labour* of his Body, and the *Work* of his Hands, we may say, are properly his.[7]

The pertinent claim here is that we are, all of us, born the sole proprietors of our bodies and, by extension, of the capacity of our bodies to labour. This is the foundation of all private property there is or can be. This is what explains how private property comes to be in the State of Nature where everything is, by assumption, held in common. The answer is *private property is there from the outset in the ownership of your body and its capacity for labour.*

And note, too, that this ownership of our bodies is perfectly democratic; it is universally true of all of us. There is a potential conflict with the Law of Nature noted earlier insofar as it is there claimed that we are the property of God, and we will get to this shortly, but let us first complete this thought by considering an example. Imagine that you are a hunter-gatherer existing in the State of Nature with your peers. You are walking along one day, feeling rather hungry when you come upon an apple tree. Reaching up, you pull an apple from the tree and, just as you are about to eat the apple, one of your fellow hunter-gatherers runs up to you and says, "Hey, give me that. I'm hungry. I want it." In relation to Locke's earlier questions we ask: What, if anything, could give you the *right* to say, "No. This apple is mine. Get your own"? In other words, what could ever give to one, in the State of Nature, the right to exclude others from access to anything? And Locke's answer is that, insofar as you are the owner of your body and its capacity for labour, you make yours whatever you acquire from the Commons by the application of your labour.

But this in turn raises another difficult question: how exactly does this happen? Well, the short answer is, you used your labour (a capacity of the body of which you are the undisputed sole proprietor) to take that apple from the Commons and this is what makes it yours. Troubling questions are raised by this, which we shall get to shortly, but it's important to emphasize that it is proprietorship of one's body that is the foundation of all private property and this ownership is there from the outset, even in the State of Nature. And, crucially, this is a perfectly and universally democratic notion of property insofar as *all* persons are *sole proprietors* of their body. It is not clear that Locke himself fully realized the sweeping implications of this premise, but that is a debate I will forgo here.

But now we must return to consider the conflict noted earlier of this premise with the Law of Nature, which for Locke is foundational to his entire enterprise. We have just noticed that Locke's third premise implies that the foundation of all private property lies in the sole proprietorship that an individual has over their body and its capacity for labour; yet in the Law of Nature, Locke claims that we are all of us God's property. How, if at all, can this conflict be reconciled? The only way to reconcile this premise with the Fundamental Law of Nature is to suppose that it is Locke's view that although we are God's creatures, we are put into the world as independent stewards of our bodies and our lives during our lifetimes. Indeed, by the Law of Nature, we are all of us stewards of humankind. So, although you have not the right to end your own life, or even put it at risk unless for some "nobler purpose," you are the independent steward of your life until God chooses the moment of your death to suit divine purposes. And just as God has provided us with a world filled with all the basic resources needed to survive and thrive, so each of us has been provided with a body capable of the labour necessary to do this. Thus, you are in this restricted sense the sole proprietor of your body and its capacity for labour, and this you share with everyone else. Looked at this way, the Law of Nature is an injunction to all of us to recognize this relationship of stewardship and the obligation it imposes on us to be stewards of the welfare of the human family in its entirety. Equally important, though, no one other than God can impose their will on another, at least not with respect to our most basic rights to life and the liberties needed to sustain this life.

With this caveat, I rephrase Locke's third premise as follows:

P3': In the State of Nature, we are, each of us, the sole proprietor of our body and its capacity for labour and at liberty to use this capacity as we see fit to survive.

As stated, this premise asserts that one lays claim to resources in the Commons by the application of one's labour, but it remains somewhat mysterious how this is supposed to work. After all, even granting that you are the sole owner of your body and, thus, of the capacity of your body to labour, what does it matter that you used your labour to take this apple from the tree? How exactly does this extend the bounds of your property to something that was, moments ago, in the Commons? And what, exactly, does "labour" mean in this context?

The answer is given by Locke's fourth premise, which is, at the same time, an inference from the first three premises. That is, it also a *sub-conclusion* of the longer extended argument we are reconstructing here (indicated here by the "/C" notation).

P4/C: Whatsoever then he removes out of the State that Nature hath provided, and left it in, he hath mixed his *Labour* with, and joined to it something that is his own, and thereby makes it his *Property*.[8]

The implicit answer is that you make the apple your property by "mixing your labour" with it. You have, quite literally on Locke's view, joined something of yourself to the apple, thereby making it yours. Thus, the reason this is called the *Labour Theory of Value*. The crucial concept being developed here is the notion of *value added*: the idea that human labour contributes value to the resources of the Commons by altering them in some fundamental way. Importantly, of course, this works only if one is applying their labour to resources in the Commons, otherwise it is theft. Returning to our example, if you step onto someone's orchard to pick an apple from a tree there, this is theft. Someone else has already applied their labour to cultivate this orchard, and the produce of it is their property for this reason. Thus, they have, by the value added by their labour, taken this bit of territory out of the Commons.

An interesting scholarly debate can be joined here, over whether Locke is actually developing two distinct arguments. By name, they are known as the *labour mixing argument* and the *value-added argument*.[9] In essence, the first argument is that Locke is implying that it is the very act of mixing one's labour with the resources of the Commons that makes that bit of the Commons yours. In "mixing" something of yourself (your labour, of which, by assumption, you are the sole owner) you make that resource yours. If this is the argument Locke intends, it raises several troubling questions. What does it mean to say that I have mixed my labour with the thing? To use an example from Lloyd-Thomas, if I do a little dance over some apple I find on the ground, does this count as labour? And have I mixed that labour in the appropriate way with the apple to lay claim to it?

The second argument is held to be making the very distinct claim that it is the altering of the resources of nature that makes something yours. The "altering" referred to here is very specifically held to result in enhanced value. You quite literally *add value* to the resources to which you apply your labour. This interpretation also raises questions, in this case about what is to count as value added to the resources of nature. Robert Nozick provides an example to underscore this puzzlement.[10] Imagine walking up to the ocean with a bottle of dye, which you subsequently pour into the ocean. As the dye disperses, a change is wrought throughout the area of dispersal, and this change is the result of you mixing your labour with it. Can we say that you thereby own the ocean or at least the part of it that is dyed? Does pouring dye into the ocean result in the kind of change that we would call enhanced value? Or, to return to the example considered earlier, if I merely dance around the "things" in the Commons I subsequently presume to own, in what sense have my actions added value?

I am going to skirt this debate because, though interesting, these subtleties are beyond my present purpose. I will suppose that there is only one argument here and it is, as indicated here, the claim that, when one applies their labour to the resources of the Commons you make it yours by the very application of your labour *because this application of your labour alters the thing in a way that enhances its value*. The "because" here is not causal, it is conceptual. In other words, it is intended to be true by definition that labour adds value to the resources to which it is applied. In response to the troubling questions raised by these examples, Locke's point is that labour is whatever alters a thing in a way that enhances the value of the thing specifically in ways that are supportive of the Fundamental Law of Nature, which requires us always to preserve life, our own and, where possible, the lives of others. So picking the apple up from the ground or from the tree counts as labour adding positive value because it makes the apple accessible to sustain life. Pouring dye into the ocean does not alter the ocean in any way that enhances value in this sense, and nor does performing a dance before picking up the apple, whatever ceremonial value the dance might have.

The case of the apple can be misleading only because it is a case where the labour required and the value added are very slight, but when we consider the case of the cultivated apple orchard, the input of labour and the resulting enhancement of value become more obvious. Locke holds the view that if we were to undertake a careful audit of the products useful to the support of life, we will find that, as an average, it is the value added by labour that makes the far greater contribution to the total value embodied in the products we actually use. Let's call these products, as they occur in the forms in which we actually use them, *end products* or *end goods*. Then his point is that the apple as an end good is the aberrant case, the vanishingly small minority of cases where labour adds little value to the end good. In the overwhelming majority of cases, and certainly as an average across the range of products as we use them, the value added by human labour accounts for at least nine-tenths, perhaps as much as 99/100ths of the value, embodied in end goods. Consider the following quote:

Nor is it so strange, as perhaps before consideration it might appear, that the *Property of Labour* should be able to over-balance the Community of Land. For 'tis *Labour* indeed that puts the *difference of value* on every thing; and let anyone consider, what the difference is between an Acre of Land planted with Tobacco, or Sugar, sown with Wheat or Barley; and an Acre of the same Land lying in common, without any Husbandry upon it, and he will find, that the improvement of *labour makes* the far greater part of *the value*. I think it will be but a very modest Computation to say, that of the *Products* of the Earth useful to the Life of Man 9/10 are the *effects of labour*. Nay, if we will rightly estimate things as they come to our use, and cast up the several Expences about them, what in them is purely owing to *Nature,* and what to *labour*, we shall find that in most of them 99/100 are wholly to be put on the account of *labour*.[11]

Certainly, we can challenge Locke on his assumptions of value here, and we will shortly, but my present point is that there are not two distinct arguments here, there is only one. I maintain this for the following reasons. First, it answers the troubling question of what shall count as labour. According to Locke, labour is what adds value. And again, to add value is to alter the resources provided by nature in ways that enhance the resulting end goods in ways that are beneficial to the project entrusted to us by the Law of Nature, the stewardship of the human family. Second, I believe this is closer to what Locke intended. Locke was doing a lot in the *Treatises*, ranging across a broad spectrum of objectives and joining ongoing debates on many issues. It is perhaps not surprising that we will find some aspects of Locke's views that fall far short of being perfectly clear, and this is one of them. So, on this point, I doubt that Locke anticipated that his argument was equivocal in this way. Third, this way of construing Locke's argument aligns it with the instrumental tone of Locke's overall position here. By that I mean, as we have seen, Locke's overarching concern, as expressed in the Fundamental Law of Nature, is with the preservation (and beyond mere preservation, the thriving) of human life as much as possible. And it is the value added by human labour that makes human thriving possible. And, lastly, this is, I believe, how the argument has been interpreted by Classical Liberals and it is our focus here to understand this tradition as it has come to us.

So, on this interpretation of Locke's intentions, I will rephrase his fourth premise as follows:

P4/C': Thus, whatever one takes from the Commons by the application of your labour adds value to that which is taken and thereby makes it your property.

Significantly, at this point in his argument, Locke has met the first of his two goals with respect to property. For notice, with this sub-conclusion, P4/C, Locke has established that there is indeed a perfectly democratic right to private property that pre-exists civilized society and formal government. It is founded on, and guaranteed by, the fact that we are all of us born in perfect equality and perfect liberty as the sole owners of our body and its capacity for labour. And it is for this reason that among the duties that properly fall on any justly constituted civil government, prominent among these will be the duty to protect the property rights of its citizens.

But what of the second goal, the problem that arises from the fact that in any civilized society based on recognition of this democratic right to own property, there arises, with apparent inevitability, a wide inequality of wealth/ownership. How does this happen? And can it be reconciled with the Fundamental Law of Nature and Locke's other premises? To see how Locke answers these questions, let us pursue our example a little further.

So imagine now that as you, hunter-gatherer living in the State of Nature, stand there eating your apple, you look about and start to formulate an ambitious plan. You think to yourself, "I could really make something of this place." You notice one lousy apple tree, existing by itself and producing a few dozen misshapen and worm-eaten apples in a typical season. But there's a lot of space and some good soil. It will take a lot of work, for sure, as there are a lot of huge Cedars and some large rocks that will have to be removed, but you set to work. You cut down the Cedars, turning them into fence posts, and you carve off a few acres of this land from the Commons behind a fence. Then you set to work to leverage out the rocks and harvest the seeds from the apple tree, which you meticulously plant in rows carefully spaced to maximize the density of viable apple trees on your orchard. These saplings you carefully cultivate over the ensuing years, watering them, fertilizing them with such compost and fish fertilizer as you can lay your hands on and carefully weeding and removing such other trees as try to make a start. In about five or six years you have a few acres of lush orchard producing, in a typical year, several thousand pounds of large, juicy, worm-free apples, where previously the same land produced a few meagre dozens of lousy, worm-eaten apples.

It is expected, of course, that your motive is to use the surplus apples you have cultivated to engage in trade with others in your community for the things you need. Similarly, others will focus their talents and energies on the opportunities presented to them to improve their lot in life. One person, who is not so good at either hunting or agriculture, finds in herself a talent for making excellent bows and arrows, which she trades with local hunters in exchange for game. Another makes tents, clothing and so on. In this way, we see a rudimentary division of labour and an exchange market based on barter emerge quite naturally from the motivation each individual has to survive and, indeed, do better than merely survive.

This example illustrates the implicit reliance on practical Utilitarian considerations in Locke's theory. For notice, consistently with the Fundamental Law of Nature, God has given us everything needed to survive and even thrive, but it will require of us that we work, that we expend our labour, and the more industrious we are, the better will be our lot.[12] Also, in addition to personal motives to improve one's own lot in life, in this case by cultivating apples to trade for other things I need, I am indirectly acting in accordance with the Fundamental Law of Nature to enhance the survivability of others generally by increasing the abundance of end goods we all depend upon. So, in fact, the result of this personal drive to flourish is that we are all made better off in the quite specific sense that there will be more of the end goods available to us than would otherwise be the case without this division of labour and simple system of market exchange. I will call this the "Rising Tide Argument." Locke did not call it that, or even make it explicit, but he anticipates Adam Smith in this respect, and such thinking will play a prominent role throughout our discussion of Classical Liberalism. The idea is that the industry of those individuals (entrepreneurs) driven by personal ambition to work hard will have the salutary effect of making all of us better off than we otherwise would be. In this sense, we are all lifted by the rising tide of the enhanced production of goods.

However, there are limits to what can be accomplished in this way. Locke claims that there are two distinct limits, imposed on us by God, which he claims are implications of the same Law of Nature that impels us to subdue the Earth for the benefit of human life. The first is that one must never take so much that it spoils in one's possession. And second, one must never take so much that you fail to leave enough for others to satisfy at least their basic needs. Now in our previous example, presumably you will trade your surplus apples for other things you need before they spoil. That's the point, after all. But if you were to take so much of the available land out of the Commons that there was not enough for others to meet their needs, you have violated the Law of Nature that God expects us to honour. In fact, Locke says that, in such a case, others have a right to so much of your surplus product as would be required to maximize the survivability of as many persons as possible. We will return to take up this point later.

These two limits place significant constraints on the possibilities for any simple barter culture to advance much further in the production of goods and the enhancement of human life. How might it be possible to transition from such a society to a full-blown capitalist economy of the sort in existence today or even in Locke's time? The answer is given by Locke's final premise:

P5: And thus *came in the use of Money*, some lasting thing that men might keep without spoiling, and that by mutual consent Men would take in exchange for the truly useful, but perishable Supports of Life.[13]

In this premise, it is revealed that money, which is actually worthless in itself, serves two valuable purposes. First, it is the universally acknowledged measure of the market value of the end goods we consume. Thus, the reason we are willing to "take [it] in exchange for the truly useful, but perishable Supports of life." To contemporize this idea, please note that this is analogous to the role played by the American dollar on exchange markets. All other currencies are relativized to the American dollar as "peg." Thus, the prices of goods in American dollars reflects their real market value and this by convention (i.e., mutual consent). Second, as we see here, money does not spoil. Thus, agreeing to a complicated system of international conventions to regard money as exchangeable for the end goods we consume, money can be stored up as a repository of capital. As such, it greatly facilitates the emergence of more complex exchange markets with further enhancements to the division of labour that will increase even further the total production of end goods we consume. And this in large part because we can now treat our capacity for labour itself as an exchange commodity and use the money we receive in wages to buy the whole range of goods we need.

Thus, we see the emergence of the synergistic interchange facilitated between those industrious individuals who work hard and save to accumulate money, working together with risk-tolerant entrepreneurs who can now raise capital for larger projects that require it.[14] And all this activity drives the virtuous cycle of enhanced value and goods production that we have agreed to call the Rising Tide. This is the great value of money as a repository of wealth: because it doesn't spoil, it can be accumulated without limit and used for such purposes. On this understanding, I rephrase Locke's fifth premise as follows:

P5': The invention of money, and the system of conventions that govern its use, serve two valuable purposes for civilized society: it serves as a universally recognized measure of the exchange value of market goods, and it serves as a repository of capital.

And so, we get Locke's Main Conclusion, stated by him as follows:

MC: And as different degrees of Industry were apt to give Men Possessions in different Proportions, so this *Invention of Money* gave them the opportunity to continue and enlarge them.[15]

Here we see Locke drawing out explicitly the conclusions alluded to earlier. Namely, and most importantly, money serves the useful purpose of supporting human industry by its ability to be accumulated and directed to specific purposes that will increase the production of end goods, thereby directly serving the Fundamental Law of Nature. And so, we see that with this, his Main Conclusion, Locke has now met the second motive of his theory of property: he has provided a philosophical justification for the inequities of wealth that seem to inevitably accompany a capitalist economy that is based on private property and free market exchange. Note the specific wording Locke uses, "different degrees of industry were apt to give men possessions in different proportions." Although we are all born in perfect equality and freedom, with sole ownership of our bodies and their capacity for labour, it seems obvious to Locke that we are not all possessed of the same degree of industry. Put positively, it is, in Locke's view, a matter of plain common sense that humans range along a continuum from more to less industrious, and that it is this quality of industriousness that explains the differences of wealth. If you work harder, you deserve more. Moreover, as we see, money provides the opportunity to "continue and enlarge" these different amounts of wealth. So, money, for the reasons explained previously, is the principal source of economic opportunity for those who are industrious and willing to work hard to accumulate. Moreover, this has the natural result of making us all better off insofar as it leads to the production of more end goods for consumption. In contemporary form, I present this Main Conclusion as follows:

MC': Thus, money serves human industry by providing capital in liquid form to be applied readily to entrepreneurial ventures that have the effect of rewarding the industrious with enhanced production and prosperity.

So here is Locke's argument as I have reconstructed it. It is presented here in what philosophers call *Standard Form*, with each premise in its logical order leading up to the Main Conclusion, and with each inference marked by drawing a line under the last premise before the conclusion that follows from the premises immediately above.

P1': Before the advent of civilized society, humankind exists in a state of perfect freedom and perfect equality.

P2': In this state, the resources of the entire world are held in common, without private ownership of any kind.

P3': In the State of Nature, we are, each of us, the sole proprietor of our body and its capacity for labour and free to use this capacity as we see fit to survive.

P4/C': Thus, whatever one takes from the Commons by the application of your labour adds value to that which is taken and thereby makes it your property.

P5': The invention of money and the system of conventions that govern its use serve two valuable purposes for civilized society: it serves as a universally recognized measure of the exchange value of market goods, and it serves as a repository of capital.

MC': Thus, money serves human industry by providing capital in liquid form to be applied readily to entrepreneurial ventures that have the effect of rewarding the industrious with enhanced production and prosperity.

It is timely to remind the reader that this particular interpretation does some violence to Locke's intentions insofar as it dispenses with all appeals to God but is as close as one can get to Locke's intentions while presenting the argument in secular form. More importantly, though, this version of the argument captures, I am convinced, the argument as it has been interpreted by Classical Liberals in recent decades.

We might ask, at this point, is this argument valid? That is, does it really accomplish what Locke takes it to accomplish? Does it really provide a compelling argument for the claims that, first, the right to own property is a perfectly democratic right enjoyed by everyone and grounded, in the first instance, in ownership of one's body and its capacity for labour, and second, the inequities that result from a capitalist economy premised on this notion of private property are perfectly justified?

The short answer is, almost certainly not. As noted, there are troubling questions that can be raised about Locke's crucial assumptions regarding the nature of labour, the notion of value added and what it means to say that when we labour upon the resources provided by nature, we mix something of ourselves into these resources, thereby making them our private property. Moreover, there are problems raised by cases of persons born disabled without the ability to labour. On Locke's assumptions, so long as they are rational, they are part of the human family and have the same worth as anyone else, yet they seem excluded from the possibility of acquiring property and in ways that have nothing to do with a lack of industriousness.

Noting these deficiencies of Locke's argument, I propose to ignore them in what follows. We will put Locke's theory to the test in a different way by contemplating its implications. Since Classical Liberals have taken the argument seriously, let us do the same and grant Locke his argument. We can trace out some of the implications of taking this argument at face value and ask what follows from this. As we will see, some of the implications of Locke's arguments raise troubling questions for Classical Liberalism and, indeed, for all of us.

Some consequences of the theory

There are four consequences, or implications, of Locke's theory that are particularly relevant to our concerns here, and these will be themes that wind their way throughout our discussion of Classical Liberalism in the chapters that follow. First is the implication that our labour is alienable and this despite the fact that our lives are inalienable. This means that, although you are the sole proprietor of your life and your body and your body's capacity for labour, you can sell your labour, that is your body's capacity for work, to another for a wage. In this way, you alienate your labour by treating it as a commodity subject to market forces.

One might raise a question here: what can it mean to say that my life is inalienable given Locke's claim that we are God's property? This returns us to the point noted earlier in connection with the second premise of his argument, namely the issue of how it can be consistent for Locke to claim that we are born the sole proprietor of our bodies *and* that we are God's property. As noted there, the only way to reconcile these claims is to argue that Locke holds the view that God grants to each of us sole stewardship of our life and our body. Locke regards it as a moral wrong to end one's life because it is for God, and God alone, to decide the moment of one's death. So one who commits suicide violates the stewardship entrusted to them by God. But, by the same token, insofar as you are the sole steward of your body and your life, it is inalienable in the sense that anyone who denies you your life or your liberty does a moral wrong. Not only do they offend God, but they also violate the terms of the trust placed in all of us to be solely responsible, in this life, for ourselves.

This idea of our right to life and liberty and the moral limits this imposes on our interactions with one another will be a central concern when we turn our attention to normative considerations in Part III. The issue of what happens to these moral limits when we secularize Locke's theory become particularly problematic. For now, I am content to merely note the point.

But now, having argued that Locke's view can be made consistent, let us confront the question of how it can be that your life is inalienable, but your labour is alienable. The answer is that it is a consequence of the division of labour that arises out of the barter system we have seen in our earlier example. Recall that God has provided for us everything needed for our sustenance, indeed our flourishing. Moreover, there is a fundamental law of nature written into the very constitution of the world that God intends us to sustain as many people as possible and that this will require of us that we apply ourselves diligently to the resources God has provided. God wants and expects us to work hard to command and steward the Earth for all humans and to flourish as a result. So imagine now that you are this hunter–gatherer/proto–orchardist with the grand scheme to take a few acres out of the local Commons and turn it into a magnificently productive orchard. In doing so, you are providing abundance (in apples at least) for yourself and the others you will trade with and, so meeting God's intentions to assist with human flourishing. But equally obviously, you cannot do this yourself. As an individual, you would never be able to pick all the apples and get them to market before they spoil. So you hire others to help you with this. You rely on their labour to help you accomplish your goals and they rely on the wage they get from you (presumably in apples at this early stage) to meet theirs. And together, your combined efforts also increase the likelihood that the locals with whom you trade will be advantaged. All of this is clearly supportive of the Fundamental Law of Nature which is the expression of God's intentions for us.

And this synergistic effect is greatly enhanced by the invention of money. As we have seen, the invention of money serves two valuable purposes. First, it becomes the universally recognized measure of the market value of the end goods we rely on.

And second, it can be accumulated without spoiling, and so it has a multiplicative effect on the benefits that result from the division of labour. Now your employees can trade their labour for money instead of apples. After all, how many apples can a single employee take as wage for their labour before the apples begin to spoil in their possession? And we all of us need more than apples to survive. If I spend all, or even most, of my available labour picking apples in exchange for apples, how will I get the other things I need? Money addresses all of these problems. It is utterly worthless in itself. Its value lies in its capacity to maximize the application of human labour to guarantee human flourishing, which it can do only if our labour is itself an alienable market commodity.

But this response is a prudential rationalization, not an explanation or a justification consistent with the premises of Locke's argument. It merely overlooks the fact that as soon as labour is regarded as a marketable commodity, it opens the door to exploitation of those selling their labour. It does not in any way address the fact that your labour, as an intimate part of you, is now under the direct control of another. It is simply what we must accept to ensure human flourishing. But we can ask, who flourishes? And who is left to merely survive?

Intimately related to this point is the *second consequence* of concern to us here. It is the claim that we live in a *meritocracy*. That is, assuming one lives in a capitalist state, with laws protecting all of us in our lives and property, then wherever one ends up in the distribution of wealth is a matter of one's personal merit. This is the real force and the implicit message of Locke's Main Conclusion, wherein he takes himself to have provided an adequate ·philosophical justification for the different degrees of wealth we see in capitalist societies. Essentially, it comes to the claim that some people are more industrious than others, possessed of a willingness to work harder and, perhaps, more willing to take risks, bargain harder and so forth. So the implication is held to follow that these persons deserve what they have acquired, so long as we assume they have played by all the relevant rules. Specifically, they have not engaged in theft (as Locke puts it, they have taken the resources from "the State that Nature hath provided and left it in") or they've purchased labour and goods from others by fair bargaining, without fraud or coercion. If we suspect that anyone has not played fairly, there are laws, courts, an entire judicial system built on the Rule of Law, to prosecute these individuals and seek remedy from the courts. Under these conditions, everyone finds themselves to be exactly where they deserve to be in society, whether you are a wealthy billionaire or a homeless person sleeping under a bridge.

I raise a couple of points about these first two consequences before pressing on. First, consider the differences of capacity that are a matter of the natural lottery, that is the arbitrary circumstances of one's birth. It is simply false, in one very obvious sense, to say that, insofar as we are all the sole proprietors of our bodies and their capacity for labour, we are to that extent born equal, as the example of the person born disabled was meant to illustrate. We are equal in the valuational sense, of course, but that is not relevant when we turn to the supposed implication of Locke's argument that we live in a meritocracy, because that has to suppose that we are fundamentally equal in the sense of our capacity for work; we differ only in industriousness, which seems to be for Locke a matter of personal character and not a capacity in the same sense as, say, your ability to lift heavy weights or solve difficult mathematical problems. Locke's argument requires that we be equal in our capacities if we are really to deserve where we end up in the distribution of goods, since this is the only way to make it the case that where you end up is a matter of your character and not some limitation you could never transcend.

Now someone wishing to defend Locke on this point might say something like the following: "Well look, it's true we differ in our capacities across a range of talents – some people are stronger than others, while others are good at solving math problems – but, as an average, we are equally blessed." Now aside from significant problems trying to understand what might be meant by "average" in this context, I submit it simply isn't true in any event. Some people, perhaps many, are born at significant disadvantage in terms of their capacity for application of their skills as labour to their own advantage. And for the moment, I'm ignoring all the circumstances of birth that bear on the nurturing and development of whatever talents one might be born with, from lack of educational opportunity to poor diet, family violence and instability, systemic poverty and all the rest. I am supposing what is probably not true, that one could factor out the pure circumstances of one's capacities at the moment of birth and quantify these as one's *potential*. And I am asserting that, even understood in these terms, it is clear we are not all born with equal potential in the sense required by Locke's argument.

Add to this the further problem of decline as one ages, in some cases a great deal, and, again, even if we suppose people are born with equal potential, this decline in later life does not strike us all equally, nor is it a matter of choice and, thus, a kind of character flaw. Interestingly, though, Locke was aware of this and made allowances for it in his theory. Thus, as pointed out earlier, Locke claims, in relation to the two limits on acquisition, that those unable to provide for themselves, or compromised in their *ability* to meet their own needs, have a right to some of the excess production of those who are able. This "right" he takes to follow from the Fundamental Law of Nature. If this law requires us to do as much as we can do to preserve the lives of as many persons as possible, then clearly someone cannot be left to starve simply because they lack the ability to provide for themselves. The obvious case here is the case of children who are dependent for some portion of their lives, but Locke believes the point generalizes. In practice, the calculations that balance the entrepreneur's rights to ownership of the product of their enterprise versus the rights of others, those more dependent, to some of the excess production of the industrious could get quite complicated. Insofar as there is an implicit imperative in Locke to incentivize the industrious to exploit their potential, there must be limits to how much can be taken from them, but Locke is clear that, in his view, it is a natural right of those unable to care for themselves that they be provided with enough of the excess production of the industrious to reasonably sustain themselves. And this view, it must be noted, is not widely shared, or even acknowledged, by Classical Liberals as they have interpreted Locke.

A further point worth considering is that Locke permits one to claim land as property. Even supposing we grant that he has established a democratic right for all to extract what they need from the Commons and make it their own by the application of their labour, it does not follow, absent additional premises, that this also applies to land. Reconsider our earlier example where you diligently work to fence off a piece of land, cutting down all the trees and extracting all the rocks, to turn it into a productive orchard. Let us further grant that this enhanced productivity is the direct result of this individual's industriousness and that this gives them a right to the produce that comes from this parcel of land, subject in the usual way to Locke's caveat that others who cannot provide for themselves will have an entitlement to some of the excess produce sufficient at least to meet their basic needs. There is no obvious reason to grant that this individual also owns the land. We could grant that there is a right to the produce of this land so long as it is continually worked by this individual, or the enterprise they organize to accomplish this fact, without granting them title to the land itself. So, absent additional premises, we would say that the land itself immediately reverts to the Commons upon cessation of the work that makes it productive.

Consider for a moment the long-term consequences of what Locke proposes. It amounts to licence for some to accumulate land, taking it out the Commons, regardless of historical developments that might significantly alter local circumstances. I am thinking in particular of population pressures that leave many now with no recourse but to sell their labour. We might say that this feature of Locke's theory has a multiplier effect on the aforementioned consequence that one's labour is an alienable commodity. Arguably, this has the effect, over time, for capitalist societies to become plutocracies. As a fact of the history of Capitalism, we know it to be the case that land becomes the repository of rentier wealth. That is, it is the accumulation of land that creates propertied classes who, although often not productive themselves, subsist on the rents extracted from the lands they own. And this kind of wealth and privilege is precisely the kind that accumulates over generations to widen even further the inequities of wealth that Locke attributes to personal character, as though it depended only on one's willingness to work hard.[16]

The *third consequence* I wish to discuss is that nature, in and of itself, is virtually worthless. Its value for us is exhausted by its capacity to meet our needs, which it does in two respects. First, it is the source of the basic resources we use to meet our needs and fashion the end goods we use to enhance our lives. And second, it is a *sink*, defined as the place where we dispose of our waste. As we saw in the earlier quote regarding the relative value of labour, in Locke's estimation the overwhelming majority of the value embodied in the end goods we use is the result of the value added by human labour. In fact, Locke estimates the value added by human labour to the basic resources provided by nature to be in the range of 90–99 percent as an average. It is perhaps not too hard to imagine how Locke might have convinced himself of the reasonableness of this point, as his comparisons of cultivated fields with those left fallow is meant to illustrate.

Additionally, though, it is worth keeping in mind that in the 1680s, the population of the entire planet is estimated to have been between 550 and 650 million. At this level of population density, there are several features that Locke probably could not have even imagined. First, he could not have imagined that we would ever be at risk of running out of basic resources or land for that matter. But also, it is unlikely he could have imagined that our waste products could accumulate to the extent that is a reality for us today. In all likelihood, he saw the resources provided by the Commons and its capacity for absorbing our waste as, for all intents and purposes, infinite. It was evident to him that God intended it to meet our needs in perpetuity. Given this context, and his view that God intends for us to labour for all we achieve, it's not hard to see why he would put the emphasis he did on the value added by labour to the basic resources provided by the Commons. But I would argue that it is this assumption of the relative worthlessness of nature, as such, that is most directly at the heart of our present predicament as we confront the very real possibility of environmental collapse as an existential threat.[17]

This assumption is the lens through which Western capitalist cultures have viewed nature, and our relationship to the natural world has been shaped by this view. We have regarded nature as a resource for us to exploit at our pleasure, as a set of resources and a sink for our waste, and it has inclined us to see ourselves as the source of all real value in the world. Moreover, insofar as it is also premised on the idea that we are to subdue and command nature, it has inclined us for centuries to regard ourselves as separate from nature. The view is that we manipulate nature to our advantage, not unlike the way we manipulate the apparatus in a laboratory, which is to say, at a distance and with a specific goal in mind. Nature is perceived to be something *other*, out there and distinct from us, that we bring under our control. This illusory view is further enhanced by the organization of our lives in cities, where the conveniences of modern life, from central heating and air conditioning, transportation devices and all the rest of it, act as a kind of prophylactic, secluding us from the ravages of nature and seducing us into believing we are immune to the effects of these ravages. Even when we go out into nature, we regard it as entertainment, often carried along in our snow mobiles, ATVs, boats and cruise ships, leaving disturbance and degradation behind us everywhere we go, before making our way home again.

It is a relatively recent revelation for capitalist societies that we are inextricably part of nature.[18] I have argued in Part I that this revelation has come principally from the development of systems theory. As a result, we are beginning to appreciate the many services provided by nature that are essential to our well-being, indeed our lives, and which have no technological substitutes (at least not at the requisite scale). Some examples include the production of oxygen, the sequestration of carbon dioxide, the regulation of climate cycles, the production and retention of topsoil, regulation of hydrologic cycles and more. These are things Locke was completely unaware of, and this lack of awareness made it possible for him to think of nature as he did. We must now confront the impacts of this legacy with real urgency.

Lastly, as our *fourth consequence*, Locke states explicitly, in relation to God's plan for us to work hard to subdue and command the Earth and improve our lot in life, that land that lies fallow is considered to be in the Commons. That is, land that lies fallow is not presently owned and, thus, is available for someone willing to take it out of the Commons and work it to enhance its productivity. This follows directly from the prime directive implicit in the Fundamental Law of Nature, insofar as the enhanced productivity of the land, wherever it occurs, enhances the survivability of more persons by providing more end goods than would otherwise be available. Thus, if one were to take a parcel of land out of the Commons, by fencing it for example, but left it fallow, this would be a kind of waste which, as we have seen, is a violation of the two basic limits that follow from the Fundamental Law of Nature.[19]

As a historical fact, this particular implication of Locke's theory has been used extensively to justify centuries of colonialism. Throughout the 16th and 17th centuries, wherever Europeans went, they claimed to find abundant land that was largely unoccupied and, thus, theirs for the taking. And even where occupied, it was often claimed that the land remained fallow. The justification for appropriating the land of indigenous societies has often been premised on the claim that the land remained fallow, thus free for the taking. More strongly, this last observation has often been conjoined with moralistic evaluations of these indigenous cultures to the effect that they are shiftless and lazy, shirking the duty entrusted to them by God to apply themselves industriously to the betterment of their condition. This, in turn, has been at the heart of the justification of indigenous relations for centuries, from confinement on reserves (to get them out of the way of those willing to work hard) to cultural assimilation (do them the favour of making them like us).[20]

One might respond that the New World really was mostly empty, after all, and settlers had no reason to think otherwise. But we know from the historical record that wherever settlers landed they quickly came into contact with native populations. And it is now indisputable that this contact often had disastrous consequences for indigenous populations, who died in the millions as a result of diseases against which they had no immunities.[21] The response to this, perhaps, is to argue that while this may be regrettable, these natives lived in a primitive state as hunter-gatherers and, thus, the land did lie fallow. So long as settlers left them enough territory to meet their needs as hunter-gatherers, whatever these settlers did take out of the Commons for cultivation was fair game. However, recent archaeological evidence indicates that this grossly oversimplifies the situation.

In a book published in 2005, Charles C. Mann summarizes this evidence and argues that indigenous populations typically lived in a state of advanced societal development, and they did not leave the land fallow.[22] While this has been known for some time in regard to, say, Mayan culture, which has left abundant structural evidence of this advanced and technological development, in regard to the indigenous cultures of North America it is only coming fully to light in the last few decades. In many cases, it was a part of their cultural heritage to regard themselves as embedded in, and a part of, the natural world. The emphasis was on living in harmony with the natural world and, thus, they deliberately built structures that were not "permanent" but would disintegrate and return to the Earth. The decimation of their populations was so sudden and disruptive that in many cases the structural evidence of this advanced state of development disintegrated before Europeans could fully acquaint themselves with it. In any case, the moralistic evaluations by Europeans of indigenous cultures failed to recognize any stance in relation to the natural world that was not premised on the idea of nature as something to be controlled.

★★★

This concludes our discussion of Locke. In Chapter 6, we turn our attention to the second foundational theory that is constitutive of Neoliberalism, Adam Smith's theory of the laissez-faire market.

Notes

1 I call this informal history insofar as it is not my concern here to engage in the kind of detailed discussion one might expect in a critical history. My concern here is merely to provide sufficient historical context for the reader to understand Locke's motives.
2 As told by Peter Laslett in his critical introduction to, *Two Treatises*, p.17.
3 Cranston (1985, p.3).
4 Locke (1988, p.286).
5 Hobbes (1968, p.186).
6 Op. cit., pp.270–271; emphasis added.
7 Op. cit., pp.287–288.
8 Op. cit., p.288.
9 For a discussion, see Lloyd-Thomas (1995).
10 The example is from *Anarchy, State, and Utopia*, (1974). See further discussion in Chapter 6.
11 Op. cit., p.296.
12 We see here implicitly the elements of the Protestant work ethic that will be articulated by Max Weber in his classic, *The Protestant Ethic and the Spirit of Capitalism*, originally published in 1905. Quoting Locke,

> God gave the World to Men in Common; but since he gave it to them for their benefit, and the greatest Conveniencies of Life they were capable to draw from it, it cannot be supposed he meant it should always remain common and uncultivated. He gave it to the use of the Industrious and Rational, (and Labour was to be his title to it;) not to the Fancy or Covetousness of the Quarrelsome and Contentious.
>
> [Locke (1988, p.291)]

13 Op. cit., pp.300–301.
14 More on this in the next chapter where we discuss the market theory of Adam Smith.
15 Op. cit., p.301.
16 On the issue of the accumulation of rentier wealth and its impacts over generations, see Picketty (2014).
17 Indeed, it must be noted that there are many scientists convinced we are already well into environmental collapse. See Donella Meadows et al. (2004).

18 Perhaps, it goes without saying that many indigenous cultures around the world have shared, as a feature of their world view, the premise that we are a part of the natural world.

19 In relation to the earlier point about land taken out of the Commons becoming private property, the apparent contradiction is resolved, for Locke, by supposing that, as a rentier, I have laboured to take that parcel of land out of the Commons (by, say, putting a fence around it, and perhaps adding some buildings and infrastructure) and if I now rent it to a farmer who leaves it fallow, the blame accrues to the farmer renting the land, not to the rentier. The rentier is blameless and continues in their property rights to the land.

20 There has been rather a lot of discussion of this in the literature in recent decades. See, for example, Ellen Meiksins-Wood (2003). In this book, Meiksins-Wood traces the use and adaptation of Locke's theory of private property to justify the colonial appropriations of imperialist states, beginning with the English colonization of Ireland. See also, Whitehead (2010).

21 Diamond (1997).

22 Mann (2005). A particularly startling revelation in regard to the claim that native cultures left the land fallow is that evidence indicates the Brazilian rainforest may be the result of deliberate cultivation by cultures long since disappeared.

6

ADAM SMITH AND THE LAISSEZ-FAIRE MARKET

Historical context and Smith's motives

The Wealth of Nations, as it is popularly known,[1] is justly deserving of its reputation as a classic. It is widely regarded as one of the principal texts in establishing economics as an autonomous discipline, and it sets the stage for much that followed in the development of economic theory and continues to be enormously influential and relevant to this day. It is also a weighty tome in the literal sense, coming in at a little over 1,200 pages in my edition. This is worth mentioning because, though often quoted approvingly by Classical Liberals, the references are almost exclusively to a couple of passages from Book I and a single passage (concerning the infamous *Invisible Hand*) from Book IV. There is little evidence that many of those who quote him or cite him as an authority have read the book in its entirety, or even much of it, and the suspicion lingers that much of the narrative surrounding the contemporary critical discussion of Smith and his theory seriously misrepresents his real views.

The book was published in 1776, a truly momentous year as it is the year of the declaration of American Independence and the year that James Watt patented the governor (a kind of speed controller) that significantly advanced the use of steam power. A great deal has changed in the 88 years since Locke published his *Two Treatises*. First, industrialization has proceeded at a rapid rate, particularly in the Netherlands and Great Britain. Steam power has been in use for several decades, but Watt's innovations will drive rapid progress in extending the application of steam power to industrial processes and to transportation. Thus, the pace of change – social, political, technological and economic – continues to accelerate.

Casting our glance back even further, to the period since the heady days immediately following Columbus's discovery of the new world, the locus of power in Europe has shifted from the wealthy city states of the Mediterranean to the North

DOI: 10.4324/9781003388555-8

Atlantic coastal states. And this is because the wealth made accessible by the new world has favoured Atlantic maritime powers. The old trade routes across the Middle East have declined in importance. Commensurate with this, there has been a continual erosion of the Mercantilism prevalent in Europe in the 16th century, and unfettered Capitalism has become the dominant economic organization in most of Western Europe. Smith himself notes there are still some Mercantilist states in existence. He mentions Spain and Portugal (in his lengthy discussion of the relative value of silver and gold in Book II), but the failure of Mercantilism is evident in Smith's observation that these two countries are, in 1776, among the poorest nations of Western Europe. It is the industrial capitalist powers that are now dominant, in Europe at least. Additionally, there has been further intensification of urbanization and development of nationalism as the imperialist ambitions of these capitalist states drive them to coalesce public opinion and purpose around the goals of the nation state.

A further feature of interest is the extent to which Locke's theory of the democratic republican state, and of the role of private property within that state, is by this time the received view. This is evident from the casual way Smith references the view and his unquestioned acceptance of the assumptions implicit in the Labour Theory of Value in particular. Contemplate, for a moment, the significance of this. In less than a century, it has gone from being a radical and highly contentious view that required personal risk and real political and intellectual struggle to establish itself, to being the received view. More than anything, this speaks to the extent to which it aligns with the interests of the capitalist class and their now well-established and undeniable influence and power.

In connection with this last point, though, is the emergence of monopolies. The connection lies in the *quid pro quo* bargains that characterized the years that Mercantilism was ascendant, between sovereigns on the one hand and wealthy capitalist entrepreneurs on the other. Columbus is himself an example of this. After being turned down by several sovereigns, he and his private backers were finally able to solicit support from King Ferdinand and Queen Isabella of Spain. Though convinced at the time that they could find an alternative route to the Asian markets that would undermine the natural advantage then enjoyed by the Mediterranean powers, events turned out very differently. In the wake of the discovery of the new world and its abundant resources, the race was on for various kingdoms to enrich and empower themselves while simultaneously emancipating themselves from the grip of the Holy Roman Empire. Given that these sovereigns were often cash-strapped, these goals were best served at the time by entering into cooperative ventures with these wealthy entrepreneurs. Simultaneously, the emerging class of wealthy entrepreneurs greatly enriched themselves and extended the range of their political influence. As the bargain between these sometimes-conflicting interests continued through the period of rapid changes that introduced the modern nation state, certain of these entrepreneurs were able to negotiate a monopoly of many markets. And, also, these monopolies sometimes served the interests of the imperialist nation states themselves. Thus, Smith notes, as one example, the state-guaranteed monopoly of the British wool industry at the expense of the smaller and less powerful Scottish wool producers.

One characteristic of Smith worth noting is his keen powers of observation. In fact, putting it that way is rather an understatement. When one reads the book, one is often astounded by his ability to observe the details of commerce and industry and deduce conclusions that link observations of present circumstances to known historical conditions into a seamless narrative. And the purpose of this narrative is to explain the differences of wealth and prosperity amongst the various nations of his day. He notes, for example, that China, though still the wealthiest nation at that time, is in doldrums, possibly in decline. Conversely, the United States (still a colony at the time he was writing) is possessed of an economy that is vibrant and rapidly growing. Relative comparisons of the European nations reveal that the Netherlands and Great Britain are the wealthiest, with (as noted earlier) Portugal and Spain among the poorest, and this despite their relative wealth and power in the 16th century at the height of the Mercantile economy. The question that drives Smith is, how are we to explain the relative differences in the wealth of nations and how these differences change over time. Thus, the title, *An Inquiry into the Nature and Causes of the Wealth of Nations*.

In particular, insofar as it seems that Capitalism is ascendant relative to Mercantilism, we can ask why this might be the case. What is it about Capitalism that might explain its superior capacity for producing wealth and prosperity? Additionally, and insofar as there are significant differences in wealth between capitalist states, what might explain this? Are some organizations of capitalist economies superior to others? This, in the end, leads to the question of whether there might be some *ideal* organization of a capitalist economy that might unlock the power of markets to deliver wealth and prosperity. Put differently, Smith asks, "What are the conditions of a market-based Capitalist economy that could be expected to maximize market efficiency?" And what would it mean in this context to speak of market efficiency? And so, with this brief description of Smith's environment, and of his theoretical motivations, let us see what conclusions he comes to.

Reconstructing Smith's theory

When trying to understand the decline of such Mercantile powers as Portugal and Spain relative to the wealthiest of the capitalist nations, Smith is quickly led to the observation that Capitalism is unparalleled in its capacity to facilitate the growth and development of the division of labour. And it is the division of labour, in turn, which is the single most important factor for enhancing productivity. This, the reader will recall from Chapter 5, was also noticed by Locke. Locke observes that it is the invention of money, as a source of liquid capital easily directed to the most efficient uses, that results in the development of the division of labour far beyond anything that might be possible in a barter economy, and that this in turn has the beneficial consequence of enhancing the production of the end goods we use. Thus, Smith's first premise:

P1: The greatest improvement in the productive powers of labour, and the greater part of the skill, dexterity and judgement with which it is anywhere directed, or applied, seems to have been the effects of the division of labour.[2]

As an example of Smith's power of observation noted here, this is the occasion in the text for the famous "pin factory" example. So consider the manufacture of even so simple an object as a straight pin (i.e., the kind used by dressmakers etc.). Smith observed a relatively small pin factory employing ten people per shift, each devoted to a distinct part of the process. Thus, one man draws out the wire that constitutes the body of the pin, another straightens it while yet a third person puts the point to it. Other individuals make the buttons and affix them to the pin. In fact, Smith notes, the production of pins can be broken down into as many as 18 distinct operations, and generally there is a positive correlation between dividing a process into distinct operations and productive efficiency. That is, all else being equal, the more distinct operations assigned to different individuals, the more efficient the production process will be relative to the manufacture of the same good using fewer distinct operations. Performing some basic calculations on the production total in a day for the pin factory he observed (about 48,000 pins in a day) he determined that each individual was responsible for a proportionate amount of 4,800 pins in a workday. Whereas a single individual on their own, even after a bit of practice, "could scarce, perhaps, with his utmost industry, make one pin in a day, and certainly could not make twenty."[3]

But why should this be the case? What is it about the division of labour that has such a multiplicative effect on the productive capacities of industry? As Smith points out, there are at least three reasons for this. First, it has the effect of increasing the dexterity of each particular workman involved in the enterprise. Here, Smith's use of "dexterity" is meant rather more broadly than the literal meaning, "skill in performing tasks." While it is certainly meant to include that, the usual understanding of "skill" would typically direct our focus to fine craftsmanship of the sort embodied in, for example, the manufacture of hand-built guitars. In many cases, the manufacture of fine instruments by a craftsman will be exactly opposite to the case involving the division of labour. The point of craftsmanship is that a single craftsperson is typically involved in the manufacture of the product. Thus, a single luthier is often responsible for the entire process of manufacturing a beautiful custom guitar, from the selection of the tone woods to the steaming and curving of the sides and the top, the meticulous crafting of the neck and insertion into the body and even the staining and finishing. This is a paradigm example of skill, or dexterity, as normally understood, yet obviously can't be what Smith is talking about since it can take an individual luthier six months or more to craft a single guitar.[4]

The sort of skill Smith is speaking of is the sort that can result only from the repetitive application of one's attention and labour to some part of the production process that, ideally, has been so simplified as to make it easy to train any labourer in a very short time to do the task efficiently. In an example from my own experience, I once had a summer job re-binding used textbooks. The basic process involved cutting off the old covers, silk screening new ones, cleaning up dog-eared edges and then binding these cleaned-up texts to the new covers. After a bit of time spent silk screening covers, I was trained to be one of the re-coverers. There were three steps to this process. First, using a one-ton press with a very sharp knife operated by a foot pedal, I cut about a quarter inch off the three outside edges of the book. This is the step that cleans up dog-eared edges. Then, using a brush and glue pot, I would spread a thin coat of glue on the outside surfaces of the book. At this stage, one must be careful to use exactly the right amount: not enough glue and the cover falls off, too much and you'll have big trouble at the last stage with the glue leaking out the edges and onto the pages. Next, I would insert the book into the new cover (being careful to put it in the right way with the spine to the back edge of the cover) and using a different one-tonne press, I would compress the book tightly for about ten seconds. At this stage, it is important to press the spine of the book firmly against the back of the press and ensure it is straight before pressing the foot pedal. And, oh yes, be sure to get your hands out before pressing the pedal. Now, I was required to keep a count of the total number of books done each day, so both my employer and I could keep track of my progress, particularly in relation to others doing the same task. It was a surprise to me how much I improved for the first week or so, as measured by my total count, before levelling off at a daily production quantity that was a close match for the other two individuals who had been doing the job for over a year. This levelling off indicated I had reached my mature skill level on this set of operations and, coincidentally, my numbers did not improve noticeably after this first week of training.

This example can be used to pull several threads out of Smith on the issue of "skill" in connection with productive efficiency. Note, first, that as a book re-coverer, I was performing three distinct operations. It would presumably have enhanced productivity further if this part of the process had been further divided and assigned to three distinct workers. One person responsible for cutting the frayed edges off the books, another to glue the resulting "cleaned up" book and a third to press the glued book into the new cover. Recall Smith on the pin factory he visited: it had divided the process into ten distinct operations but, as Smith notes, it could have been (and, in some cases, was) divided into as many as 18 distinct steps. But we must note this need not be the case; that is, it is not *necessarily* the case that more steps assigned to distinct individuals will *always* have the effect of further enhancing productivity. It would depend very much on context, in particular on available space, organization of that space and machinery available. So, for example, in some cases it might have a negative impact on productivity to have one individual cutting off frayed edges, another to apply glue and then yet a third person for pressing the finished text. We could easily imagine a bottleneck if one operation was simply quicker to do. Ultimately, it will be a decision to be made based on experience. However, this example also illustrates the sense meant by Smith when he speaks of dexterity. Attention is required to do the job properly, and to make sure you don't maim yourself, and we can see that there is a period of acclimatization leading to a levelling off of productivity growth that typically indicates maturity in the process has been accomplished. But as such it need not require anything like the skill employed by a fine luthier manufacturing a custom guitar.[5]

The second reason that explains the multiplicative effect of the division of labour on productivity is the time saved by not requiring the worker to be continually moving from one process to another. Again, speaking to my own experience at the book re-binding factory, I am comfortable claiming that if I had been responsible for the entire process for each book, from cutting off the original cover to the final pressing of the renewed book into its new cover, it's unlikely that I could have renewed more than two dozen, perhaps three dozen on a good day. (Incidentally, these processes were, for a variety of reasons, located in two distinct buildings, which we can safely assume is not unusual.) My mature totals at the re-binding station were in the range of 950–1,100 books per day. There were three people at re-binding stations, so we were refinishing something in the range of 2,800–3,300 books a day with about a dozen employees on site. Using a straightforward proportional calculation like Smith, we can assume that working individually, our production totals would have been in the range of 240–360 books a day, and this is without taking into consideration the inputs of the other employees doing different things (e.g., accountants, sales personnel etc.) but still necessary to the end result.

Third, the division of labour, in addition to the relatively straightforward sense in which it clearly enhances the productivity of individual workers, also leads to "the invention of a great number of machines which facilitate and abridge labour, and enable one man to do the work of many."[6] In other words, the division of labour also enhances technological innovation. There are several reasons for this. First, it is a well-established fact that innovations in machinery are often discovered by workers themselves, focused as they are on some distinct part of the production process. We can speculate on whether the drive to explore and continually tinker are universal features of human nature, or perhaps it is the result of the incentives built into capitalist systems, but it is an undeniable fact of the historical record. As Smith himself says, "Men are much more likely to discover easier and readier methods of attaining any object, when the whole attention of their minds is directed towards that single object, than when it is dissipated among a great variety of things."[7] Additionally, though, Capitalism encourages others to devote themselves to the challenge presented by the drive to be continually increasing productivity by means of invention, either of machines or of enhancements to processes or supply chains. Again, quoting Smith:

> In the progress of society, philosophy or speculation becomes, like every other employment, the principal or sole trade and occupation of a particular class of citizens. Like every other employment too, it is subdivided into a great number of different branches, each of which affords occupation to a peculiar tribe or class

of philosophers; and this subdivision of employment in philosophy, as well as in every other business, improves dexterity, and saves time. Each individual becomes more expert in his own peculiar branch, more work is done upon the whole, and the quantity of science is considerably increased by it.[8]

This last observation by Smith also illuminates a further feature of his first premise. In addition to the claim that it is the division of labour, more than anything else, which has been responsible for enhancements to the productivity of labour, Smith also claims that it has been responsible for "the greater part of the skill, dexterity, and judgment with which it is anywhere directed, or applied." We get some insight into his meaning on this point in his discussion of invention in relation to the enhancement of productivity. His point is that it is the division of labour itself which unleashes the innovation that drives further productivity, which in turn incentivizes the investment of time and energy into further innovation and so on in a beneficial cycle. The rewards of enhanced productivity incentivize efforts to find more opportunities for enhancing productivity, including further refinements to the division of labour itself, that is, finding ways to further divide production processes, supply chains and so forth. And it is noteworthy that it is often the supports provided by technological innovation which make it possible to further subdivide production processes where it might not have been possible previously. Lastly, as we've seen, it is *the focus of attention* that is the result of the division of labour itself which is the principal cause of this innovation in technique and technological supports. A truly wondrous and beneficial cycle which, when unleashed, can drive growth in productivity potentially indefinitely.

So here is Smith's first premise rephrased here as two distinct claims to capture the essence of this discussion:

P1': It is the division of labour which has made the greatest contribution to the enhancement of productivity.
P2': This enhanced productivity in turn results in, and incentivizes, a beneficial cycle of innovation in production leading, in turn, to further enhancements of productivity.

And here is Smith's third premise, which is, I submit, best seen as a sub-conclusion that follows from the first two premises:

P3/C: It is the great multiplications of the productions of all the different arts, in consequence of the division of labour, which occasions, in a well-governed society, that universal opulence which extends itself to the lowest ranks of the people.[9]

Here, Smith is making explicit a claim we found to be only implicit in Locke. In my discussion of Locke, I called this the Rising Tide Argument as there was an implicit attempt to persuade the reader that if society should undertake to structure private property relations according to the premises of Locke's theory, and thereby unleash the possibility of the market exchange of goods under these conditions, it would have the overall effect of making us all better off than we otherwise would be. Metaphorically, it would act as a rising tide that lifts all boats. We see this explicitly in Smith when he claims that the universal opulence "extends itself [even] to the lowest ranks of the people," (qualifier added for emphasis). Classical Liberals are keen to point to this feature of liberalized markets and you may have heard it referred to as the *Trickle-Down Effect*. That is, if we liberalize markets according to the principles alleged to follow as prescriptions from Smith's theory, we can expect the wealth, created primarily by wealthy capitalists, to trickle down to the lowest economic brackets of society. This latter label (i.e., trickle down) was embraced enthusiastically by the administration of Ronald Reagan where it was used essentially as a kind of promissory note as they began to undertake the dismantling of the welfare state and the disempowerment of the unions that had endured to that point since the administration of Franklin Roosevelt. The message to the public, in essence, was, "Trust us, and you will see that, in the fullness of time, we shall all be better off as a result." Later, we will turn our attention to the empirical question of whether, and to what extent, this promise has been realized.

In any event, it is clear, I think, that Smith intends this claim to follow from his first two premises; that is why I have labelled it a sub-conclusion. And this because, as he inserts into this premise, this universal opulence is a direct *consequence* of the division of labour discussed in what I have reconstructed here as his first two premises. Now, it is true that this is intended by Smith to be a causal consequence (i.e., the division of labour is the cause of the enhanced productivity that results), but in Smith's mind the inevitability of this causation, at least under the ideal circumstances he sets out to describe, licenses the inference of P3 from P1' and P2'. The underlying idea is that the enhancement of productivity, which, as we've seen, is a direct consequence of the division of labour, results in more end goods, more of the things we need and want, and even if the distribution of these goods is not perfectly equal, there will still be more for everyone. A bigger pie, as it were, means more pie for everyone.

It is often said, in connection with this, that in a capitalist society, even a beggar lives better than the king of an uncivilized society. Of course, stated in this way, the phrase is merely rhetorical and one might reasonably ask in what sense it might be true. When Europeans first encountered Polynesians, they found a society that, by some measures, could be said to live very well. They hardly worked, able to sustain themselves on the natural produce available to them, had abundant time for play and recreation and seemed very content and unbothered by the sorts of stresses and strains that inevitably accompany the ambitions of more materialist societies.[10] Not surprisingly, they were characterized by Europeans as lazy, incontinent and dishonest. But being charitable, the intended claim is that, in a primitive society, even a king is in a state of continual risk and want in a way that goes beyond the risk and want that would characterize the life of a street person in, say, Toronto, who can avail themselves, if they wish, of meals and a bed at the shelter. The assumption is that the homeless person in contemporary civilized society has access to supports that ameliorate the worst risks faced as a daily reality by persons living in what Smith would have referred to as primitive cultures.[11]

Lastly, I wish to comment on the phrase, "in a well-governed society," which points to the ideal circumstances required for the division of labour to bring about its beneficial consequences. This has a very specific meaning for Smith, and we will explore the subtleties as we develop the argument, but we can anticipate the broad outlines here. A well-governed society would respect the property relations set out in Locke's theory, including the claim that our labour is an alienable commodity.[12] But also, a well-governed society is one where government undertakes to ensure that all transactions in the marketplace are fair and free in the sense that there is no fraud, coercion or withholding of relevant information, and the Rule of Law guarantees legal remedies available to anyone who has been a victim of fraudulent or coercive activity. And, as a related point, all negotiations that conclude with an agreement are regarded as binding contracts that must be honoured by all parties to the contract. This respect for contracts is essential to fortifying the trust that is absolutely necessary for the market to function optimally.

Rephrasing Smith's third premise, we get the following:

P3/C': As a consequence of this enhanced productivity that results directly from the division of labour, there will be (in a well-governed society) an increase in the production of goods which can be expected to raise the living standards of all members of that society.

And here is the fourth premise, which requires a somewhat lengthy quote to provide the relevant context:

P4: This division of labour, from which so many advantages are derived, is not originally the effect of any human wisdom, which foresees and intends that general opulence to which it gives occasion. It is the necessary, though very slow and gradual, consequence of a certain propensity in human nature which has in view no such extensive utility; the propensity to truck, barter and exchange one thing for another.[13]

The essence of this premise is the claim that the division of labour, already established to be so beneficial to us, is the consequence of a natural propensity in human beings to engage one another in trade. We characterize it as a propensity rather than an "instinct" to avoid any implications that it might be a kind of fixed action pattern, an innate ability as it were. Smith's point is not that trade is an instinctual activity in the sense that would make our trading behaviour predictable in ways similar to that in which the blink response or "fight or flight" behaviour might be predictable.

Rather, the propensity in us is a natural tendency to *engage* one another, and this can, and often does, include engaging one another in trade. Put somewhat differently, I think it's fair to say that Smith regards humans as, by nature, social beings. As social beings, we are prone to a variety of behaviours, quite plastic in the range of possibilities they open, for collecting together into groups of various sorts and engaging one another in various ways. Sometimes we fight, often we assist one another, and sometimes we engage one another in trade. All of this is natural given the kinds of beings we are, but none of it results in the sorts of fixed action patterns that characterize what are usually referred to as instincts.

I think it's fair to say of Locke that he was also convinced that we are, by nature, social beings, but there is one big difference between Locke and Smith on this point, hinted at when Smith says that the division of labour, and the advantages that result from it, "is not originally the effect of any human wisdom, which foresees and intends that general opulence to which it gives occasion." Smith is asserting that the division of labour and the markets that can result from this division of labour are not the result of deliberate planning but are a natural consequence of the propensity to engage one another, whereas Locke would insist that markets, indeed all matters of civil life, must be carefully and deliberately constructed, according to carefully worked-out principles and maintained by the coercive powers of government.

Returning to a point raised briefly in Chapter 4, Smith's view is typical of philosophers of the Scottish Enlightenment, which includes David Hume, Francis Hutcheson, Thomas Reid and, of course, Smith. The central ideas of this movement are that humans, as individuals, are possessed of a common sense best characterized as a sound practical judgement concerning matters of everyday affairs and need only to be left alone to act on their sound judgements. As such, there is an emphasis on individual liberty in this philosophy. Moreover, this sound practical judgement is rooted in an instinct for self-preservation that typically manifests itself as a tendency to judge for oneself from the perspective of self-interest. This is not to say that we always act exclusively from motives of self-interest, and Smith and Hume are quick to point out that a natural propensity for empathy plays a profound role in determining our behaviour, particularly in relation to others. Nor is it to deny, or in any way contradict, the claim that we are social beings in the sense noted here. It is merely to emphasize what Smith takes to be an obvious statement of fact, namely that we orient ourselves to the world and our interactions with the world from the perspective of self-interest first, and we are capable of sound practical judgement in most cases as to what will have the desired result; that is, the result that can be reasonably expected to maximize survival and self-interest. Moreover, this sound practical judgement is focused on the immediate facts of the case before us, and we judge on that basis, with less attention to future consequences. Indeed, our insight into future consequences diminishes exponentially the further out into the future they lie. That is, our ability to accurately foresee the future drops off very rapidly.

If we grant this to Smith, at least for the sake of argument, then we are in a position to better understand Smith when he says, in the premise now under consideration, that the division of labour and its benefits are not the result of any deliberate foresight, or wisdom, or intention to achieve that result. Rather, it is "the necessary, though very slow and gradual, consequence" of our propensity, as social beings, to engage one another in trade. This is sometimes referred to as "emergence," which in this context means that, if we leave people alone to judge for themselves in all their transactions with one another, then the beneficial cycle of the division of labour leading to enhanced productivity, leading to further innovations in the division of labour and the supporting technology will simply arise without requiring design or deliberate planning of any kind. More strongly, Smith claims that when we attempt to intrude deliberate design into markets, we usually have a negative impact on productivity rather than the reverse, and this in large part is because of our poor ability to foresee the longer-term consequences of our policies and plans.

I think Smith would want to avoid confusion here by pointing out that, when we leave people free to act on their practical common sense, their focus is on their immediate needs and how "this" exchange can best serve these immediate needs, and this focus on the immediate outcome of each transaction acts as a kind of containment on unforeseen consequences. To repeat, this marks out Smith's view as being quite distinct from Locke's, where Locke would insist that getting things right, concerning government, markets and all matters of social life, requires rational planning and deliberate foresight. Notice, also, the emergence of a *sophisticated* division of labour and efficient markets is slow and gradual. It simply *emerges*, or happens on its own, if market participants are left free, but it happens incrementally, which acts as another check on negative unforeseen consequences. There is always time to respond and adapt. Here is Smith, from one of the most oft-quoted passages that speaks to some of these themes:

> [M]an has almost constant occasion for the help of his brethren, and it is vain for him to expect it from their benevolence only. He will be more likely to prevail if he can interest their self-love in his favour, and shew them that it is for their own advantage to do for him what he requires of them. Whoever offers to another a bargain of any kind, proposes to do this. Give me that which I want, and you shall have this which you want, is the meaning of every such offer; and it is in this manner that we obtain from one another the far greater part of those good offices which we stand in need of. It is not from the benevolence of the butcher, the brewer, or the baker, that we expect our dinner, but from their regard to their own interest. We address ourselves, not to their humanity but to their self-love, and never talk to them of our own necessities but of their advantages.[14]

Summing up then, Smith's fourth premise is best regarded as making a claim that has both positive and negative connotations. Positively, he is claiming that the division of labour is the incremental result of a propensity for humans to engage one another in trade (which is just to limit our focus to one kind of engagement that is part of a larger propensity to engage one another as social beings). And put negatively, it is the claim that this division of labour is best left to the incremental advances made possible by the self-interest of market participants and *not* to the strivings of deliberate design. With this understanding of Smith's meaning, I rephrase this premise as follows:[15]

P4': The division of labour is the result of our natural tendencies as social beings and, in particular, the propensity to engage one another in trade. (That is, the markets and enhanced productivity that result from this division of labour advance incrementally when participants are left to act on these natural propensities from motives of self-interest and are often impacted negatively when we attempt to interfere from intentional design or foresight.)

So, as hinted here, the term "market" has a very precise meaning for Smith. A well-governed society will be one that guarantees the integrity of its markets, so that market participants will be able to act on their propensity to engage one another in trade, each pursuing the transaction from the perspective of immediate self-interest. This means that such a society will guarantee the trust that is essential to preserving the market, which requires that there are guarantees – social and legal – that contracts duly negotiated will be honoured. Specifically, the Rule of Law will guarantee punishments for violators and legal remedies for plaintiffs. Thus, negotiations will be free of fraud, coercion and withholding of relevant information. Other than these formal guarantees, there is to be, as much as is possible or advisable under the circumstances, a restraint on imposing other constraints that might interfere with the motive of self-interest on the part of market participants.[16] Let us call markets in this sense, laissez-faire markets, or, as they are sometimes referred to, Free Markets.[17]

With this as context, here is Smith's final premise:

P5: As it is the power of exchanging that gives occasion to the division of labour, so the extent of this division must always be limited by the extent of that power, or, in other words, by the extent of the market.

Smith's point here is that the division of labour, and the enhanced productivity that is the result, can flourish only in circumstances where there are laissez-faire markets that support exchanges contemplated and negotiated exclusively from the motive of self-interest (for all market participants). Thus, if we are to unleash the full potential of the division of labour and the enhanced productivity that results from this, we must do as much as possible to realize the fullest extent of these laissez-faire markets. Put slightly differently, we want our economic activity to fall, as much as possible, within the scope of laissez-faire markets. To the extent that our economic activity falls outside the scope of free markets, there is every expectation that the productivity will fall below what it otherwise would be and, thus, we will all be impacted negatively to some extent (the inverse of the Rising Tide, as it were).

So putting this premise into our contemporary language, we have:

P5': The extent and scope of the division of labour is directly proportional to the extent and scope of the free markets on which it depends. (That is, the capacity of a society to realize the potential of the division of labour will correlate directly with the extent and scope of the free market in that society.)

And from these premises, we are now in a position to make the conclusion of Smith's argument explicit (Smith leaves it unstated, probably to incline the reader to draw it for themselves):

C: (All else being equal) the wealth of nations will depend directly on the scope and extent of the free markets they encourage and sustain.

The idea being expressed here gets to the nub of things for Smith. When we enquire into the wealth of nations, its causes and the reasons for the wide disparities between nations in this regard, we find that it is related directly to the scope given to free markets in relation to commercial activity: the more commercial activity that falls within the scope of these free markets, the more likely it is that the potential of the division of labour will be realized. Now recall that this means we move incrementally in the direction of the beneficial cycle of refinements to the division of labour itself, which drives technological innovation that drives further refinements of the division of labour itself and so on. In this way, productivity is continually, if incrementally, increasing, resulting in more of the end goods we all desire relative to the inputs required to produce them. Thus, we are all made better off; again, the rising tide that lifts all boats.

But this simply raises another question: what is it about laissez-faire markets that leads to the enhancement of productivity? Why should it be that laissez-faire markets have this remarkable potential that exceeds the capacity of the best minds to intentionally design a market system? The short answer is that, in a laissez-faire market, the system responds exclusively to the forces of supply and demand, which are internal to the market system itself. In such a system, there will be competition between the various market participants that drives the innovation that in turn drives the enhancement of productivity. That is to say, it is this *perfect* and *fair* competition between market participants that is the motive force, the impetus, that drives the enhancement of productivity. And this perfectly free and fair competition is possible only under the conditions of a free market.

So consider the following example. Imagine that you are about to enter the market as a private entrepreneur. You are keen to better your condition, you are willing to work hard, and, so the process begins with a careful examination of the options before you. Mindful of your particular skill set and the resources available to you, the issue turns principally upon where you might best direct your energies. Let us imagine further that, at present, there is a shortage of bread. That is, the supply of bread is at present significantly below what the local market demands. Moreover, being a bread maker would be a good match for your skill set and resources, so you enter the market as an independent baker. Having a lot of liberty to set prices fairly high, and being industrious and hard-working, you quickly grow your business. In time, though, other entrepreneurs have the same thought as you. Those already in the business are also growing their businesses, producing more bread to meet demand, and others enter the market from time to time as new producers. Thus, demand drops off as supply rises. At this point, competition between bakers forces them to find ways to secure their market share at the expense of other competitors.

Now assuming the conditions of a free market obtain, this competition will be perfectly free insofar as all competitors are free to act on their self-interest, but fair because all business conduct will be utterly without fraud or coercion of any kind. A range of outcomes is possible here. Some market participants may decide that the bread market is no longer sufficiently profitable and they may elect to exit that particular market. Others will double down, relying on their skills to offer consumers a better product at attractive prices. They may, for example, get into producing a range of alternative bread products such as Ezekial, sourdough and various sprouted grain bread products. In time, the market adjusts again, without any direct intervention by planners or bureaucracy of any kind, leaving in place only those producers who remain competitive at existing levels of demand and competition. And, importantly, consumers are made better off by being offered better products at better prices. And this example is meant to generalize to any individual market we wish to consider (e.g., automobiles, housing, air conditioners) or, at a more abstract level, markets in the more general sense (e.g., the labour market, the energy sector).

In this way, then, free markets demonstrate the sense in which they are maximally efficient: they allocate resources to where they will be most efficiently used to enhance productivity, and they achieve this without the deliberate intervention of planners or bureaucrats.

It is also worth noting the important role played by *willingness to pay*.[18] In this regard, it is said that the price signal is the information that drives movement in the market. If the price is below what consumers are willing to pay, demand will spike, tightening supply, which signals the opportunity to increase production and/or raise the price. This happens until the price rises above what market buyers are willing to pay which will be signalled by a drop in demand, resulting in over-supply, which in turn signals the need to lower prices and/or reduce production. Thus, prices hover around the point where supply and demand balance each other, and the market produces only those goods that will be consumed and produces them in the amount that will see all of them consumed (in principle). And all of this happens without direct intervention, "as though directed by an invisible hand."[19]

This phrase is quoted from Part IV of Smith's text where he discusses the tendency for market participants to favour domestic products over imports, but it is usually quoted absent all context and taken to imply that the free market as articulated by Smith is maximally efficient at allocating resources where they will find their most productive use. More strongly, it is usually held that any attempts to intervene into the market is likely to have the effect of undermining market efficiency, a cautionary tale for governments everywhere.

While I think it's clear that Smith thought something like this to be true, it's also clear he did not think it to be true in the unqualified sense typically advocated by Classical Liberals (more on this later). In any event, the best way to interpret the notion of the invisible hand, I think, is as follows. If you imagine aliens observing us undetected from outer space, and focused on observing our market behaviour in particular, then the conclusion they would draw is that this behaviour must be directed from behind the scenes to have this remarkable effect of maximizing the efficiency with which resources are directed to where they have the most beneficial outcomes. Though this direction would not be visible to them from their vantage point, it would beggar belief to suppose that such beneficial outcomes could be accomplished without explicit direction and planning. How else to explain the remarkable tendency of this behaviour to synchronize continuously and flawlessly with best possible outcomes so as to continually make these communities better off, at least materially?

A final comment about Smith's argument as reconstructed here concerns the qualification implied by the phrase in brackets of Smith's conclusion, "all else being equal." Strictly speaking, the presumed direct correlation between the scope of free markets and the wealth of nations holds only in cases where we compare nations that are otherwise exactly similar with respect to such factors as, for example, level of industrialization, types of industries, skills of the labour sector and such. More importantly, though, Smith is clear that he's *not* to be taken as advocating the view that there is a simple equation: more scope given to free markets is always better. He is willing to contemplate cases where the state would have good reasons for limiting the scope of the market to prevent, for example, the emergence of corporate monopolies and cases where the profit motive might come into conflict with the public interest.

We will consider this further later when we turn our attention to the implications of Smith's view. But even at this stage it will be helpful, for purposes of clarity, to have some examples of commercial activity that fall outside the scope of free markets. One such example would be commercial activity within the public sector, such as the activity of food inspectors, police, military personnel, policy analysts and in some jurisdictions, educators. Also included, though, is the non-governmental sector (NGOs), comprising those organizations that involve themselves in activities that very often supplement the services ordinarily supplied by governments but in cases where governments may be constrained by a lack of resources. Disaster relief organizations are obvious examples, but this category can also include various "Think Tanks," organizations devoted to analysis of public policy for example. NGOs are outside the market when they depend for their funding on charitable donations or fundraising activities (i.e., they are run as not-for-profit organizations).

With this understanding of Smith's argument, I bring the reconstructed version together here in standard form as we did for Locke's argument:

P1': It is the division of labour which has made the greatest contribution to the enhancement of productivity.

P2': This enhanced productivity in turn results in, and incentivizes, a beneficial cycle of innovation in production leading, in turn, to further enhancements of *productivity*.

P3/C': As a consequence of this enhanced productivity that results directly from the division of labour, there will be (in a well-governed society) an increase in the production of goods which can be expected to raise the living standards of all members of that society.

P4': The division of labour is the result of our natural tendencies as social beings and, in particular, the propensity to engage one another in trade. (That is, the markets and enhanced productivity that result from this division of labour advance incrementally when participants are left to act on these natural propensities from self-interest and are often impacted negatively when we attempt to interfere from intentional design or foresight.)

P5': The extent and scope of the division of labour is directly proportional to the extent of the free markets on which they depend. (That is, the capacity of a society to realize the potential of the division of labour will correlate directly with the extent *of the free market in that society*.)

C: (All else being equal) the wealth of nations will depend directly on the scope and extent of the free markets they encourage and sustain.

Let us now turn our attention to some of the implications of Smith's theory, particularly as it has been interpreted by Classical Liberals.

Some consequences of the theory

There are four implications of Smith's theory as Classical Liberalism has interpreted it that I wish to focus on here. First, is the implication that, in a genuinely free market system, all participants must be left to act exclusively from purely selfish motives. Second, in relation to extending the scope of free markets as much as possible is the prescription for social policy to privatize as much of the commercial activity as possible. That is, to move as much of the commercial activity being carried out by the public sector into the private sector as possible. Third is the prescription to minimize the influence of government in the private sector as much as possible, principally by means of deregulation. Lastly, and as a consequence of the successful implementation of the second and third of these prescriptions, to minimize taxes as much as possible.

Turning our attention to the first of these implications, namely the tendency to read Smith as insisting that a laissez-faire market is one which guarantees that all market participants are left to act exclusively from *selfish* motives, let us begin by getting clear on the difference between selfishness and self-interest, as these terms pertain to market behaviour. *Selfish* is usually defined as an adjective descriptive of a person, or their motives, or as an adverb descriptive of their behaviour (i.e., he behaved selfishly) that is lacking in consideration for others. Someone who acts from selfish motives exclusively is acting egotistically, and they care nothing about the consequences of their actions for others, except insofar as this may rebound on them, by virtue of impacting negatively on their reputation, for example. *Self-interested* is usually defined as an adjective descriptive of actions, motives or behaviour that focus on one's personal advantage but not to the exclusion of all concern for others. An individual can act from motives of self-interest while simultaneously taking into consideration the interests of others and, on that basis, be motivated to negotiate a win–win deal. Smith would be concerned to point out, I think, that we usually act from motives of self-interest because it is our nature to do so, and it is our nature because this is what preserves us and is the means by which we secure our own happiness. But, as we've seen, Smith also argues for the claim that we are, by nature, social beings. Thus, acting from self-interest cannot be incompatible with our social propensities.

Now thus far, this might sound like a subtle difference, a mere matter of degree, but I think that is mistaken. The difference is revealed by asking what *would* happen in a variety of circumstances. For example, we can imagine circumstances where a person would be required to choose between acting exclusively from selfish motives, knowing that their actions will impact negatively on others but in ways that will never rebound on them, versus acting self-interestedly but avoiding harm to others even though this might reduce the benefit to themselves. Under such circumstances we can ask if there would be a difference in behaviour between the pure egoist, the individual motivated exclusively by selfishness, and the ordinary person acting on their sound practical judgement from the perspective of self-interest. It's clear, I think, that the answer is "yes." The egoist, the purely selfish person, will always act to maximize the benefit to themselves, regardless of the consequences for others (again, presuming that this can't rebound on themselves, presumably because it remains undiscovered). Whereas the ordinary individual would stop short of acting in ways that would visit harm upon others, even if they thought this could never reflect on themselves, they would accept less benefit to themselves (possibly even no benefit to themselves) just to avoid harming others.[20]

In part, such behaviour is motivated by relations of reciprocity. Reciprocity is a fragile thing that depends on trust and knowing that purely egotistical behaviour is not the norm. But for Smith it goes further than that. Smith, together with Francis Hutcheson, David Hume and the other members of the Scottish enlightenment made a point of emphasizing that our social natures are grounded in our moral sentiments, which include feelings of honour and pride on the one side, that can drive us to combative behaviour on occasion, but also feelings of sympathy and empathy that can incline us to be cooperative and helpful of others in need or distress, even at some cost to ourselves.[21] In fact, honour can often motivate behaviour that is self-sacrificing because this is, in the context, seen as the honourable thing to do.[22] We are equipped with a sound judgement in practical matters, wherein we judge the situation immediately before us on the basis of self-interest but where this self-interest is always measured against the feelings aroused by the situation before us. Thus, Smith's conclusion is that, left to themselves to act on their sound practical judgement, most people will act in ways that sustain the trust and the reciprocity on which free markets depend.

Moreover, recent experience tells us that encouraging purely selfish behaviour risks damaging this trust. The mistake of the Classical Liberals is that they have urged that, in a truly free market, all participants must be left free to act exclusively from selfish motives while reassuring us that, on Smith's argument, this will make us all better off. The example of Martin Shkreli is a cautionary tale for Classical Liberals, as the predictable outcome of a free market that undermines the trust and the relationships of reciprocity that are required to sustain the system. Shkreli is an American businessman who founded the pharmaceutical firm Turing Pharmaceuticals, among others, which bought patents for several drugs that were for the treatment of life threatening but relatively rare diseases, promptly raising the price of these drugs significantly, making them unaffordable for many. In one case, for a drug known as Daraprim, the price went from US$13.50 per pill to US$750.00 per pill.[23] The ramifications of Shkreli's behaviour continue to negatively impact public trust.[24]

It is best to consider the next three implications of Smith's theory together because of their interconnectedness. Looking at Smith's conclusion, it is clear that the major consequence to come out of this will be a general directive for governments to do what they can to extend the scope of free markets to cover as much of the productive and commercial activity as possible. In practice, this has been taken to prescribe three directives as matters of social policy that have become something of a rallying cry for Classical Liberals: deregulate, privatize and reduce taxes (and the size of government).

Beginning with deregulation, the general idea is to deregulate productive and commercial activity as much as possible. By this is meant that governments ought to minimize the regulations that apply to various economic sectors and greatly streamline and simplify what remains. In fact, some go further and insist that regulation ought to be outright eliminated where possible. Some examples of the sort of regulation include consumer protection laws that require manufacturers of consumer products to meet regulatory standards with respect to product safety, including, in the case of pharmaceuticals, standards for the production site, adequate testing and so forth. Other regulatory controls relate to the requirements to meet environmental standards, including the responsibility to undertake adequate environmental impact assessments, treatment of effluents or even regulations that protect employees from harm.

Now one might wonder, "Why deregulate? How is this relevant to Smith's goals?" Well, the first thing to notice is that regulation imposes costs onto business and these costs leave less money available for reinvestment into growing the business, so productivity is necessarily less than it otherwise would be. This is what is meant by the phrase "regulatory burden": the burden that requires businesses to devote resources of time and money to meeting the regulatory hurdles imposed by government. And, insofar as businesses will be required to pass these costs along to their customers, the public generally will consume less. Thus, there is less overall production, less consumption, fewer jobs and all the rest that is presumed to flow from this.

Moreover, as we saw earlier, Smith's point is that, in the case of a free market, the prices of goods respond to supply and demand exclusively (at least in the ideal case) which, under the conditions of a free market, balance each other because of the perfectly free and fair competition that results. So any exogenous conditions that require prices to respond to conditions other than supply and demand will be expected to have the effect of undermining competition, thus leaving markets less extensive and less efficient at allocating capital and resources than they otherwise would be. Regulations are one very obvious example of such an exogenous condition. With regulatory burden, prices are being distorted by costs that have nothing to do with supply and demand; they are being imposed by government from "outside" the market.

The negative impacts on competition are obvious when we recognize that different jurisdictions often have different regulatory burdens. Thus, Company A will be at a disadvantage relative to Company B to offer comparable goods at competitive prices if the regulatory burden on A is significantly higher than it is for B. The amount of money required for this purpose can be extensive in cases where the regulatory burden is itself extensive, and these costs must be passed along to consumers, which can have the effect of making these products less competitive in cases where foreign competitors meet a smaller regulatory burden. The bottom line here is that regulation can impose many conditions that undermine the perfectly free and fair competition on which the free market depends.

In relation to taxes, it is clear that this regulatory activity requires an enormous bureaucracy that includes qualified policy analysts, lawyers, drafters of legislation and inspectors at the very least. As a result, it is easy to show that, in addition to the specific costs noted earlier that result in less money in the hands of capitalists for investment in growing the economy, there will also be less money in the hands of consumers for the discretionary purchases that stimulate demand. After all, policy must be clear, and often involves coordination with various jurisdictions (federal and state/provincial for example). Teams of lawyers, policy analysts and, often, external advisors are required to draft effective legislation that will meet the goals and objectives set out by policy. Inspectors are required to ensure that manufacturing sites and retail sites comply with regulations. Typically, products must not only meet stringent testing to be licensed for the market, but ongoing testing will also be required to ensure the products continue to meet relevant safety standards. For products coming in from foreign manufacturers, border officials will play a prominent role, often in conjunction with inspectors. Regulations require implementation and enforcement, so there will also be some imposition on the resources of the justice system to deal with recalcitrant offenders. It must be seen that the justice system will backstop the regulations that are in place. So the bureaucracy of regulation will require a lot of resources on the side of regulators (i.e., government) and this must be paid for by taxes additional to the regulatory burden imposed on businesses directly. Thus, insofar as Smith's overarching concern is with the wealth of *nations*, the social costs, imposed as the tax burden required to support this government infrastructure, will also have the effect of making the free market less extensive than it otherwise would be. Put bluntly, the money available for discretionary purchases will be lessened in more-or-less direct proportion to the tax burden; every dollar spent on government is no longer available for other goods or services. So less consumption, less demand, which in turn means supply must shrink. Thus, lower employment levels, in the view of some, puts in place a kind of vicious cycle of economic decline. The remedy, it is clear, is to stimulate economic growth by leaving more money in the hands of consumers.

Which brings us, quite naturally, to the issue of privatization. As discussed earlier, there is an important assumption at work here, namely that the private sector is always more efficient than the public sector. And this, the reader will recall, means that the private sector will always allocate resources to exactly where there is demand and in exactly the amounts needed, and will always do so more effectively than the public sector, for comparable goods and/or services. This implies that the private sector can always supply more of the same good at a lower price than the public sector, across the board for all goods and services.

A central part of the narrative on this point holds that the public sector, as a part of government, is by necessity bureaucratic, and bureaucracy is inevitably wasteful. Consider this argument as applied to a specific example, the case of privatized prisons.[25] Here it is argued that moving the operation of prisons from the public sector into the private sector can save rather a lot of money, though we presumably accomplish the same civic goal, which is the punishment of offenders and in ways that support the goals of deterrence and public protection from criminal activity. Any decision to keep this service in the public sector, extracts more money from society than it would otherwise require, and this is not only wasteful but is a kind of punishment delivered on the innocent public. By the same reasoning, we are urged to privatize utilities, airports, transportation (rail, air, public transit), communications, education and hospitals. Indeed, in principle, there is nothing that should remain in the public sector: even policing and military services can be privatized, at least in part, as indeed is being experimented with in some jurisdictions.

Having done my best to make the positive case for these prescriptions, let us now consider how they might be viewed differently from a critical perspective. Certainly, regulation can be burdensome, and any regulation must be designed carefully to meet public goals with as little burden as possible. It is a recipe for disaster, though, to suggest that we should do away with regulation altogether. Even the more limited prescription to minimize regulation must be approached with caution. It must be remembered that regulation is intended to protect the public interest by, for example, ensuring consumers are protected from foreseeable harms, workers are protected from exploitation, and we are all protected from environmental toxins that could threaten our health, even our lives. Thus, though the narrative for deregulation seems persuasive on its face, it is worth pointing out that it is misleading in at least two respects.

First, it is often suggested that, in many cases, we need not worry about regulation to protect the public welfare because corporations can be trusted to do the right thing simply because it's in their self-interest to do so. We could, for example, deregulate the forestry industry and trust the industry to take care in developing the requisite network of access roads, to not over-harvest, to clean up after themselves responsibly with minimal damage to the harvest area and to replant trees as required to both restore forests and prevent topsoil erosion. After all, their corporate interests depend upon the careful management of this resource. But this is simply naïve. Whether, and to what extent, a deregulated industry or economic sector will proceed in ways that would protect the public interest will depend a lot on other circumstances, not least of which is their competition. This will be particularly prominent in cases where this competition is from other jurisdictions where different regulations will likely apply. The first interest of business is profit and the public interest will almost always be sacrificed for the sake of profit whenever the two come into conflict.

Second, another thread to the narrative is to use extreme examples of bad regulation, flawed in various ways, inviting the public to draw the conclusion that all regulation is bad. For example, badly written policy can make it difficult for businesses to understand what compliance will require of them or may put enormous bureaucratic hurdles in the way of compliance. Certainly, we want well-crafted regulatory policy, but most of it is well-crafted, at least in jurisdictions where there are low levels of corruption and the Rule of Law prevails and where all stakeholders are given the opportunity to have inputs to the process. In this context, "well-crafted" means well-written to achieve policy goals, clarity around what compliance will require and doing so while simultaneously imposing the smallest burden possible. Those who object to regulation on principle will always insist regulation is bad, but this is to put the public interest in a compromised position.

In any event, the historical record is clear: in nearly every case where there has been widespread deregulation of a sector, harm to the public interest has followed and has sometimes been quite severe. A compelling argument has been made that it was the deregulation of the financial sector in the United States, in particular the repeal of the Glass-Steagall Act by President Clinton in 1999, that led ultimately to the financial crash of 2008/2009.[26] We take up this point again, with an extended discussion of this example and others in Chapter 8.

Privatization can also be entirely appropriate, even salutary, but it's wise to have fulsome public debate on each individual case. It is not in the public interest to assume we must always privatize and do so everywhere. As an example, consider telecommunications. There was a time, not so long ago, when phone service (as it was then called) was in the hands of public utilities in many jurisdictions. Selling off the public infrastructure and opening the sector to private service providers has certainly resulted in a much richer set of devices and services available to consumers and possibly even at comparable or better values (though I don't have data to support this). Let's say, for the sake of argument, that this has been a good thing for the public interest. It is worth noting, though, that in the early 20th century, as phone service was beginning to enjoy widespread diffusion, government involvement was needed to build the infrastructure. No private entrepreneur, or even a group of investors, would have considered building the infrastructure required without government involvement. It was simply too expensive and too risky. The same point could be made about transportation services. In Canada, for example, the development of the rail network that linked the country from Atlantic coast to Pacific coast was a central part of nation building. It served the public interest to have a nationwide rail system to deliver goods and people and was, moreover, a *sine quo non* of national unity.[27] Again, no private entrepreneur, or group of investors, could have undertaken a project of this scope with so little guarantee of short-term returns on investment.

But there is a further reason, a matter of principle, for keeping some goods and services in the public sector. The principle might be expressed as follows: whenever a good or service is required for the public interest, it ought to be protected from the intrusion of the profit motive. Assuming fulsome public debate, we can agree that in some cases a service might no longer be so vital to the public interest that it must be protected from the profit motive.[28] Or, alternatively, public choice becomes more important than guaranteeing the good or service. The example of telecommunications might be such a case. There was a time when development of the infrastructure and providing services at affordable rates to ensure widespread adoption made it necessary to rely on the public sector. It seems, though, that under present circumstances protection of the public interest can be achieved by sensible regulation while leaving delivery of the service to private providers. But other goods and services are arguably so vital to the public interest and sensitive to distortions by the profit motive that they must be retained within the public sector. The example of privatized prisons is one which, it seems to me, is such a case. The administration of justice generally, from courts to prisons, is the foundation of the trust requisite for societies based on the Rule of Law. Intrusions of the private sector put this trust at enormous risk. Other examples that occur to me are education and basic healthcare and possibly basic utilities.[29]

One reason for asserting this claim is that, in the case of basic education, for example, it is required to guarantee that, upon reaching maturity, individuals can enter the market with the basic skill set needed to be competitive; to have a fair chance at finding opportunities to survive and thrive. Put somewhat differently, it is about guaranteeing *equality of opportunity*. Anyone willing to accept the premises of a capitalist market society will be willing to accept inequality of outcome but not inequality of opportunity. Inequality of outcome becomes a problem if it becomes so extreme that it begins to undermine the legitimacy of the political system. For example, when wealth accumulates and facilitates the emergence of a kleptocracy that has the power to undermine democracy, resulting in the "capture" of government, there are real reasons to be concerned.[30] But the real problem is when people perceive that they no longer have genuine equality of opportunity, when they cease to believe that they can improve their position. Once this happens, the majority of citizens lose faith in government, and it becomes impossible to coalesce public opinion around a set of shared goals. People perceive, quite rightly, that the system is rigged to favour the interests of the wealthiest. We take up this discussion again in greater detail in Chapter 7 when we compare the views of Robert Nozick and John Rawls on the nature of the ideal polity.

In any event, we have been assuming till now what is manifestly untrue, namely the Classical Liberal assumption that the private sector is always more efficient at allocating resources. The available data consistently challenge the claim that the private sector always provides equivalent services more efficiently. Recent examples of privatization of public utilities in California and Ontario speak for themselves: blackouts, brownouts, disruptions of service and escalation of costs to consumers have created political firestorms in these jurisdictions. Additionally, insufficient resources are dedicated to maintaining the infrastructure, leaving us increasingly vulnerable to potentially catastrophic failures.

A particularly illustrative example is the American healthcare system, which is structured in such a way that health insurance and, in many cases, healthcare, is provided by the private sector but backstopped by government. In other words, the system is designed to be the perfect compromise: we leave delivery of the service (in this case, health insurance and healthcare) to the private sector but guarantee accessibility by having government pay for the services for those who qualify for government support. This arrangement inevitably results in what are called "perverse incentives," the idea being that the structure of the arrangement guarantees that the profit motive operative in the private sector will incentivize the providers to act in ways that will undermine the principles we are trying to protect by having government involved in the first place.

A private healthcare provider will be strongly incentivized to charge as much as possible, to schedule unnecessary procedures and so on, because government is acting as the guarantor of payment. Simultaneously, these same incentives motivate the insurer to deny claims whenever it can be expected to result in a loss, by citing pre-existing conditions or any number of other reasons. The patients are compromised and costs inflate uncontrollably, adding fuel to the arguments on the part of those who would do away with all government-guaranteed healthcare.[31] More telling, though, is the data comparing cross-national healthcare outcomes. Though it is notoriously difficult to settle on a universally acknowledged definition of "efficiency" in this context, one result is to compare national life expectancies against costs per person. Such rankings, calculated by the World Health Organization (WHO), consistently place the American system among the least efficient of the Organization for Economic Cooperation and Development (OECD) member nations.[32]

★★★

This concludes our discussion of the foundational theories of Classical Liberalism. The goal has been to carefully reconstruct the theories and the supporting arguments of Locke on private property and Smith on the free market. As the conceptual foundations of Classical Liberalism, the ultimate purpose is to articulate the underlying narrative of this ideology by making the assumptions explicit and subjecting them to critical scrutiny and tracing out the central implications of these theories. We have surveyed at least some of the ways in which these implications may cause concern for some.

In the next two chapters, I want to explore further these implications as applied to the domains of the polity and corporate governance. To provide a frame of reference for this discussion, I will introduce the alternative ideology that has historically been the main competitor to Classical Liberalism, the view I will refer to as Progressive Liberalism. And so, in Chapter 7 this exploration is carried out vis-à-vis a comparison of the Classical Liberal conception of the ideal polity, as envisioned and articulated by Robert Nozick, with the Progressive Liberal conception of the polity embodied in the work of John Rawls. And in Chapter 8, the focus is on corporate governance and corporate social responsibility as these have been conceived under the ideology of Classical Liberalism, articulated by Milton Friedman, and comparing this with the Progressive conception articulated by Edward R. Freeman.

Notes

1 The full title is *An Inquiry into the Nature and Causes of the Wealth of Nations.*
2 Smith (2003, p.9).
3 Op. cit., p.10.
4 Though, to be profitable, they often work simultaneously on more than one guitar, typically with each instrument at a different stage. See Brookes (2005).
5 The reader can explore the process of mass-producing guitars to get an idea of how the craftsmanship of the master luthier gets transformed into an industrial process that still produces fine instruments. Visit, for example, the website for Taylor guitars: www.taylor guitars.com (retrieved October 26, 2022).
6 Smith (2003, p.14).
7 Op. cit., p.16.
8 Op. cit., p.18. Note that here Smith is using the term "philosophy" in a sense that will strike modern readers as a bit peculiar, to refer to anyone working in speculative, imaginative disciplines, and technological invention and innovation would be the relevant example here.
9 Op. cit., pp.18–19; underlining added.
10 This is not to imply that it was a perfect society. It was, by all reports, a very sexist society, for example.

11 I can only assert here that it is my opinion that this cavalier claim greatly underestimates the risk, existential peril really, faced as a fact of daily life by homeless persons.

12 As mentioned at the start of the chapter, Smith takes Locke's theory to be understood and accepted as the received view of property relations by his readers, at least for the most part.

13 Op. cit., p.22.

14 Op. cit., pp.23–24.

15 For simplicity, I rephrase the claim here as a single premise. I take the part in brackets to be clarification, not a distinct premise. Some might parse it differently but I am confident this does no violence to Smith's intentions.

16 We will consider shortly the sorts of cases where Smith entertains the idea that it might be advisable for government to impose additional constraints on the market.

17 I will use both terms as stylistic variants.

18 We will take up the concept of *willingness to pay* again in Chapter 9 when we discuss Preference Utilitarianism.

19 Op. cit., p.572.

20 This assumes it would not result in harm to themselves or their family and friends. Some persons will even accept harm to themselves to avoid harming others, or to protect others from harm, and this often results in what we call heroic behaviour. What's important, for present purposes, is to note that heroic behaviour is actually not a rare thing.

21 We consider this approach to ethics, the Theory of Moral Sentiments as it is usually known, in greater detail in Chapter 9.

22 For more on this point, and the results of recent research into the extent to which people are willing to engage in behaviour that is self-sacrificing for the sake of the "team," see Haidt (2012).

23 https://en.wikipedia.org/wiki/Martin_Shkreli (retrieved October 27, 2022).

24 https://www.cnbc.com/2018/03/09/martin-shkrelis-legacy-shaping-the-drug-pricing-debate.html (retrieved October 27, 2022).

25 I consider the case of privatized prisons at greater length in Chapter 9.

26 In particular, see Stiglitz (2010).

27 Berton (1971).

28 What can cause real resentment is when a government privatizes public goods, lands or services without public debate. When deals are concluded behind closed doors, it undermines trust in government. An example here would be the case of BC Rail: https://thetyee.ca/Opinion/2011/12/27/BCRail/ (retrieved October 27, 2022).

29 Thanks to my student Serena Gearey who brought to my attention Smith's vigorous defence of public services, basic education in particular, in her unpublished paper, "Freedom to Sell: A Laissez-Faire Free Market Economy, as Imagined Under the Trump Administration," presented January 20, 2017 at Thompson Rivers University's 10th annual, *Philosophy, History, and Politics*, undergraduate conference, under the category of "Trump, Democracy & North American Politics." Here is the link for the programme schedule: https://www.tru.ca/__shared/assets/2017_program39871.pdf (retrieved October 27, 2019).

30 For a detailed and systematic exploration of the ways Classical Liberalism has put adverse pressures on democratic institutions, the following are invaluable: Mayer (2016) and Jacob S. Hacker and Paul Pierson (2016).

31 The best discussion I have seen of the intrusion of perverse incentives is to be found in Kuttner (1996). Chapter 4 is explicitly concerned with the American healthcare system.

32 Many sources are available. Here is one: https://bmchealthservres.biomedcentral.com/articles/10.1186/s12913-015-1084-9 (retrieved October 27, 2022).

7

CONTRASTING VISIONS

Classical versus Progressive Liberalism and the ideal state

Having set out the conceptual foundations of Classical Liberalism in Chapters 5 and 6, we now turn our attention to how this ideology manifests itself in contemporary debate in relation to two issues: conceptions of the ideal state or polity and conceptions of corporate governance and corporate social responsibility. In this chapter, we will investigate and compare Classical Liberal and Progressive Liberal conceptions of the state with an eye to how they differ and what lessons, if any, this might have for us as we turn our attention in Part III to reflections on the moral evaluation of Consumer Capitalism and its supporting ideology. As we shall see, despite sharing many assumptions, these two conceptions of the state are at odds on many details. As Nobel laureate economist Joseph Stiglitz has claimed, the debate regarding these conceptions is at the heart of the central ideological battle of our times,

> [O]n one side [are] those who believe that markets *mostly* work well on their own and that most market failures are in fact government failures. On the other side are those who are less sanguine about markets and who argue for an important role for government. These two camps define the major ideological battle of our time. It is an *ideological* battle, because economic science – both theory and history – provides a quite nuanced set of answers.[1]

We begin with a basic understanding of these distinct visions of what the state should be like, and the supporting arguments each offers, drawing on representative examples from the literature. Our goal, ultimately, is to explore some of the implications and test these against our moral intuitions, as we have done in previous chapters with the views of Locke and Smith. We start with the Classical Liberal conception, defended most ably by Robert Nozick.

DOI: 10.4324/9781003388555-9

Robert Nozick: the Night Watchman State

The view we are about to consider is articulated and defended by Nozick in his classic, *Anarchy, State, and Utopia*, published in 1974. In this book, he refers to his conception of the ideal state as the Night Watchman State because it conceives of the ideal state as absolutely minimal in something rather like the way we might say the single night watchman who guards a condominium, or other facility, is the minimum protection one could have to adequately protect the facility. The suggestion is that the protective apparatus of the state should be as minimal as possible, and there are few, if any, roles the state ought to undertake beyond protective functions. Nozick's argument proceeds in two stages. In the first, he gives the reader an invisible hand argument for the emergence of the minimal state. Though we will get to the details shortly, the implicit idea is that it is possible to give a plausible explanation of how the state *could* emerge from the State of Nature by the ordinary activities of persons in that condition without intending, or directly planning, to have that outcome. In the second stage, he presents a very compelling argument for the claim that this minimal state is the most that can be given adequate justification. More specifically, any state more intrusive than the minimal state can be shown to violate the rights of its citizens.

Stage 1: leaving anarchy behind

We begin with a few remarks to set the context. First, on the point of the invisible hand argument at the heart of Stage 1, it begins by imagining life in the State of Nature and proceeds to show that the civil state *could* have emerged from this condition by a process wherein none of the participants had that particular goal in mind. Questions arise as to the relevance of beginning from an imagined vision of humanity's pre-history. Can we know, with any reasonable assurance, that the imagined state from which we begin bears any resemblance to the actual state of affairs that prevailed? And if not, can it be relevant to build a theory of the state on such an imagined scenario? Such questions have been at the heart of critical discussions of every thinker who has relied on this device to arrive at a theory of the state, including Hobbes, Locke and John Rawls (whose views will be examined next). Nozick's response is that we do not need to suppose that the imagined scenario bears any resemblance to the actual historical facts to be useful. As he says,

> State-of-nature explanations of the political realm *are* fundamental potential explanations of this realm and pack explanatory punch and illumination, even if incorrect. We learn much by seeing how the state could have arisen, even if it didn't arise that way. If it didn't arise that way, we also would learn much by determining why it didn't; by trying to explain why the particular bit of the real world that diverges from the state-of-nature model is as it is.[2]

Nozick begins from an imagined position that is rather like Locke's State of Nature, discussed at length in Chapter 5. It is from this existence in anarchy that the state will emerge. For Nozick, this is properly called anarchy precisely because there is not yet anything like the apparatus of the state in existence. Colloquially "anarchy" carries connotations of being a state of chaos, but this is not how Nozick is using the term. For Nozick, it is simply the term that properly designates the condition of human existence in the absence of a state apparatus. The crucial feature of the state, in this regard at least, is that the state has the exclusive right to use coercion to achieve its aims. So in the absence of this centralized bureaucracy that claims exclusive rights to the use of coercion, people live in a state of anarchy in the sense that each person is left to defend themselves. The right to use force to defend oneself, or one's property, or to use force to seek remedies for wrongful actions on the part of others, resides in each individual, and that is anarchy, a state of completely independent self-rule.

Notice, though, that for Nozick, as for Locke, individuals in the State of Nature have basic rights. These are rights to protect oneself and one's family from harm, to protect one's property from damage or theft and to determine for oneself how one shall live, at least within the limit that, in doing so, one does not intrude into the similar rights of others. This is not a trivial assumption and there are many who would refuse to accept the existence of rights in such a state.

Thomas Hobbes is a prominent example, and he claims that it makes no sense to speak of rights in the absence of the state. For Hobbes, rights come into existence only with the state because rights depend for their existence on the centralized locus of power that resides in the state and gives it the right to lay claim to the exclusive use of coercion to implement laws and protect the interests and rights of its citizens. In the absence of the state and its protective apparatus, there can be no such thing as rights and, in fact, it makes no sense to even speak of rights in this context. There is only the behaviour of individuals acting under the imperatives of the instinct to survive and doing so within the constraints of limited resources and competition with others. Now it is true that Hobbes's conception of human nature is rather dark. Thus, he is inclined to portray the State of Nature in dismal terms. Nozick is at pains to point out that he is quite deliberately assuming a State of Nature (and the corresponding conception of human nature) that aligns most closely with Locke's precisely because Locke's conception, coming as it does between the extremes of Hobbesian brutality and Godwin's pastoral, bucolic utopia,[3] is for this reason likely to be the best one could reasonably hope for in the State of Nature. As the *best one could hope for* in the State of Nature, it follows that if we could show that a state would be (morally) superior even to this best-case scenario of anarchy, then this would in itself "provide a rationale for the state's existence; it would justify the state."[4]

But where do these rights come from? Nozick's answer is that these basic rights are *moral rights*, that is rights that we are entitled to because we are moral beings. We have the moral right to not be harmed and to defend ourselves from harm and seek reparations for harms. Nozick is quite clear that such moral rights *logically precede* and assume precedence over any political rights there may be. Thus, any state that enforces political rights must do so within the limits of these basic moral rights; any political right that violates or supersedes a basic moral right will be unjustifiable.[5]

Our starting point then, though nonpolitical, is by intention far from nonmoral. Moral philosophy sets the background for, and boundaries of, political philosophy. What persons may and may not do to one another limits what they may do through the apparatus of a state, or do to establish such an apparatus. The moral prohibitions it is permissible to enforce are the source of whatever legitimacy the state's fundamental coercive power has.[6]

So persons have basic moral rights in Nozick's State of Nature, just as they do in Locke's State of Nature, but whereas these rights in Locke's conception are guaranteed by God, in Nozick's they are features of our status as persons. In this regard, Nozick cites Kant with approval.[7] In somewhat more detail, the argument is as follows. Locke supposes that in the State of Nature, people are mostly reasonable, though they can certainly act unreasonably from time to time.[8] But the level of reasonableness we can assume in people is sufficient to claim that even in this state, we are persons, that is, moral beings. And this because the fact that we *can* act reasonably much of the time is sufficient to guarantee that *we know when we are acting unreasonably*. Otherwise put, we know the difference between right and wrong, and we can, as Kant points out, voluntarily decide to act on imperatives that reflect this knowledge. So even in the State of Nature we exist as moral beings, and this is enough to guarantee that we are entitled to the basic moral rights to self-protection.

Now, if we put these last two points together, namely that an adequate justification of the state would have to show that it is morally superior to the best we could hope for in a state of anarchy, and that, as persons, we have these basic moral rights that must be protected, then we are driven to the conclusion that the moral superiority of the state, vis-à-vis anarchy, will reside in the superior ability of the state to protect these rights. The goal for Nozick is to show that such a state could have arisen from a state of anarchy by a sequence of steps, each of which is morally permissible and without the actors having intended the state as outcome. Rather, they sought merely to defend their rights. How is it imagined this happened?

Well, as Locke pointed out, even presuming that people can be expected to behave reasonably most of the time, it is still true that this leaves it to each individual to act as juror in deciding what defensive force might be appropriate and to decide what shall count as an appropriate remedy for previous wrongs. Given that people can be expected to favour their own interests, there is every reason to expect anarchy to lead inevitably to the abuse of power to extract vengeance beyond what a disinterested party would regard as fair. This is perhaps the strongest motive people might have for leaving this state of anarchy. In Nozick's invisible hand explanation, the next step is the emergence of mutual-protection associations. That is, people will quite naturally bargain with each other to form groups for their mutual protection, recognizing that, in doing so, they can be safer than they would be if left to be self-reliant for all protection of their interests. Moreover, they will bargain for such a kind of mutual-protection association precisely to have all cases judged by some common standard. And so, a natural market, involving entrepreneurs selling protection packages, individuals being hired into a variety of roles including security personnel, sales personnel, policy wonks and so forth emerges naturally.

However, problems ramify one level up. Specifically, claims between individuals represented by different security agencies and competing claims between individuals falling within the protective domain of the same agency will have the effect of driving further competition and innovation. And so, over time imbalances between competing protective agencies, particularly those smaller agencies that cannot offer adequate protection to their clients when in conflict with a more powerful agency, will drive the emergence of a Dominant Protective Agency (DPA) or perhaps several of these representing areas sufficiently far apart that conflict between them would be rare. And all of this has arisen as a result of persons seeking protection of their rights by surrendering the right to self-protection in favour of a system that would ensure that all claims would be adjudicated and settled by a common set of policies and norms. That is, every step in the sequence has involved the voluntary actions of the participants, and no step has involved the violation of anyone's rights,

> In each of these cases, almost all the persons in a geographical area are under some common system that judges between their competing claims and *enforces* their rights. Out of anarchy, pressed by spontaneous groupings, mutual-protection associations, division of labour, market pressures, economies of scale, and rational self-interest there arises something very much resembling a minimal state or a group of geographically distinct minimal states.[9]

Now this DPA is not actually the minimal state for at least two reasons. First, all protective arrangements are private, that is, a matter of contractual agreement between private individuals and the DPA. Thus, the situation is one that still leaves, or permits, some individuals to enforce their own rights, namely all those residing within the domain of the DPA yet unwilling to enter into a contractual agreement with it. Second, it leaves some persons within its domain without protection though they might want it, namely all those unable to afford protection services from the DPA.

Recall that it is a necessary and distinctive condition of statehood that the state claims exclusive right to the use of coercive force within its domain and the right to punish all who violate this monopoly. If I understand Nozick correctly, it is this monopoly on the right to use coercion that confers on the state a legitimacy no purely private protective agency can possibly enjoy. The DPA, as such, lacks the power and the legitimacy to insist that everyone within its domain purchase its services (this would turn it into a protection racket and violate the rights of those individuals who purchase protection services only under duress). Nor does it have the power or the legitimacy to demand of some of its clients that they overpay for protection services to permit the DPA to extend the umbrella of coverage to those within its domain that cannot afford to pay for themselves. This would violate the rights of those forced to pay more than would otherwise be required to protect themselves.

To see our way out of this difficulty, consider a kind of state intermediate between anarchy and the minimal state, what Nozick calls the *ultra-minimal state*. Like the minimal state, the ultra-minimal state claims exclusive rights to the use of coercive force within its domain, but unlike the minimal state, it does not offer protection services or enforcement of rights to everyone within its domain. It offers these services only to those willing to pay for them and sets the price at the level that reflects personal protection of the purchaser and their immediate family only. So, unlike the DPA, the ultra-minimal state will not permit private acts of defence or retribution within its domain by those individuals unwilling to enter into contractual arrangements with the state (but who could afford to do so) and will act on its exclusive right to the use of coercion to punish any it is aware of. But unlike the minimal state, it leaves some individuals within its domain without protection at all (and, hence without rights since they are prohibited from defending themselves), namely those that are unable to afford its services.

This ultra-minimal state, as a *logical* predecessor of the minimal state, raises a dilemma for the latter insofar as it appears, on its face, to threaten the possibility of an invisible hand explanation of the emergence of the minimal state. After all, the point of this invisible hand explanation is to show how such a state *could* emerge from anarchy and could do so without at any step violating the basic moral rights of its citizens. But now we learn that the minimal state will be such that it offers its basic protective services equally to all within its domain (this is what distinguishes it from the ultra-minimal state), but this fact, in itself, perforce requires that it violates the rights of at least some of its citizens. And this because the minimal state, unlike the ultra-minimal state, will be *redistributive*. It will be required to force those who can afford services to pay for them even if they otherwise would choose not to do so, and it will extract additional payments from those able to afford the protective services so as to enable the state to provide its services even to those who cannot afford these services themselves.

Here is the dilemma stated succinctly: this redistribution will be enforced to ensure that everyone within the domain of the state is covered by its protection services, and it is this universal coverage of protective services that confers legitimacy on the minimal state, yet this redistribution comes at the price of violating the rights of some of its citizens, which simultaneously undermines this legitimacy. Put somewhat differently, either we settle for the ultra-minimal state and accept that some persons will be left unprotected or we must find some way to justify the minimal state consistently with an invisible hand explanation.

Nozick will take the latter option arguing that the sacrifice of the rights of some individuals – essentially making them the means to the ends of others – is unacceptable in any framework that, like his, is built on Lockean/Kantian principles of the basic moral rights of persons. The problem for Nozick, then, is to argue that the minimal state can be shown to emerge from the ultra-minimal state without *actually* violating anyone's rights. More strongly, he argues that the ultra-minimal state *must* take the steps necessary to become a minimal state. As Nozick says, "The operators of the ultraminimal state are morally obligated to produce the minimal state."[10]

The key to Nozick's argument for this conclusion is that the minimal state is not really redistributive but merely appears to be so. Whether a situation really is a case of redistribution (which really would be a violation of the basic rights of at least some citizens) will depend on the reasons that explain and justify this "redistribution." Quoting Nozick,

> We might elliptically call an arrangement "redistributive" if its major (only possible) supporting reasons are themselves redistributive. Finding compelling nonredistributive reasons would cause us to drop this label. Whether we say an institution that takes money from some and gives it to others is redistributive will depend upon *why* we think it does so. Returning stolen money or compensating for violations of rights are *not* redistributive reasons.[11]

This quote has the hint to resolving Nozick's dilemma: the minimal state is not redistributive in relation to the ultra-minimal state because the money it takes from some that is additional to what they would be required to pay *ceteris paribus* for the same level of service in the ultra-minimal state is required to *compensate* others for the violation of their rights. In particular, insofar as the ultra-minimal state, in claiming exclusive monopoly on the right to use force prohibits these others from defending themselves, and threatens to punish them for doing so, their most basic rights are violated. To leave these persons uncompensated would be a violation of rights that would threaten the very legitimacy of the state. Thus, the ultra-minimal state is obliged to extend protection services and enforcement of *basic rights* to all citizens within its domain, thus becoming the minimal state.[12]

Stage 2: limits of the state

Having arrived at the emergence of the minimal state by a sequence of steps, none of which violate the rights of its citizens, and without the stages being informed specifically by any intentions to have the state as outcome, we ask, "Why stop there? Why not have a more extensive state that offers more than the minimal protection services to its citizens?" In this stage of the argument, Nozick purports to show that the minimal state is the most that can be adequately justified because any state more extensive than this can be shown to violate inescapably the basic rights of at least some of its citizens. In other words, any attempt by the state to use its monopoly on coercion to enforce additional services would be (really) redistributive and would sacrifice its moral legitimacy. It is to this argument that we now turn.

Any discussion of the issue of the state in relation to the goods and services provided to its citizens is a discussion of Distributive Justice. Looked at from this perspective, the question is whether there are plausible principles of justice that justify (i.e., require as a matter of justice) the state in delivering services additional to the protection of the basic rights of its citizens. For example, if one is a radical egalitarian, one might suppose that justice requires the state to ensure a completely equal distribution of the goods and services of the society it governs, and this will inevitably require that the state engages in more or less continuous efforts to redistribute goods as inequalities develop. In a word, the state will be required to use its coercive powers to take from some and give to others. We can regard the minimal state as occupying the opposite extreme, wherein the state uses its monopoly on force to guarantee the basic rights of its citizens but otherwise lets the distribution of goods and services arise as it will from the productive activities and free exchanges of its citizens. This is, according to Nozick, the distribution that is most just.

To understand his reasoning consider the distribution of goods[13] in a society at any particular moment in time. Call this distribution, D. For Nozick, the justness or otherwise of D is determined by his Entitlement Theory. According to this theory, the justness of a distribution in a particular society at a moment in time is determined by three principles. First is the Principle of Acquisition, which is the principle that determines whether the initial acquisition of the goods in question is just. This will be the case if, and only if, these goods are acquired justly, that is, without in any way violating anyone's rights. Now according to Nozick, this is essentially given by Locke's theory of acquisition: a good is acquired justly if it is taken from the commons (i.e., it is not presently held by anyone else) by the application of their labour.[14] Next is the Principle of Transfer, which concerns the transfer of goods from the present holder(s) to another (or others). This principle says that a transfer will be considered just if, and only if, the present holder of the good in question is entitled to the good, and it is transferred justly. And a transfer will be just so long as doing so violates no one's rights, and this will be the case so long as transfers occur within a market system of free exchange, with appropriate safeguards in place to ensure all participants are free to act exclusively from their (self-interested) desires. That is, the safeguards ensure there is no coercion, fraudulent activity or manipulation, and there is a system of jurisprudence to ensure that all freely negotiated contracts are honoured. In other words, transfers are just if they happen within a Smith-style free market. And last is the Principle of Closure, which says that a holding is just only so long as it is the result of repeated applications of acquisition and transfer.

So the overarching Principle of Distributive Justice says that a present distribution of goods, say D, is just if everyone is entitled to their holdings under this distribution, and they will be entitled to their holdings so long as the distribution satisfies the terms of Entitlement Theory. For clarity, we can exhaustively specify *justice in holdings* with the following inductive definition:

1 A person who acquires a holding in accordance with the principle of justice in acquisition is entitled to that holding.

2 A person who acquires a holding in accordance with the principle of justice in transfer, from someone entitled to the holding, is entitled to the holding.
3 No one is entitled to a holding except by (repeated) applications of 1 and 2.

Thus, as Nozick says, "Whatever arises from a just situation by just steps is itself just. The means of change [i.e., from one distribution to another – WIH] specified by the principle of justice in transfer preserve justice."[15]

The thing to notice is that this Entitlement Theory is historical. The justness or otherwise of a present distribution will be a matter of how it has come to be. By contrast, other theories of just distribution are ahistorical; the historical facts are irrelevant to the issue of whether the present distribution of holdings is just. Instead, the issue is to be decided by a principle, determined in advance, that fixes what the distribution must be. To return to the example of radical egalitarianism, the principle is Perfect Equality, and we ignore historical facts looking only to the actual distribution of holdings to see if it meets our criterion of perfect equality. If it doesn't, a redistribution must occur to establish our goal state of perfect equality. Nozick calls such principles *end-state principles* or, alternatively, *end-goal principles*. The principle is justified by some theory of justice and then is appealed to in determining what our goal should be with respect to the distribution of goods. Such redistribution as will be required to meet our goals will be considered the rectification of an unjust distribution.

Now redistribution may be required in the minimal state too, and for this purpose Nozick introduces what he calls the Principle of Rectification, which becomes necessary only if it should be discovered that D is not just with reference to Entitlement Theory. The difference, though, is that, for Nozick, any rectification must be sensitive to the historical facts,

> This principle uses historical information about previous situations and injustices done in them (as defined by the first two principles of justice and rights against interference), and information about the actual course of events that flowed from these injustices, until the present, and it yields a description (or descriptions) of the holdings in the society. The principle of rectification presumably will make use of its best estimate of subjunctive information about what would have occurred (or a probability distribution over what might have occurred, using the expected value). If the actual description of holdings turns out not to be one of the descriptions yielded by the principle, then one of the descriptions yielded by the principle must be realized.[16]

The central point about all ahistorical, end-goal schemes of distribution that reveals why they are unjust is that they require the distribution of goods in a society at any moment in time to conform to a pattern, but individuals left free to choose for themselves on the basis of their desires and act on them will inevitably disrupt any such pattern. Thus, the redistribution required to bring the distribution into line with the requisite pattern will require the state to violate the freedoms of its citizens, which is to violate their most basic rights, and to do so continuously.

Nozick introduces a unique thought experiment to reinforce this point, the Wilt Chamberlain example. So consider Wilt Chamberlain, a star basketball player of his era, working in the NBA as a free agent. He negotiates a deal with the owners of a team to play for them for a year at some fixed wage, plus an additional surcharge of $2.50 per ticket to all of his games. The idea is that, because of his drawing power as a star player, fans will be required to pay an additional $2.50 per ticket for the privilege of seeing him play. Moreover, and this is important to the example, the surcharge is public; the fans know in advance that they are being charged this extra fee and that it will go to Chamberlain in its entirety. At the end of the year let us say that Chamberlain is richer by $250,000 additional to his wage because of the surcharge. If one is committed to some patterned distribution, you might think this additional money should be reclaimed from Chamberlain by taxation and returned, through redistribution, to the patrons that paid the additional charge. But how is this to be justified? They knowingly and willingly paid the fee for the privilege of seeing Chamberlain play. It was on this understanding that they attended his games, in record numbers, indicating it was their freely chosen desire to see Chamberlain play, even at the additional price. Why should this freedom be interfered with?

It's helpful to recall at this point that Nozick's central assumption is that any social policies or legislation will be constrained by his moral side constraints. In other words, the basic moral rights take priority, and any laws or policies designed to meet social goals will be constrained to not violate these basic rights, the most important of which is the libertarian freedom of self-determination. And this includes the freedom to dispose of any holdings to which one has just entitlement. Thus, if any patterned distribution is to be justified, it must be shown to be compatible with these basic rights, which is arguably impossible.[17]

> If entitlements to holdings are rights to dispose of them, then social choice must take place *within* the constraints of how people choose to exercise these rights. If any patterning is legitimate, it falls within the domain of social choice, and hence is constrained by people's rights.[18]

On this basis, Nozick is driven to conclude that any taxation that goes beyond the minimum required to sustain the minimal state is theft. It is, as he puts it, a form of forced labour. The state uses their monopoly on coercion to force people to work for purposes they did not themselves voluntarily consent to, to redistribute the products of this labour to social goals the state has determined to be just. This is redistribution, unlike the compensation that might be exacted to ensure some citizens are not left unprotected in their basic rights. There are, in this case, no compelling non-redistributive reasons that can be appealed to. Thus, it cannot be justified and is to be regarded as theft; the deliberate taking of property to which these citizens are justly entitled. And this is tantamount to giving some persons (the recipients of redistribution) a claim of ownership to the personhood of those forced to pay,

> End-state and most patterned principles of distributive justice institute (partial) ownership by others of people and their actions and labor. These principles involve a shift from the classic liberals' notion of self-ownership to a notion of (partial) property rights in *other* people.[19]

Rhetorically at least, Nozick is suggesting that nothing could be more unjust than this: having the state use its coercive powers to grant to some of its citizens property rights to the personhood of various other citizens. This is the most heinous violation of rights Kant could contemplate: the state facilitating the use of some of its citizens as the means to the ends of its other citizens.

I will leave the critical discussion of Nozick's theory until after we have considered the alternative Progressive Liberal conception of the state, but it will be helpful here to sketch out some of the implications of this theory. The perfect Night Watchman State is unlikely to be realized anywhere, though we could expect some states to approximate more closely than others to the ideal it implies. So, we ask, what would this ideal look like?

Well, insofar as the state must be absolutely minimal in extent, the provision of all goods and services additional to the basic protective services will be provided by the private sector. This includes educational institutions at all levels, hospital and care facilities and even infrastructure such as roads, bridges, sewage and water treatment and so forth. Unless an argument is forthcoming that these services and infrastructures are part of, or are essential to, the delivery of basic protection services and, thus, their provision by the state would count as compensatory rather than redistributive, it must be that they are provided by the private sector exclusively. The state will be involved in the provision of such basic protection services as will be required to guarantee that its citizens are secure in their lives and their property, and that all contracts are negotiated under conditions of perfect freedom with an absence of fraud and deceit, and that any failures of market participants to honour the terms of a contract so negotiated will be remedied by a civil court system. So the state will be required to provide protection services in the form of police for internal protection and armed forces to protect against foreign aggressors. But even here, there will be opportunities to allow private sector players to provide some of these services, though under the terms and legal constraints set by the state. Examples include private security firms that provide additional protections to those clients willing to pay for these additional services, but also supplementations to regular police services, privately run prisons and supplementations to armed forces personnel.

Insofar as these services are privately provided, they will be made available exclusively to those willing and able to pay for them. It's somewhat difficult to imagine in detail what this might look like in its fullest realization, but it would be an extension of such divisions as one may find presently in those places wherein the overwhelming majority of persons, those unable to pay, live in poverty and without basic education or medical care, segregated into communities such as slums or favelas, with minimal infrastructure services such as plumbing, water treatment and so forth. Meanwhile, those able to pay live in walled sectors, or secluded areas, patrolled by private security firms (i.e., providing security additional to the basic security provided to all), their children attend the best private schools and universities, enjoy the best medical care available and enjoy the benefits of well-maintained roads, bridges, plumbing and sewage treatment and so on.

Let us now consider the Progressive Liberal conception of the state and the extent to which it might seem to offer a genuine alternative to the Night Watchman State.

John Rawls: *justice as fairness*

It is widely agreed that the finest articulation and defence of the Progressive conception of the state is the one given by John Rawls. Developed over a period of decades in a series of published papers, the view is presented in definitive form in his book, *A Theory of Justice*, published in 1971. Upon his death in 2002, it was said in one obituary that this work "was perceived as a watershed moment in modern philosophy."[20] And of Rawls, it has been said that, "John Rawls was widely recognized as the greatest political philosopher of the 20th century."[21] In this book, Rawls calls his view "justice as fairness" and the central insight is that for a political system to have legitimacy, it must bring forth the willing participation of its citizens, and this will require that they perceive it to be just in the sense of being fair.

As with Nozick's view, we begin with some contextual remarks. Interestingly, Rawls begins from the assertion of the inviolability of the basic rights and freedoms of individuals, and he remarks that this sets it apart from Utilitarianism.[22] He goes on to draw connections repeatedly between his view and the contractarian view of Kant. I say "interestingly" because this is essentially the same starting place as Nozick. As such, this points to a central narrative thread in my reconstruction suggested by the question, "How do they end up at such different conclusions?" More generally, we might wonder how different their views are in fact. I shall be suggesting the difference is not so much as many have supposed but is probably best explained by how they differ regarding what shall count as coercion (more on this point in the concluding remarks).

Now the subject of the theory, justice, is concerned with the principles that structure the basic institutions of society (to be defined shortly) because these institutions determine in turn the distribution of both rights and duties *and* the advantages that flow from the social cooperation facilitated by these basic institutions.[23] As such, these principles determine and regulate all subsequent agreements negotiated under their terms, thus specifying the forms of social cooperation that are possible and the general form that government shall take. So the goal of Rawls's theory is to discover, articulate and defend the principles of justice that persons would agree to as the ones to determine the structure of their relations to one another. As such, we make some assumptions about these individuals and the circumstances under which the principles are chosen. First, we assume that these individuals choose freely, which requires that they choose under circumstances that guarantee their fundamental equality with one another. There is no coercion and no imbalance of power that could conceal or mask coercion. Second, we assume them to be rational, in the minimalist sense employed by economics. That is, they are concerned to advance their own interests and they can effectively calculate the best means to achieve such ends as they set for themselves. They are otherwise mutually disinterested. That is, they take no particular concern to interfere in the affairs of others except when these others present obstacles to the pursuit of their own interests. They act from self-interested but not selfish motives. Under these conditions, any principles mutually agreed to, assuming such principles can be found, would characterize justice as fairness and would for this reason bring forth the willing participation of the citizens living under their terms.

The theory and its supporting arguments are developed in two stages. In my reconstruction, the first stage will articulate and explain the principles that would constitute the foundations of a just society. And in the second stage, we characterize the conditions under which these principles would be chosen, called the Original Position, and provide a plausibility argument that the principles of Stage 1 would *always* be chosen under such conditions.[24] In Rawls's theory, the Original Position functions as the analogue of the State of Nature in Contractarian theories of the state. Specifically, the Original Position requires us to carefully delineate the circumstances under which persons *could* come together to negotiate the principles of justice that would determine the basic institutions under which they would subsequently live with one another. The theory is not falsified by the fact that no one has ever actually come together under such conditions for such purposes. As noted in our discussion of Nozick, the point is to delineate the circumstances that *would* be required, in this case to guarantee perfect equality amongst the participants for the duration of the negotiations that would have the principles of justice as their output.[25]

In the phrase used earlier, "provide a plausibility argument," what is meant is that it is not supposed that this result is *proven* in the sense meant by that term when we speak, for example, of a deductive geometric proof. There can be no such proofs in the realm of politics and value theory. Rather, these principles are made plausible by being shown to be the result of Reflective Equilibrium. By this, Rawls means that it is not to be presumed that these principles are the ones that persons would agree to on first meeting them, as their first and unreflective response, so to speak. Rather, it is plausible to believe that all participants in the original position *would come to agreement* on these principles after careful reflection and debate with one another. The equilibrium achieved is the best balance between one's initial reaction as measured off against contemplation of the consequences of adopting these principles.

The idea is that one tests the adequacy of the theory itself by tracing out the consequences implied by the principles of the theory and asking how these consequences *align* with those positions where one's views are already settled; that is, those cases where one's moral convictions are firm. If it's possible to find agreement on these settled cases, at least in large measure, one is encouraged to explore the implications that concern those issues where one's views remain unsettled, with an eye to asking how plausible the guidance provided by the theory seems, on balance, concerning these more challenging cases. Assuming the theory under contemplation passes the first test of aligning reasonably closely with the cases where one's convictions are settled, the process typically goes back and forth for several iterations at least, between realignment of one's basic convictions and/or adaptations of the theory being considered to bring the two into closer alignment with what seems reasonable on such reflection. In fact, there is no reason the process should come to a final conclusion. In my own experience, it is a lifelong project of continual rebalancing of one's settled moral convictions against the implications of moral theorizing. So as Rawls says in relation specifically to the principles of justice,

> [J]ustice as fairness can be understood as saying that the two principles . . . would be chosen in the original position in preference to other traditional conceptions of justice, for example, those of utility and perfection; and that these principles give a better match with our considered judgments on reflection than these recognized alternatives.[26]

I think it is implicit in this quote that Rawls is suggesting that the principles of justice are chosen over the alternatives considered *because* they better match our considered judgements on reflection.

And so, with these contextual remarks in place, we proceed to reconstruct Rawls's theory.

Stage 1: the Principles of Justice

The Principles of Justice are said to determine the *basic institutions* of society. An *institution* is here conceived of abstractly as a public system of rules which collectively define offices and positions, together with the rights and duties, powers and immunities of those offices and positions.[27] And an institution is *basic* to the extent that it is determinative of the rights and duties of the citizens living under these terms and also of the distribution of the advantages of the forms of social cooperation facilitated by these institutions. As such, these basic institutions determine and regulate the kinds of social cooperation that are possible and the form that government will take (in broad outline) under the circumstances.

It is important to emphasize the public character of the rules that determine these basic institutions.[28] The underlying idea is that it is the *public* character of the rules underlying the foundations of society that ensure the trust essential to the sustainability of a just society. For a society to be genuinely just, it must be perceived to be so by the citizens of that society, and this will be because they trust that the system is fair. It is this trust in the fairness of the system, grounded in the publicity of the rules, that serves over time to cultivate the sentiment of justice itself, the commitment of the public to a sense of fair play, and has the consequence over time of aligning the self-interest of citizens with the socially desirable behaviours that encourage the like behaviour in others. The system becomes self-reinforcing, as it were.[29]

There is an interesting parallel here to the discussion in Mill's Utilitarianism, wherein Mill argues that the success of Utilitarianism, as a moral and political scheme to underwrite society, requires for its success that we undertake to arrange the laws, policies, jurisprudence and educational resources of society to bring them in line with Utilitarian values so that, over time, citizens would regard their self-interest as indistinct from what would be in the best interest of the collective good.[30] The difference lies in the fact that Mill requires society to undertake a deliberate programme to shape the views of its citizens so as to have this effect, whereas in Rawls's case, it just happens as a matter of course if we imagine a society founded on the principles of justice and full transparency about the rules. As Nozick might say, the emergence of the social values of justice, reciprocity and trust are explained in Rawls's case by appeal to a kind of invisible hand argument, whereas in the Utilitarian case, active intervention on the part of the state is required to have this effect. For Rawls, these values are the ones that citizens will freely choose for themselves under the conditions of a just state as defined by his principles. This is, in part at least, what Rawls means when he asserts of the principles that "they bring forth the willing participation" of the citizens living under their terms.

Here, then, is a first statement of the principles:

First: each person is to have an equal right to the most extensive scheme of equal basic liberties compatible with a similar scheme for others.

Second: social and economic inequalities are to be arranged so that they are both (a) reasonably expected to be to everyone's advantage and (b) attached to positions and offices open to all.[31]

Implicit in this statement of the principles is the idea that, in looking at the resulting society through the lens of justice, this society will be constituted of two parts that are, analytically at least, quite distinct. The first part defines and secures the basic rights and liberties (entitlements) of citizens. And the second part, established by the second two-part principle, which Rawls calls the Difference Principle, specifies and regulates the social and economic inequalities that result. Now these principles are, to use Rawls's phrase, "lexically ordered" by which he means to underscore the logical and valuational priority of the basic rights and freedoms of citizens in any society that presumes to call itself just. So the first principle is logically prior in that the second principle cannot even be understood as a principle constitutive of justice unless it is first understood that the basic rights and freedoms secured by the first principle have been itemized and secured. And it is valuationally prior in the sense that the rights and freedoms secured by the first principle set a floor, as it were, a lower limit in our treatment of others. However, the social and economic inequalities may be arranged to meet the conditions specified by the second principle, they can never be arranged so as to have the effect of impacting negatively on these basic rights and freedoms. The basic rights and freedoms, in other words, are not negotiable at the margins for the possibility of some greater social good.

There is an unmistakable Kantian quality to these arguments that give priority to the intrinsic dignity of persons (as Kant would put it). And in this sense, Rawls's view is again distinguished from a Utilitarian or Consequentialist view. This is made clear by the following general principle. As Rawls puts it, this principle reflects the most general conception of justice and the aforementioned principles of justice are to be thought of as a special case of this more general principle:

> All social values – liberty and opportunity, income and wealth, and the social bases of self-respect – are to be distributed equally unless an unequal distribution of any, or all, of these values is to everyone's advantage.[32]

The point to notice in this general principle is that the equality of distribution is to be departed from only if it can be shown that doing so will be to *everyone's* advantage and not merely the aggregate advantage of the majority (at the expense of some). It is my view that, insofar as Rawls is committed in this way to the Kantian view that gives logical and valuational priority to the basic rights and freedoms of individuals, he is much closer to Nozick than many have recognized.

In any event, for Rawls, these principles define the structure of a society he characterizes as Democratic Equality as opposed to, for example, Socialist Equality. Specifically, the kind of equality to be guaranteed is *equality of opportunity*. Citizens will be willing to accept inequality of outcome as long as they believe the system to be fair, and their perception of fairness will depend in large measure on being reassured that there is genuine transparency and equality of opportunity. If this is guaranteed by the structure of the society, and people can perceive this to be so, they can accept an unequal distribution of goods as reflective (within limits) of differences in individual contributions.

As noted earlier, transparency is guaranteed by the full publicity of the rules. But how is equality of opportunity to be guaranteed? Is "guarantee" too strong in this context? According to Rawls, the guarantee is a consequence of the fact that the system has been set up and is administered as an instance of Pure Procedural Justice (hereafter: PPJ). He defines this notion as follows,

> PPJ obtains when there is no independent criterion for the right result: instead there is a correct or fair procedure such that the outcome is likewise correct or fair, *whatever it is*, provided that the procedure has been properly followed.[33]

PPJ and equality of opportunity are mutually reinforcing here: keeping the principle of fair equality of opportunity in mind is what ensures that the principles of the system must be chosen under conditions characterized as PPJ, and choosing the principles under conditions of PPJ is what ensures that the outcome will guarantee equality of opportunity.[34] The implication is that, if we have taken care to set up the basic institutions of society with complete impartiality, then we accept the outcome, whatever it may be with respect to the distribution of goods, *because* we will know the system guarantees equality of opportunity.

Now when Rawls speaks of there being "no independent criterion for the right result," he is again rejecting Consequentialist visions of society. Consequentialist theories start from the presumption that we have a criterion for the kind of outcomes we wish to have, say maximization of aggregate welfare. Thus, if we fail to achieve these outcomes, we must intervene to redistribute. In other words, Rawls is, again like Nozick, rejecting patterned outcomes. We don't decide in advance what outcome we wish to achieve. Rather we decide in advance how the basic institutions of society are to be structured and then accept the results that flow from that arrangement. Moreover, it is crucial to keep in mind here that the lexical ordering of the two principles guarantees that any resulting distribution of goods must be such that it never requires the infringement of the basic rights and freedom of its citizens.[35]

Another detail to be considered, though, is the nature of the goods under contemplation. What are the goods to be distributed under the Difference Principle? They are conceived to be basic, primary goods. Basic, primary goods are those goods that every rational individual can be presumed to want, in such abundance as may be possible, because they are essential to enacting a life plan for oneself. These are such things as rights, liberties, opportunities, self-respect and income. Without access to these goods in sufficient quantity (we will not consider how to interpret "sufficient" in this context), no one can undertake to live their life freely according to their own plan. And so, we arrive at the following refined version of the Difference Principle:

> Social and economic inequalities are to be arranged so that they are both (a) to the greatest expected benefit of the least advantaged and (b) attached to offices and positions open to all under conditions of fair equality of opportunity.[36]

Having considered the principles of justice, we now turn our attention to how they are arrived at and justified as plausibly the ones to be chosen by rational beings concerned to advance their interests in a position of bargaining equality.

Stage 2: the Original Position

The task of Stage 2 is to show that the principles of Stage 1 are "those which rational persons concerned to advance their interests would accept in this position of equality to settle the basic terms of their association."[37] This task will be accomplished to the extent that it can be argued plausibly that any agreements reached in the Original Position (hereafter: OP) will be fair, and this, in turn, depends on demonstrating the OP is a position of perfect equality. This is not the equality of Locke's State of

Nature, wherein each individual is equal in the limited sense of being equally free to defend themselves and their property and seek remedies as they think fit for perceived wrongs. For Rawls, the relevant notion of equality is that which comes out of procedural justice. That is, care is taken to set up the decision procedure of the OP in such a way that arbitrary contingencies of any kind are made irrelevant. Arbitrary contingencies include such things as one's position in society, one's gender or religion and the particular arrangement of power and class relations as they have evolved in the society and so on. Such facts are arbitrary, thus irrelevant to the decision, precisely because they are accidental features of historical circumstance and/or the natural lottery and not in any way reflective of moral desert.

But how is this equality to be accomplished? The answer is the *veil of ignorance*. That is, we impose on the decision-makers the requirement that their deliberations and, ultimately, the decision itself be made from behind the veil of ignorance. This is Rawls's term of art for the situation in which we imagine the participants know nothing particular about themselves or their position in the society they are to bring forth. In particular, we are to imagine they do not know their gender or their race. They do not know anything specific about the position they *will* occupy, nor about the life plans they *will* seek to enact. Importantly, they do not know where they *will* end up in the class structure that determines the resources one will have available for those life plans. And everyone behind the veil of ignorance is in the same epistemic position of knowing nothing particular about themselves or their future situation in the society their choices will bring forth. It is the veil of ignorance that guarantees the equality of all participants and the impartiality of the choices made; no one is privileged in any way concerning knowledge of facts about the future society they are to bring into existence. They know only certain general facts about human society. They know, for example, that they will be embedded in a network of relations that will depend for its success on reciprocal interactions with others, and they know that they will have certain values, desires and goals, though they don't know anything about the specific character of these values, goals and desires.

It is assumed, then, that the participants are rational and concerned to advance their own interests, though they don't know what these interests will be. They are also mutually disinterested. That is, they take no interest in each other's welfare. Under these circumstances, the argument goes, participants would, under Reflective Equilibrium, always choose the principles of justice articulated in Stage 1 as setting the basic terms of their association. And this because, from behind the veil of ignorance, the decision is made in reference to all and only the information that is relevant, *and* this information is always the same because it expunges all information concerning particulars. Thus, as Rawls says, the OP provides us with a unique viewpoint to assess principles of justice. Anyone may, at any time, adopt the position of the veil of ignorance and, from that perspective, adjudicate justice as fairness,

[T]he original position must be interpreted so that one can at any time adopt its perspective. It must make no difference when one takes up this viewpoint, or who does so: the restrictions must be such that the same principles are always chosen. The veil of ignorance is a key condition in meeting this requirement. It insures not only that the information available is relevant, but that it is at all times the same.[38]

As Rawls notes, the veil of ignorance is implicit in Kant's notion of the categorical imperative, and the condition of universality in particular.[39] The principles that a rational person would choose from behind the veil of ignorance will be the principles that will apply to all persons universally because they will apply regardless of what one's particular aims are. When we turn our attention to Kant's theory of moral duty in Chapter 10, we will see that, for Kant, to act on an imperative reflective of one's particular objectives is to act from a *hypothetical imperative*, and this is not to act morally at all. Morality requires of us that we act on what Kant calls a *categorical imperative* and that these apply to us unconditionally, regardless of what we might want for ourselves. Additionally, Kant emphasizes the requirement of *reversibility*, namely that we must act only on those maxims we would be willing to accept even if situations were reversed and you were the moral patient. The underlying idea is that, if you do not know that you will not end a woman in the future society, you will not choose principles that guarantee the systematic exclusion of women from positions of power and authority; if you don't know that you will not end up a person of colour, you will not choose principles that will subjugate people of colour to the lower economic class and deny them the equality of opportunity needed to improve their condition.[40]

A big part of Rawls's reasoning for the preferability of the principles of justice vis-à-vis alternative conceptions of justice consists in demonstrating of the most plausible alternatives (e.g., Utilitarianism in several versions and Perfectionism, in particular) that they would *not* be chosen under Reflective Equilibrium. It would be a lengthy digression for us to consider these arguments in detail, but it is worth pointing out that one prominent consideration in favour of justice as fairness is the weakness of Rawls's assumptions. By contrast, Utilitarianism, in all its versions, must assume universal benevolence on behalf of those choosing the basic principles. They must be willing to choose principles that might consign themselves and others to subordinate positions enjoying less freedoms and rights, and fewer of the basic goods, just because this arrangement has the consequence of raising overall benefits for society as an average or as an aggregate quantity.[41]

Summarizing, then, justice as fairness has the following features in its favour. First, and given the finality of the agreement (i.e., it sets the basic terms for their association and for the adjudication of all claims going forward) the participants will never choose principles that have consequences they can't possibly accept. And so, because of this, they have chosen principles that guarantee their basic rights and freedoms and insure themselves against the worst losses. This latter follows from the equality of opportunity guaranteed by the Difference Principle. Second, there is the publicity of the principles: the principles are public and known by all members of society as setting the basic terms of their association. And this feature has the tendency to nurture and enhance the values and behaviours conducive to justice, the conditions of trust and reciprocity on which a just society depends. Lastly, it also has the consequence of enhancing the self-respect of individuals which, in turn, increases the effectiveness of social cooperation. If I know that, regardless of my position in society, I enjoy the same rights and freedoms as others, and whatever my level of income and the prestige of my position, I know that the basic institutions of society are arranged in such a way as to guarantee that I shall never be permitted to be used as mere means by another, then my self-respect is as secure as it can be, contingencies aside. To put the point into Kantian terms again, it is to build into the basic structure of society itself the conditions that incline us all to treat one another as ends-in-themselves and not merely as means.[42] To anticipate the discussion to follow in the next section, this is, I believe, a big part of what the Sociologist Robert Putnam refers to as social capital.

We ask, what kind of society might this be? What would a society structured by the principles of justice as fairness look like? Politically, it will certainly be a democracy of some kind. Only a democracy could be compatible with the first principle that guarantees the basic rights and freedoms of the citizens. Compatible with being a democracy, though, a rather wide variety of actual implementations are possible. Economically too, the principles of justice are compatible with a wide range of options, from socialism wherein the workplaces are organized under worker collectives, to various degrees of free market Capitalism. The only economic arrangement clearly excluded would be any kind of central command economy as it would be incompatible with the guarantee of basic rights and freedoms, in particular the right to choose one's employment.[43]

The most natural expectation would seem to be a democratic free market society probably not so different from what was in place in most Western democracies in the decades immediately following the Second World War, the period roughly 1945–1975, called by the French, *Le Trente Glorieuses*. One might object that there was rather a lot of injustice in this period, in all of the countries that might be captured by this designation.[44] While this is true, Rawls does not demand that a society founded on his principles be perfectly just; indeed, such a thing is impossible. Even in a society founded on these principles, governing is, in practice, a case of Imperfect Procedural Justice. The principles themselves are chosen under the perfect equality of the OP, thus guaranteeing Pure Procedural Justice (i.e., strict impartiality) at the point of choosing the basic principles themselves. But the construction of a constitution based on these principles, and the business of governing itself, will be affected by contingent facts pertaining to, for example, historical circumstances, a lack of knowledge of all the relevant facts, short-sightedness and the biases of bureaucrats in these positions. Thus, they will always be subject to the possibility of injustice.[45]

In this respect, the principles of justice are analogous to the operation of jurisprudence. We choose the principles of jurisprudence, the Rule of Law, innocence until guilt is proven by a jury of one's peers, strict procedures for jury selection and so on under conditions of Pure Procedural Justice. But in practice, the law can make egregious mistakes, miscarriages of justice. In practice, it too is an example of Imperfect Procedural Justice. So what Rawls would say is that we should expect a society founded on these principles to be nearly just or as close as is possible under the circumstance.

It is in light of this that he goes on to develop a theory of civil disobedience that articulates the circumstances under which civil disobedience can be justified in a state that is founded on the principles of justice. Civil disobedience serves the interests of a just society by functioning as its conscience, as it were,

> By resisting injustice within the limits of fidelity to law, it [i.e., civil disobedience – WIH] serves to inhibit departures from justice and to correct them when they occur. A general disposition to engage in justified civil disobedience introduces stability into a well-ordered society, or one that is nearly just.[46]

So a society that is just, or nearly so, in Rawls's sense, will be one that respects the basic rights and freedoms of its citizens and guarantees these in law, leaves people free to choose their employment and the plan of life that seems to them to be best. It will guarantee equality of opportunity by, for example, ensuring that all positions of power and authority are open to all (in principle). Moreover, the economic arrangements, including taxation, will be arranged to enhance equality of opportunity by arranging things to ensure that such inequalities as may persist will be to the greatest benefit of the least advantaged. Plausibly, this will require the state to guarantee a basic education for all children at the very least.[47] It is hard to imagine how equality of opportunity could be guaranteed unless people can be expected to have the basic skill set required to be minimally competitive upon entering the working world and/or to pursue further educational opportunities that would enhance these basic skill sets should they choose to do so. Additionally, it seems likely other basic goods and services will have to be guaranteed, at least in quantities sufficient to ensure that citizens can start their working lives in a condition that makes it possible for them to even imagine life plans that they would value. I am thinking of such things as basic healthcare, and safety from abject and debilitating poverty, the sorts of conditions that damage body and mind and are utterly ruinous of one's self-respect. Under these circumstances, and knowing that the basic arrangements are fair, citizens are willing to accept such inequalities of outcome that result.

The two views compared

Rawls's theory has been the focus of critical discussion for some 50 years or so now, and some of these criticisms pose deep problems for the view. I will mention two of these. First, the social contract applies strictly to only those who could, in principle, be signatories to the agreement. This means it does not apply to non-human animals, or even humans who fall below the threshold of rational comprehension of the terms of the contract. Any protections for these individuals will depend on the generosity of actual members of the contract, their willingness to extend the umbrella of protection to such individuals.[48] Moreover, though, it has been argued that the OP is impossible in principle insofar as it is not possible to imagine being without gender, for example, or without class membership, or any knowledge of one's interests and values beyond knowing that one will have values and interests. As some feminists have pressed this concern, it is to assume an ideal of objectivity that is, in fact, quintessentially Eurocentric and masculine; it is to assume that this perspective is one that can "see the world as it is." But, the objection is pressed, this is to assume what is manifestly false, namely that one can set aside one's identity as a concrete and situated person, and that this would be to see the world much as "I" do when I, for the moment, look at the situation "from out there." As embodied, concrete beings, we are gendered, we inescapably have values, shaped in part by our feelings of membership in various groups, including socio-economic class. To ignore these facts about ourselves has only seemed possible because Rawls and others in the tradition of male-dominated Western philosophy have assumed that being objective, seeing the world from out there, is a matter of seeing the world pretty much as they do. It is, to appropriate a phrase introduced by Thomas Nagel, what might be called "the view from nowhere."[49]

These are hard problems that deserve to be acknowledged, but it would be beyond my purposes to attempt to deal with them here. I would suggest, though, that they threaten to distract us from what is really at issue here, which is the difference between these two conceptions of the state, Classical Liberal on the one hand, and Progressive Liberal on the other; the central ideological battle of our times as Stiglitz characterized it in the quote that opened this chapter. I have been at pains to argue that both theories start from much the same place: a Kantian notion of the inviolability of the individual. As such, both are concerned to guarantee the basic rights and freedoms of individuals. They are, both of them, emblematic of their time, committed as they are to the individualism that is, historically, a very recent arrival and perhaps an aberration. Having said that, though, there is this difference: Rawls focuses our attention on the very salient fact that any civilized society that expects the willing engagement of its citizens must establish and maintain the conditions for people to perceive it to be fair, and for this to be the case, they must feel that equality of opportunity is genuinely possible.[50] Under these circumstances, people will be willing to accept inequality of outcome.

My goal in what follows is to evaluate these alternatives in relation to two closely related issues, equality of opportunity and social capital. The reader is invited to compare the following observations with the discussion of these topics in Chapter 4, where the discussion proceeded under the terms of the narrative we are calling Biosphere Consciousness, the world view that underwrites Natural Capitalism.

Joseph Stiglitz and the price of inequality

We take up the discussion of equality of opportunity with a more careful examination of the recent writings of Joseph Stiglitz. In several recent books, he has explored the economic data in relation to equality of opportunity and the consequences of recent changes in its extent.[51] Although he focuses on the United States because of the wealth of data available and the extent of the change in a mere 40 years or so, he makes it clear the same points apply in varying degrees to almost all advanced industrial countries.[52] For the United States, the data are unequivocal: there has been a stark decline in equality of opportunity since the mid-1970s, a decline that has gained momentum since the early 1980s. As recently as the early 1970s, the United States was consistently among the best of the advanced industrial democracies for equality of opportunity, as reflected, for example, in social mobility, the likelihood that someone born into a family among the lower-income brackets could move upwards into even the highest-income brackets and into positions of prestige and authority. Since about 1975 or so, and accelerating with the election of Ronald Reagan in 1980, the United States has declined to the point where it has the highest levels of inequality among the relevant comparative nations.[53]

A useful measure of inequality for comparative purposes is the Gini coefficient. This is a statistical measure of income distribution, and also wealth distribution, that provides a comparative measure of inequality between various income percentiles in a population or comparisons across populations.[54] Though it is strictly a measure of inequality of income or wealth distribution, its relevance to inequality of opportunity is secured by the fact that, as the evidence clearly indicates, where there is significant inequality of income and wealth, there is also inequality of opportunity. When reading figures on Gini coefficients, keep in mind that it is an indicator that ranges between 0 and 1. Specifically, a Gini coefficient of 0 for a population would indicate a state of perfectly equal distribution for that population, and a Gini coefficient of 1 would be the distribution where one individual has all the income and everybody else nothing; a state of perfect inequality if you like. Keep in mind, too, that small differences in the Gini index can indicate significant differences in inequality. For the United States, in 2018, the Gini coefficient for income distribution was 0.49; in 1990, it was 0.43; and in 1974, it was 0.35; reflecting a steady progression of increasing inequality since the mid-1970s. As a comparison, in 2017, the figure for Denmark was 0.29 and for France it was 0.32.[55]

In a careful and systematic analysis of the data, Stiglitz traces out the negative consequences of this growing inequality for productivity, market efficiency, democratic and legal institutions, the foundations of trust and, perhaps most importantly of all, the nation's very sense of its identity,

Of all the costs imposed on our society by the top 1 percent, perhaps the greatest is this: the erosion of our sense of identity in which fair play, equality of opportunity, and a sense of community are so important. America has long prided itself on being a fair society, where everyone has an equal chance of getting ahead, but the statistics today, as we've seen, suggest otherwise: the chances that a poor or even a middle-class American will make it to the top in America are smaller than many countries of Europe.[56]

Implicit in this quote, the growing inequality of income and wealth has been overwhelmingly to the benefit of the top 1 percent of the income distribution, the smallest fraction of American society.[57] While I would not want to assume that Stiglitz takes fairness to be synonymous with equality of opportunity, it is clear in this quote that, at the very least, he certainly holds the latter to be the most prominent indicator of the former. Moreover, and directly relevant to our present concerns, this growing inequality has been the result of a programme to deregulate the economy, privatize as much of the public service as possible, and by these two means, shrink the government to the smallest size possible to facilitate reductions in taxes, under the promise that this would stimulate the economy and benefit everyone. Ronald Reagan actually went so far as to claim that his tax cuts would stimulate the economy to such an extent ultimately that government revenues would increase making it possible to restore public services.[58] None of this turned out to be true, yet the administration of George W. Bush passed further tax cuts during his term as president, and more recently we saw Donald Trump proclaim his round of tax cuts to be his greatest legislative accomplishment. And again, it is worth pointing out that these tax cuts have been overwhelmingly to the benefit of the richest of Americans, disproportionately favouring the top 1 percent (whether as a measure of income or of wealth). And all of this is perfectly consistent with Nozick's vision of the Night Watchman State and the incremental movement towards the aspirational goal of the minimal state.

Thus far, it is clear from the data that the promise of a Rising Tide that would follow these initiatives and lift all boats has failed to happen. A response one often hears to this, though, is that the programme of deregulation, privatization and tax cuts has not yet proceeded far enough; the problem is a failure of government, not a failure of markets.[59] Keeping Nozick's aspirational goal in front of mind, the closer we can get to the full realization of the minimal state, the more likely we are to enjoy the predicted benefits. This, in fact, was implicit in the messaging from George W. Bush and his administration when his two rounds of tax cuts were before Congress in 2001 and 2003. But, in fact, the political and social divisions that have accompanied the growing inequality have in the United States reached a state of virtual political paralysis[60] and apparently intractable social upheaval and violence, as evidenced in the assault on the US Capital on January 6, 2021.

As a general trend, this outcome was entirely predictable. In fact, it was predicted by many. Consider, as we approach the minimal state, and the wealthiest among us retreat into their walled villas and secured high-rise condos, with access to the best in privatized medicine and private schools for their children, what incentives are left to press upon them the need to also support public education, health services and other services upon which so many depend? If I am paying to send my child to a private school, why should I also pay taxes to support the public education system? And so, we see the divisions exacerbated, with little chance of rapprochement. As Stiglitz puts the point,

> The more divided a society becomes in terms of wealth, the more reluctant the wealthy are to spend money on common needs. The rich don't need to rely on government for parks or education or medical care or personal security. They can buy all these things for themselves. In the process, they become more distant from ordinary people.[61]

In any event, and returning momentarily to the Rising Tide Argument, the evidence clearly shows that productivity is enhanced by increased public investment, a point emphasized decades ago by John Kenneth Galbraith in a series of books.[62] The reasoning is straightforward. It is the public investment into infrastructure, healthcare and education that supports the development of commerce that grows an economy. This point was made in our discussion of Smith (Chapter 6) in relation to the growth of telecommunications in the early decades of the 20th century. No private entrepreneur, or even a consortium of entrepreneurs, could have undertaken to build the infrastructure needed to make telephone service both affordable and sufficiently wide in coverage to become attractive to the general public, insofar as it would have taken far too long to realize a return on capital investments. We could repeat examples *ad nauseum* in relation to power grids, roads, bridges, satellites, basic research and more. As Stiglitz, puts it, "If the government doesn't provide roads, ports, education, or basic research – or see to it that someone else does, or at least provides the conditions under which someone else could – then ordinary business cannot flourish."[63]

It is worth noting again the point made by Rawls that the citizenry must perceive the system to be fair if they are to be expected to give their willing support, a point with which Nozick apparently agrees. As he says, "No doubt people will not long accept a distribution they believe is *unjust*. People want their society to be and to look just."[64] It is not entirely clear to me how or why Nozick would have thought that people could possibly perceive the minimal state to be just. In Chapter 10 of his book, titled, *A Framework for Utopia*, he seems to be suggesting that people would find it inspiring because of the maximal liberty it permits citizens to organize their lives as they see fit.[65] I think that Rawls would reply to this by saying that one's freedoms count for little in the absence of enough of the basic goods to live a decent human life and to enjoy self-respect. This last point about self-respect is crucial because it is self-respect that is so dependent upon perceiving the system to be fair by virtue of guaranteeing equality of opportunity. As Rawls says, self-respect is the most important of the basic goods.[66] So rhetorically one might ask,

> What does my freedom count for if there is not even the remotest opportunity for me or my children to ever hold positions of prestige and authority, or enjoy

the educational opportunities that make such advancement possible? And how could self-respect be possible under such circumstances?

Robert Putnam and social capital

Which raises the issue of social capital and the erosion of trust. In a groundbreaking book published in 2000 titled, *Bowling Alone*, political scientist and sociologist Robert D. Putnam explores the decline in the United States of social capital.[67] What is social capital? It is defined in analogy with physical capital and human capital. Thus, just as physical capital is defined as the tools, machinery, and physical assets that contribute to productivity, and human capital is defined in terms of the education and training that enhance the skills and thereby the productivity of workers, so social capital is defined as those features of our social networks that enhance productivity. In this sense, the market is itself a network of social relations that contribute enormously to human productivity, a fact agreed to on all sides.

Additionally, though, the network of personal relations that help an individual locate a job is an example of social capital that is more individualistic in nature, though not exclusively so because, while it certainly enhances the productivity of that individual, it also contributes indirectly to overall productivity. The point is that we all of us form, or become embedded in, networks of social connections, extending outwards in complex relations, from the family as foundation, sometimes overlapping, sometimes contiguous to one another and sometimes opposed to one another – a complex structure of networks within networks, if you will. Some of the benefits, and in some cases the costs, of the activities conducted within these various networks of social connections are externalized beyond the individuals and groups immediately benefitting from them. This is capital, an investment that pays dividends, though it is of necessity an investment that is social in nature. Quoting Putnam,

> Social capital can thus be simultaneously a "private good" and a "public good." Some of the benefit from an investment in social capital goes to bystanders, while some of the benefit rebounds to the immediate interest of the person making the investment. For example, service clubs, like Rotary or Lions, mobilize local energies to raise scholarships or fight diseases at the same time that they provide members with friendships and business connections that pay off personally.[68]

Putnam examines a range of phenomena under the rubric of social capital that includes political participation (e.g., likelihood of voting, assisting or joining a political party), civic participation (e.g., serving on a civic committee, writing a letter to the editor of a local newspaper), religious participation (e.g., likelihood of attending church regularly), social connections in relation to work (e.g., everything from union membership to meeting friends after work for a beer), informal connections (e.g., such things as bridge clubs, likelihood of having friends over for dinner), altruistic activity and volunteerism (i.e., donating blood, or volunteering/donating to a charitable organization) and even trends in relation to reciprocity, honesty and trust. And the salient fact to emerge from the data is that there has been a decline across all of these measures, in some cases quite steep, that correlates perfectly with the decline of equality of opportunity that began in the mid-1970s and gained momentum in the early 1980s. And there is solid evidence that this correlation is not mere coincidence.[69] Summarizing these figures in relation to league bowling, which functions as both an indicator and a metaphor for these trends, Putnam says,

> [L]eague bowling has plummeted in the last ten to fifteen years . . . a profile that precisely matches the trends in other forms of social capital that we have already examined – steady growth from the beginning of the century (except during the Depression and World War II), explosive growth between 1945 and 1965, stagnation until the late 1970s, and then a precipitous plunge over the last two decades of the century.[70]

Attempting to determine causal relations is enormously difficult. Utilizing various sophisticated statistical techniques, Putnam examines some of the most likely candidates that have been proposed to explain this measured decline in social capital: increasing demands on people's time as families manage two careers, lengthy commutes and the impacts of urban sprawl on everything we do; the tendency for technology and mass media to capture our attention[71] and intergenerational differences, among others. It turns out that, even when correcting for education, income levels, gender and region among others, consumption of television as passive entertainment is "the single most consistent predictor" of civic disengagement Putnam has discovered.[72] Even here, though, conclusions about causation remain inconclusive. The best that can be said is that the lines of causation run between various factors in a way that is often mutually reinforcing (not unlike positive feedback loops in relation to climate change), a typical example of what might be called a downward spiral. As television consumption, in combination with urban isolation, and a host of other factors encourage values associated with individualism and consumption, one is inclined to watch even more TV, bowl alone (rather than join a bowling league) and so on, in a perpetual self-reinforcing cycle.

But here is a crucial fact about social capital: it depends for its very lifeblood on trust. Trust is the lubricant that underwrites the reciprocity that is at the heart of social capital. The reason is easily explained. Trust reduces transaction costs on market exchanges. If we meet to negotiate a deal embedded in a network of trustworthiness and reciprocity, we do not need to expend time and energy ensuring the details will unfold as agreed. We do not need to pay extortion costs to corrupt officials.[73] But the trust that is essential to reciprocity, and the social capital it facilitates, is directly eroded by widening the gap of inequality, particularly equality of opportunity. When people perceive the system to be rigged, to be unfair, to be arranged in ways that guarantee many positions and opportunities will be systemically inaccessible to them, there is no room any longer for trust or reciprocity. Reciprocity becomes the prerogative of the sucker. As Stiglitz says,

> An economy with more "social capital" is more productive, just like an economy with more human or physical capital. . . . But the idea of trust underlies all notions of social capital; people can feel confident that they will be treated well, with dignity, fairly. And they reciprocate. . . . Social capital is the glue that holds societies together. If individuals believe the economic and political system is unfair, the glue doesn't work and societies don't function well.[74]

We can now make explicit the connection between the earlier point about equality of opportunity and the present point about social capital. The explicit reference to "dignity" in this quote recalls the earlier discussion in regard to Rawls's claim that self-respect is the most important of the basic goods that must be guaranteed by a just (fair) society. The chain of consequences is clear. People cannot feel a genuine sense of self-respect, or intrinsic dignity, in cases where they are treated unfairly and, as a result, there is an erosion of trust. Reciprocity collapses, divisions amongst the population grow, social capital declines, and we are all left worse off by the dramatic increase in transaction costs that inevitably result. Even the legitimacy of government is called into question as a result.

Before rushing to embrace these conclusions, though, I must address the criticism expected from the defender of the minimal state to the effect that, in any event, the redistribution required to guarantee equality of opportunity and facilitate social capital amounts to theft and, thus, a violation of the rights of some citizens. The answer, I think, is that a case can be made that parallels Nozick's argument for the move from the ultra-minimal state to the minimal state. That is, it is *not redistribution* because the justification is that it is required to prevent the even worse violation of the rights of the potential victims of inequality and the erosion of trust, and this answer is consistent with Nozick's theoretical commitments. Moreover, I believe the suggestion is implicit in Rawls. In speaking of the second part of the Difference Principle (i.e., the part of this principle intended to guarantee equality of opportunity), he says,

> It expresses the conviction that if some places were not open to all, those kept out would be right in feeling unjustly treated even though they benefitted from the greater efforts of those who were allowed to hold them. They would be justified in their complaint not only because they were excluded from certain external rewards of office but because they were debarred from experiencing the realization of self which comes from a skillful and devoted exercise of social duties. They would be deprived of one of the main forms of human good.[75]

The deprivation, then, that results from inequality of opportunity is a violation of the basic rights of those deprived because it systematically denies them access to a basic good accessible to others. It is, moreover, one of the main forms of human good. Such a deprivation, insofar as it is a violation of basic rights deserves compensation. But the only compensation here that can remedy the situation requires the state (presumably the minimal state) to repair the features of their social arrangements in a way that will restore equality of opportunity.

But what, precisely, is this basic good that some citizens are systematically deprived of in a situation of unequal opportunity? It is the right, the freedom, to live one's life freely according to one's own plan, the most basic freedom/right of all, the very foundation of the legitimacy of the minimal state. For if I am to be truly free to live my life according to a plan that I adopt for myself, I must at least have the real opportunity to actualize that plan. It is one thing for me to try to actualize my ambitions and fail for reasons that concern only myself, or, at least, where this failure is not because of any obstructions systemic to the organization of society itself. For such a failing I can hold only myself responsible, and thus I willingly accept the unequal outcomes of society secure in the knowledge that I am as well off as I could be. But if the arrangements of civil society are such that I am excluded from the opportunities open to others and that are essential to my self-actualization, then a wrong has been done; a wrong that must be compensated.

And to return to the aforementioned point about the centrality of self-respect to the goodness of one's life, we see that self-respect depends on one's ability to self-actualize. Where one has no chance to realize the self that one freely chooses, it is one's dignity that suffers. This is particularly the case where the obstructions to one's self-actualization are premised on arbitrary discriminations of one's membership in a particular class of persons. One is excluded, say, from certain opportunities because one is a person of colour, or poor or a woman. And a primary opportunity that functions to exclude many persons in the United States by virtue of being poor is (the lack of) educational opportunity. Educational opportunity is central because it is the primary means by which one enhances their skill set to open other career and positional opportunities. Speaking of this Stiglitz says,

> [G]overnment alters the dynamics of wealth [and thereby the equality of opportunity – WIH] by, for instance, taxing inheritances and providing free public education. Inequality is determined not just by how much the market pays a skilled worker relative to an unskilled worker, but also by the level of skills that an individual has acquired. In the absence of government support, many children of the poor would not be able to afford basic health care and nutrition, let alone the education required to acquire the skills necessary for enhanced productivity and high wages. Government can affect the extent to which an individual's education and inherited wealth depends on that of his parents. More formally, economists say that inequality depends on the distribution of "endowments," of financial and human capital.[76]

What, then, explains Nozick's resistance to this conclusion? The answer, anticipated in the introduction to this chapter, is explained by the differences between Nozick and Rawls with respect to how they conceive of coercion. Or, better, differences regarding what they *recognize* as coercion. As a species of Libertarianism, Classical Liberalism is inclined to recognize only negative rights, the traditional entitlements as "freedoms to do" certain things. Thus, coercion can consist only of deliberate interference with these rights. I deliberately interfere with your freedom of movement, say, or intentionally prevent you from expressing yourself. Progressive Liberals, though, acknowledge positive rights in addition to these negative rights (liberties), positive rights being entitlements to such things as will require the active intervention of the state to guarantee fair access – education and adequate healthcare being primary examples – which the Classical Liberal refuses to acknowledge. Thus, for the Classical Liberal, inequalities of opportunity that may result from limited access to such goods, and even where such limited access is a systemic function of the legal and structural features of the society, are not the result of deliberate interference. Rather, they are merely the outcome of the market doing its thing. No one has specifically undertaken to prevent me from having access to these opportunities, so it's not any kind of coercion. If my actions are "forced" by limited choices, this is not force as coercion. It is merely a function of the situation for which I alone must be held responsible.[77]

In conclusion, we have seen that although they begin in a similar place, with the assumption of the primacy of the basic rights and freedoms of citizens, Nozick and Rawls end up in a very different place. I have been at pains to argue that this difference can only be understood by taking account of the very different understanding they have of what basic rights consist of, and how this in turn informs their views of what shall count as coercion. For Rawls, and the Progressive Liberals generally, basic rights include the right to dignity and equality of opportunity, and these can be guaranteed only by the active involvement of the state. To abrogate its duty to protect these basic rights would have the state leave its citizens open to unfair coercion, and the steady erosion of trust required for political legitimacy is the predictable outcome.

★★★

In the next chapter, we will undertake an examination and comparison of these ideologies in relation to corporate governance and corporate social responsibility.

Notes

1 Stiglitz (2012, p.172).
2 Nozick (1974, pp.8–9); underlining added, italics original.
3 In Chapter 5, I used Rousseau as the exemplar of what Nozick here refers to as the Godwinian conception of human nature. Presumably, Nozick is referring to William Godwin (1756–1836), a journalist and anarchist who defended anarchism on the grounds that people left to themselves are perfectly reasonable (or very nearly so) and, thus, do not need the paternalistic state to defend their interests.
4 Op. cit., p.5.
5 Nozick refers to these basic moral rights as *moral side constraints* on the origins and operation of the state.

6 Op. cit., p.6.
7 For example, see p.31, *et passim.*
8 See discussion of Locke on this point in Chapter 5.
9 Op. cit., pp.16–17; italics in original.
10 Op. cit., p.52.
11 Op. cit., p.27.
12 In regard to how we should understand the relation between the DPA and the ultra-minimal state, Nozick is somewhat equivocal. Sometimes, he seems to be suggesting that the DPA *is* the ultra-minimal state. At other times, and in my view more plausibly, he seems to suggest that the DPA will transition naturally to the ultra-minimal state under competitive market pressures that will require it to address such issues as risk and free riders. In doing so, it will perforce claim a monopoly on the right to coercion and, thus, transition naturally to the minimal state. See, in particular, p.109 on the DPA as *de facto* monopoly.
13 Whenever speaking of the distribution of goods, I shall mean to speak of the total sum of goods and services available for distribution at that time (something like the GDP at a moment in time) but for convenience will shorten this to "goods."
14 Nozick expresses some caveats about this. In a very subtle discussion of Lockean acquisition, he sets out a number of difficulties for the theory as normally interpreted (some of which we considered in Chapter 5) but thinks there is some way of addressing these issues.
15 Op. cit., p.151.
16 Op. cit., pp.152–153.
17 Though, as Nozick points out, it may be logically possible that there might be some patterned distribution that is compatible with the basic rights, it would have to be so weak that it would be uninteresting. Certainly, any patterned distribution committed to egalitarianism will be thwarted (p.164).
18 Op. cit., p.166; italics original.
19 Op. cit., p.173; italics original.
20 Douglas Martin writing in *the New York Times*, November 26, 2002.
21 Thomas Scanlon, quoted in *Prospect*, by James Garvey, November 13, 2017. Retrieved October 29, 2022, from: https://www.prospectmagazine.co.uk/philosophy/john-rawls-died-15-years-ago-heres-why-you-should-read-a-theory-of-justice.
22 Rawls (1971/1999). The reference to basic rights is from p.3, and for the remark about Utilitarianism, see p.13, *et passim.*
23 Op. cit., pp.6–7.
24 Op. cit., p.13 and p.43, bottom.
25 Op. cit., p.11.
26 Op. cit., p.43.
27 Op. cit., p.47.
28 Op. cit., pp.48–49.
29 Op. cit., p.49; see also Chapter 8, "The Sense of Justice."
30 See Mill (1962), Chapter III, pp.10–11. See also the discussion in Rawls at pp.439–440, where Rawls seems to suggest that Mill's position is inconsistent or, at the very least, in tension with a pure Utilitarian view insofar as Mill seems to accept the view that the advance of civilization will depend on citizens accepting that there can never be grounds for setting aside the claims of some, not even to enhance the aggregate good of society as a whole.
31 This first presentation of the principles undergoes refinement as the details of the theory unfold. We will consider some but not all of these reformulations. This statement of the principles is from Rawls (1971/1999, p.53).
32 Op. cit., p.54.
33 Op. cit., p.75; italics mine.
34 Op. cit., p.76.

35 Rawls notes two kinds of cases that may justify the restriction of basic liberties, though in both cases this can only be "for the sake of liberty itself." See pp.214–215.

36 Op. cit., p.72.

37 Op. cit., p.102.

38 Op. cit., p.120.

39 In particular, see Rawls's comments at pp.221–227.

40 The double negative formulation used here, while somewhat awkward, is to emphasize the fact that, from behind the veil of ignorance, it is a matter of what one does not know.

41 Op. cit., p.111 and pp. 160–168.

42 Op. cit., pp.156–157.

43 Op. cit., p.241.

44 Additionally would be the injustice externalized onto various colonies, former colonies and even trading partners.

45 Op. cit., pp.311–312.

46 Op. cit., p.336.

47 Smith (2003) argues for this very point in Book V of *Wealth of Nations*. See, in particular, Book V, Chapter 1, pp.986–988.

48 Rawls deals with this by observing that his theory is not meant to be a complete theory of morality. It would need to be supplemented by such other theories as may be required to comprehend the full range of issues (see, in particular, pp.448–449). This still leaves it to signatories of the basic agreement, though, to decide how these other issues shall be dealt with. After all, the principles of justice set the basic terms for our association and all that flows from that. Also of interest in connection with this point is *Zoopolis: A Political Theory of Animal Rights*, by Will Kymlicka and Sue Donaldson (2013).

49 Nagel (1986).

50 Within limits. Everyone knows there will be violations of principle in any society, however just. As Rawls puts it, government in practice is a case of Imperfect Procedural Justice (see earlier discussion).

51 See, for example, *The Price of Inequality: How Today's Divided Society Endangers Our Future (2012)* and, *The Great Divide: Unequal Societies and What We Can Do About Them* (2015).

52 And because of globalization and externalities, the impacts of these changes have been spread worldwide. An insightful discussion of the impacts of financial globalization in shaping the economic crisis of 2007/2008 is found in Adam Tooze (2018) *Crashed: How a Decade of Financial Crises Changed the World*.

53 See, in particular, Stiglitz (2012, pp.20–24).

54 Wealth is measured separately from income and is a measure of total equity or net worth, as given by property, investments and so forth, for various percentiles of a population. For example, a measure of total accumulated wealth for the top 1 percent of the US population exceeds $25 trillion, which is more than the combined wealth of the bottom 80 percent of the population. See: https://www.brookings.edu/blog/up-front/2019/06/25/six-facts-about-wealth-in-the-united-states/ (retrieved October 29, 2022).

55 https://data.worldbank.org/indicator/SI.POV.GINI?locations=US (retrieved October 29, 2022).

56 Stiglitz (2012, p.117).

57 The reader will recall the slogan of the Occupy movement of 2011–2012, "We are the 99%."

58 Op. cit., p.71.

59 Former US Speaker of the House, Paul Ryan, for example, has suggested this on many occasions.

60 On the point about political paralysis, I am reminded of the speech by John McCain given to the American Senate shortly before his death, on July 25, 2017. You can read the text of this stirring speech at: https://www.npr.org/2017/07/25/539323689/watch-sen-mccain-calls-for-compromise-in-return-to-senate-floor, (retrieved October 29, 2022).

61 Op. cit., p.93.
62 See, for example, *The Affluent Society* (1958) and *American Capitalism* (1952).
63 Op. cit., pp.92–93.
64 Nozick (1974, p.158); italics original.
65 In particular, see his comments on p.332, where he says, "the Minimal State is Utopia."
66 Rawls (1971/1999, p.348).
67 The full title is, *Bowling Alone: The Collapse and Revival of American Community*.
68 Putnam (2000, p.20).
69 See, in particular, the discussion on pp.359–360.
70 Op. cit., p.112.
71 For more on this point, a book I highly recommend is Matthew Crawford (2015) *The World Beyond Your Head: On Becoming an Individual in an Age of Distraction*.
72 Op. cit., pp.230–231.
73 Op. cit., pp.134–137.
74 Stiglitz (2012, p.122).
75 Rawls (1971/1999, p.73).
76 Stiglitz (2012, pp.30–31). Cf., Picketty (2014).
77 It must be noted that, for Nozick, the only rights which there are may be property rights. This is strongly suggested by the language on p.238.

8

CORPORATE GOVERNANCE AND THE LIMITS OF CORPORATE SOCIAL RESPONSIBILITY

In this chapter, the focus will be on differing visions of the nature and extent of corporate social responsibility (hereafter CSR), and we will examine this topic from the perspectives of representative Classical and Progressive Liberal writers. Let us begin, though, by getting clear on the concepts involved.

A *corporation* is defined as an organized body of individuals authorized by the laws of the state where domiciled to act as a single entity, usually referred to as a legal person and recognized as such by the relevant statutes and usually authorized for specific purposes often defined by a corporate charter. The principal idea is that the membership of the persons organized as a corporation will change over time, but the artificial person that is the corporation itself exists in law *in perpetuity*.

The focus here will be on for-profit publicly traded companies. It is worth pointing out that the two terms "company" and "corporation," though their referents overlap, are not synonyms. There are many organizations that are not companies that can become incorporated, for example, cities and other jurisdictions, hospitals and others. Moreover, not all companies are incorporated, for example, privately held companies. In regard to those corporations that are publicly traded for-profit companies, the owners are the shareholders (or stockholders), and these individuals own a proportion of the company that is determined by the number of shares in their possession. Often, the number of shareholders can be vast and distributed globally and can include persons invested indirectly by virtue of being members of, say, a pension fund that owns shares directly. These owners do not, as a rule, manage the corporation themselves but elect a board of governors to do so, though it is common for some shareholders to sit as board members. These boards typically meet regularly to debate the overall direction of the firm and to vote on resolutions and other matters of business related to the firm, including the election of new board members. In this sense, it is the board that directs the corporation, though usually the day-to-day

DOI: 10.4324/9781003388555-10

management of the firm is left to in-house management (i.e., the executive team and mid-level managers), whose members are bound by a fiduciary duty to manage the firm in the interests of the owners. It is not unusual for members of the executive to also sit on the board, sometimes as voting members, sometimes not.

Now when we speak of CSR, the presumption seems to be that corporations, in their activities, have got a responsibility to act in ways that will contribute to various social goals, whatever these may be. We may suppose the social goals themselves can change over time, or from one jurisdiction to another, but some likely candidates might be charitable contributions to the community, perhaps assistance with the building of schools or hospitals or contributions to their maintenance; voluntarily engaging with community activists in tree planting activities at former logging sites or even organizing employees in various charitable activities, including assisting with emergency response.[1] A question one might ask is, why do we make this presumption? From where is it supposed this duty to social responsibility derives? To understand the answer to this question, it is helpful to consider a brief and informal history of the corporation and corporate governance.[2]

The first corporations in the sense with which we are concerned emerged in the Mercantile era, examples being the Dutch East India Company, the Hudson's Bay Company and the (English) East India Company. These companies were created under charter by governments as enterprises to advance the imperial and Mercantile interests of these nations and for this reason were usually granted monopolies with respect to both geographical areas and commercial markets. For example, the Hudson's Bay Company was initially chartered principally to secure access to furs and for this purpose was given a monopoly on the fur trade for northern North America, the basin of Hudson's Bay (i.e., what is now Canada, more or less). In this way, there was a kind of a *quid pro quo* bargain between the interests of the private investors on the one hand, who gained the advantage of a government-guaranteed monopoly on a potentially profitable venture (and often the military protection of the state was included implicitly in this guarantee), and the imperial interests of the state, on the other hand, to secure their access to territory and resources.[3] The *quid pro quo* nature of this relationship and the monopolies that were the legacy of this period were the focus of Smith's critical analysis, as discussed in Chapter 6.

In any event, these charters implied that the companies themselves were under the control of the state, and their charters were subject to revocation if the corporation failed to meet the purpose for which it had been chartered. Thus, we see that the idea that corporations are granted permission by the state where domiciled to undertake their activities in exchange for meeting certain obligations is implicit from the outset. However, the existence of these very monopolies, and the intrusion of the state into the affairs of these corporations, became increasingly problematic, in part at least, because of the critical investigations of Adam Smith.

As discussed, Smith argues very effectively that the wealth of nations is enhanced by the liberalization of trade and leaving market participants to engage one another exclusively from motives of self-interest. And so, as European economies transitioned from Mercantilism to Capitalism and proceeding into the 19th century, there emerged a more laissez-faire approach that tended to eschew direct government control of corporations as instruments of colonial policy, and the modern form of the private sector company came into being as a result of a series of legal reforms. Prominent among these, the Joint Stock Companies Act (enacted in United Kingdom, 1844)[4] greatly facilitated the process of incorporation by eliminating the need for direct government charter in favour of a process of legal registration. This had the effect of making it possible for individuals to incorporate by a relatively simple and transparent process of registration, a process that created the corporation as a *legal entity* and thereby simplified the rights and duties of investors. Notably, the control of companies had passed from government to the courts. Coordinated with this was a period of significant deregulation and the emergence in law of the notion of *limited liability*, an innovation that allowed investors to limit their liability in the event of business failure to the amount they invested in the company.[5] In Britain, this began in 1855 with the passing into law of the Limited Liability Act. This was effected, in part, to encourage private investment. The underlying reasoning is that corporations are granted permission to operate because it is presumed they will be of benefit to society and, thus, we should encourage such private investment.

The last feature of interest here is the emergence of the notion of corporations as legal persons. The principal idea is that, insofar as the corporation exists in perpetuity as a legal entity that is independent of the members that constitute it at any moment in time, it should enjoy at least some of the rights and responsibilities of natural persons. In part, at least, this is done to facilitate corporations entering into contractual agreements with other parties and makes it possible for them to bring civil suit or to be sued by other parties. This not only provides the legal assurance that is essential to the trust on which business is conducted but also significantly reduces the transaction costs of conducting business. In the United States, an important event in connection with this was the result of a legal ruling in 1866, *Santa Clara v. Southern Pacific.*[6] In this decision, the courts affirmed that the Fourteenth Amendment applies to corporations as much as to natural persons. The Fourteenth Amendment grants citizenship to all persons born or naturalized in the United States and guarantees them equal protection of the laws on this basis. While the implications of this ruling remain controversial and interpretation continues to evolve in the courts, in a recent ruling, *Citizens United v. Federal Election Commission,*[7] the Supreme Court affirmed the right of corporations to unlimited contributions to independent political broadcasts on the basis of an appeal to the First Amendment, essentially arguing that to limit such corporate spending would be a violation of their right to freedom of expression.

And so, in response to our earlier question, history suggests that corporations have a social responsibility that derives from the fact that they are granted permission to act in jurisdictions where registered, and they are given a number of freedoms and rights that attach to this privilege. All rights come with attendant duties and the case is no different for corporations as legal persons. So, having established the claim that corporations do indeed have social responsibilities, the real question becomes, "What is the nature and extent of CSR?" Or, to put the question somewhat differently, when wondering precisely who should benefit directly from CSR, we ask, "In whose interests are corporations to serve?" I will argue for the view that CSR implies a commitment to benefit civilized society as a whole; that is, in the terminology central to this debate, they are to act in ways that benefit all stakeholders (not just all stockholders), or, at the very least, to do no intentional harm to any stakeholders. In fact, this commitment is implicit in one of the arguments central to the Classical Liberal position, the Rising Tide Argument, already touched upon earlier. We revisit that argument here to note further implications, but moreover, as we shall see, other arguments can be brought to bear in defending this view. Particularly relevant to our concerns is the argument that the Stakeholder view of corporate governance and CSR is the only one compatible with the narrative of Biosphere Consciousness sketched in Chapter 4.

The Shareholder Model (Milton Friedman)

We begin with the Shareholder Model (aka Stockholder Model) defended most conspicuously and ably by Nobel laureate economist Milton Friedman. Friedman is, in my view, often mischaracterized as an unrepentant and doctrinaire libertarian capitalist, closer in spirit to Ayn Rand than the facts will bear. Reading his book, one is struck by his reasonableness and his sincerity.[8] While a missionary zeal is undeniable, his concerns for freedom are equally apparent, and his enthusiastic defence of free market Capitalism might in large part be attributed to historical circumstances.

He commenced his career as an academic economist in the immediate post-war years. As a young academic at the University of Chicago in the 1950s, his views are taking shape at the height of the Cold War. Paranoia and fear were ubiquitous, as evidenced by the McCarthy hearings and the House Un-American Activities Committee. The Soviet Union and communism were perceived in some quarters as constituting a genuinely existential threat to Western democracies, not least because of their avowed mission to bring the communist revolution to the world but also because of the allure communism held for many citizens of these very democracies.[9] This allure was based on the promise (whether real or illusory) of employment for life for everyone, universal pensions and benefits and perhaps also as a reaction to what some felt was a tendency for capitalist societies to force a conformity based on values of consumerism, commodification of all things and the resulting (Marxian) alienation. More pointedly, there was a narrative that emerged in the years immediately following the Roaring Twenties and the subsequent stock market crash that the overwhelming majority of citizens were hapless victims of corporate excess and that such benefits as were extended during the Roosevelt New Deal era were extracted grudgingly from the captains of industry to quell unrest.[10]

Friedman sees the views he defends as the necessary corrective to these trends, an attempt to restore faith in the feature of capitalist democracies that is the best they have to offer, namely freedom, both in markets and politically. It is this freedom he seeks to defend, but it begins with the freedom of markets. And, whereas Rand is frequently guilty of obvious fallacies in her reasoning and shameless caricature of those she perceives to be opponents of her views,[11] Friedman's work is carefully argued and attentive to the available evidence. I do think he makes some mistakes in both logic and fact, as noted later, but his arguments deserve careful attention.

I will be relying on two principal sources for the following exposition. First, his book, *Capitalism and Freedom*, originally published in 1962, and an op-ed piece published in *the New York Times Magazine* titled, "The Social Responsibility of Business is to Increase its Profits," published in 1970.

Let's start by getting as clear as we can on Friedman's thesis,

[T]here is one and only one social responsibility of business – to use its resources and engage in activities designed to increase its profits so long as it stays within the rules of the game, which is to say, engages in open and free competition, without deception or fraud.[12]

In one respect, at least, this is very clear: the increase of profits is the *exclusive* social responsibility of business. One might wonder how this is a *social* responsibility at all, but this is to be explained by appeal to the Rising Tide Argument introduced in Chapter 5 – we, all of us, benefit from the positive externalities (or "neighborhood effects" as Friedman prefers to call them) that result from business attending to what it does best, and which it should do exclusively, which is to focus on the increase of profits. It becomes clear subsequently that this is a directive to business to *maximize* profits, so long as it does so within the "rules of the game." However, this latter qualification is somewhat less clear, though we're told that this means, in part at least, that it must do so while engaging in free and open competition and without resorting to deception or fraud. Later, in the op-ed piece, we're told that this requires of business that it observe the laws of the land and ethical customs. As he says there, the responsibility of the executives "is to conduct the business in accordance with their desires [i.e., the desires of the shareholders – WIH] while conforming to their [sic] basic rules of the society, both those embodied in law and those embodied in ethical custom."[13]

A lot of questions can be raised about how the executives are to know what the relevant ethical customs are, particularly when operating in foreign jurisdictions, and how to operationalize these as constraints on the conduct of business, but I pass over this in what follows since the directive to strictly observe the laws of the land seems sufficient to capture Friedman's intent here. It must be noted, though, that elsewhere Friedman makes it clear that this requirement to observe the strict letter of the law is consistent with strategizing to maximize profits by making sure to never exceed the minimum required by law. For example, considering laws to limit carbon emissions, should the firm voluntarily decide to spend money from profits to reduce its emissions even further than required by law, it would be violating its strict responsibility to maximize profits within the limits of the law. So it is the laws of the land that set the "rules of the game."

The choice of metaphor, to see the conduct of business as a game, is telling in this context and a host of troubling questions emerges in contemplation of this. Given the free movement of capital under various trade agreements negotiated in the last few decades, corporations are enticed to maximize profits by moving capital investments when feasible to jurisdictions where the regulatory burden imposes fewer constraints. What are we to say about corporations that maximize profits within the strict limits of the law by moving production facilities to locations where, for example, there are relatively few laws to protect the health and safety of workers? Or moving to locations where there are few environmental protection laws? Or fewer safety regulations resulting in accidents impacting even those non-employees living within the zone impacted by the activities of the facility? One is reminded of the citizens of Bhopal impacted by the explosion of the Union Carbide facility on the night of December 2 – 3, 1984, which spewed toxins into the local environment. Setting aside Union Carbide's behaviour to avoid legal penalties following the disaster, it would seem that their original decision to locate their facilities in Bhopal and conduct their activities in the manner that led to the disaster was consistent with strategies to maximize profits, as Friedman's directive would seem to require. And if business is a game, with the object of winning, and where winning is defined as the maximization of profits consistently with the rules of the game, this is exactly what they are *required* to do.

The matter is made even more complicated when we consider that corporations are almost always involved in the negotiation of international trade agreements, and it is consistent with Friedman's mandate of profit maximization that they should be trying very hard to get regulatory burdens reduced and movement of capital resources liberalized as much as possible. In doing so, they are directing their energy and resources to profit maximization. In practice, this often involves seeking to have clauses inserted into these agreements that have the effect of undermining open and free competition, in ways injurious to smaller, local industries and businesses. Additionally, these trade agreements often have the effect of diminishing the capacity of nation state governments to act in the interests of their citizens when these interests come into conflict with such agreements. It is not just that these results are the frequent outcome of multilateral trade negotiations, but more significantly, they seem to be exactly what is required if profit maximization is the sole directive for a corporation. The point to be emphasized here is that it is difficult to understand how Friedman's constraints on profit maximization could be framed consistently so as to actually be any kind of meaningful restraint on corporate behaviour at all.

A further aspect of the game metaphor, though, is that it has the effect, whether intended or not, of implying there can be no such thing as victims in a free market system that adheres to the rules. There can be winners and losers certainly but not victims and oppressors. This is reinforced by the view inherited from Locke that, in a free market society, where one ends up is a matter of personal merit. And note the role played by freedom in all of this. The very point of free markets is that one is free, in principle, to walk away from any deal under negotiation. There is no coercion, there is just a game that one is free to join or not.

And lastly, the game metaphor encourages a view I believe to be false and injurious to meaningful progress in realizing a sustainable form of Capitalism, and that is the view that business is a zero-sum game. Specifically, and insofar as games are by nature won or lost, to see business from this perspective is to embrace the view that for me to win, the other party (or parties) must lose. And this despite the fact that this is at odds with the fundamental premise at the heart of the Rising Tide Argument, that market exchanges undertaken by individuals acting on their self-interest are undertaken because they are mutually beneficial. At the very least there is a tension here if not an outright contradiction. In Part I, it was urged that a sustainable form of Capitalism will require us to see the goal of all negotiations as intended to facilitate an outcome that is win–win. But let us set such concerns aside for the moment to consider Friedman's position and his arguments in detail.

In what follows, I will set out *five distinct lines of argument* for Friedman's thesis and for ease of reference, I will give them names. I will refer to the arguments in the following order and using these names: the Private Property Argument; the Reward for Risk Argument; the Rising Tide Argument; the Sanctity of Voluntary Contracts Argument and the Freedom Argument. Two caveats here. First, the reader should be aware that these names are never used by Friedman (at least not to the best of my knowledge). Moreover, though for clarity we treat these arguments as analytically distinct, they are in fact deeply intertwined and mutually supportive. In particular, freedom does a lot of work in Friedman's texts. It functions both as a powerful metaphor and rhetorical device and as a conceptual device at the heart of several of his arguments.[14]

Turning, then, to the first argument, the Private Property Argument, Friedman says, "The corporation is an instrument of the stockholders who own it."[15] The premises are that the corporation is the (exclusive) property of the shareholders and that the rights and freedoms attached to proprietorship under law give these owners exclusive prerogative (within the limits of the law) to direct how this property shall be used. And, so, the conclusion follows that, unless directed otherwise, the executives of the corporation have a fiduciary duty to maximize profits for these owners. There is an assumption at work here as an unstated premise that the shareholders will ordinarily want that their property to be used in a way that maximizes their self-interest which, in the case of capital interests, is usually taken to mean used in such a way as to maximize returns.[16] Friedman might say to us,

Consider your property – whatever you wish, perhaps your home, your car, your favourite guitar – you wouldn't tolerate someone else telling you what is to be done with it. You would insist that you are at liberty to use it exclusively in your own interest; you have the right to exclude others from enjoying the use and benefits of this property. So why would we willingly tolerate the owners of corporations being told what to do with theirs?

Here we see the first use of freedom in the context of an argument: the meaning in this case being the legal rights and freedoms defined by the property laws of capitalist societies (ignoring subtleties of how these laws might differ from one jurisdiction to another).[17] The real insight here, I believe, is that, for Freidman, property rights are exclusive in two distinct senses. First, as noted earlier, my ownership of anything gives me exclusive disposition over that thing. But, and perhaps even more importantly, Friedman seems to imply that property rights are all the rights there are. Or, at least, the only rights that really need to be enshrined in law and defended by the state. This is why, in his view, the state has few legitimate roles beyond those required to accomplish this. As already hinted in Chapter 7, this point is made more clearly and starkly by Nozick. Speaking of rights generally, Nozick remarks that rights to equality of opportunity, and even the right to life, can be realized only in a society with the appropriate "substructure of things and materials and actions; and *other* people may have rights and entitlements over these."[18] Thus, all rights additional to property rights are derivative of property rights, which include rights to one's personhood and, thus, to self-protection. He concludes,

> The particular rights over things fill the space of rights, leaving for general rights to be in a certain material condition. The reverse theory would place only such universally held general "rights to" achieve goals or to be in a certain material condition into its substructure so as to determine all else; to my knowledge no serious attempt has been made to state this "reverse" theory.[19]

If this is indeed what Friedman assumes, then we can easily make sense of his claim that the fiduciary duties of corporate executives are owed exclusively to the shareholders.

The next argument, implicit in Friedman's discussion of patents, I shall call the Reward for Risk Argument.[20] The argument starts from the observation that there is always an enormous risk involved in the investment of capital. It's not just that investments can occasionally fail. It's that when they do, the failure can be egregious and investors can lose their entire investment. Moreover, it is claimed that it behoves capitalist societies to encourage these capital investments, to do what is possible within the limits of the law and prudent policy to stimulate capital investment as much as possible. One very obvious way to encourage investment in the face of such risk is to reward investors by permitting all accumulation of profit not needed for reinvestment into the business to keep it viable and growing to be returned to investors as dividends.[21] Thus, the conclusion follows that maximization of profits for return to stockholders is the justified reward that offsets the risk of investment.

But now we ask, "Why would society wish to encourage, stimulate, such investment by giving it these extraordinary rewards?" The answer is given by the familiar

Rising Tide Argument: by stimulating such investment, we directly enhance productivity, grow the economy and produce jobs. Thus, we all benefit. Ronald Reagan used this argument to great rhetorical effect in his frequent pronouncements that entrepreneurs are the true heroes of capitalist societies.[22] The emphasis in Friedman's writings is on freedom, and appeal to the Rising Tide Argument is less prominent there than it would become by the time Reagan took office, but it is there, running throughout his work in an important supportive role.

It is interesting to speculate on the causes underlying the change in the temper and tone of public discourse in the United States over this time frame, from the early 1950s to the late 1980s. Friedman's concern for freedom amounts almost to an obsession. Capitalist economies were flourishing and, thus, attention was riveted by the existential threat posed by the communist bloc, as noted earlier. By the time Reagan takes office in 1981, capitalist economies are stagnant, and the threat posed by the communist bloc declines steadily through the 1980s. This perception that the threat was in decline was partly the result of negotiations to limit armaments and partly because of Gorbachev's initiatives to introduce *perestroika* (restructuring) and *glasnost* (openness/transparency) into Soviet society, initiatives broadly seen to be liberalizing and to constitute a dialling back of the Soviet Union's imperialist ambitions and posture as defiant enemy of the West.[23] Particularly after the collapse of the Soviet Union in 1991, and the triumphalism of the Washington consensus that followed,[24] the focus of American discourse shifted from freedom to prosperity. In any event, the Rising Tide Argument plays an important supportive role in Friedman's model, in particular as this is related to risk for other stakeholders.

To see this, consider the flip side of this risk for investors: the risk for all other stakeholders. These stakeholders include at the very least such groups as employees, the executives, the suppliers, the clients and members of the local community. These groups are subjected to risks that, on the face of it, seem to offset to some degree the benefits alluded to by the Rising Tide Argument. Depending on the conduct of the corporation, there can be lay-offs of employees, sinking revenues for suppliers on reduced demand or even failure to make payment. In contemplation of such risks, we might ask, "What protections ought there be to off-set these?" If investors deserve to have their risks offset by the possibility of enhanced profits, what might be done for these other stakeholders? Enter the fourth argument, the Sanctity of Voluntary Contracts Argument.

That's a bit of a mouthful, but bear with me for a moment and the significance of the name will become clear.[25] The argument begins with the assertion that other stakeholders enter into a relationship with the firm on the basis of a contract, whether it be a contract of sale between supplier and the corporation or a contract of employment between employees and the firm. In this latter case, the contract may be a collective agreement between the firm and employee bargaining unit or an individual contract of employment negotiated by an individual with the firm. And in cases where there is not an explicit contract between the firm and the stakeholder under contemplation, it is usually supposed that there is an implicit contract. For example, in the case of customers (i.e., end users purchasing a product or service), the claim is that there is an implicit contract between vendors on the one hand and clients on the other, as implied by the legal term *caveat emptor*. While different jurisdictions will circumscribe the application of the legal principle differently, the idea is that all parties are bound by law to the terms of an implicit contract that typically places the burden of due diligence on the consumer before making a purchase, though the producer and seller are bound to not misrepresent the product and/or service.

Now if we generously grant this assumption that *all* stakeholders other than shareholders have their relationship with the firm and, thus, their risks in relation to that relationship, circumscribed by the terms of a contract, either explicit or implicit, then these contracts have two features directly relevant to our present concerns. First, they are entered into voluntarily (that is, freely), and second, in capitalist societies they have a special status before the law that I would characterize as the secular equivalent of being sanctified. Specifically, the trust that is essential to the proper functioning of the market requires that the law guarantee that contracts freely negotiated must be honoured and significant penalties are attached to the violation of contracts for this reason. Thus, the conclusion follows that all stakeholders other than the owners have their risks adequately ameliorated by contractual agreements. Further, it is supposed that the security provided by contractual protections of their interests is all that the situation requires. Thus, these stakeholders are not subject to risk in the same way as shareholders for whom there cannot possibly be a contract that would protect them from the potential loss of their investments (though, of course, limited liability laws are in place to limit the losses to those direct investments only).

Once again, freedom plays an important role in this argument, but here the reference is to the freedom of agents to negotiate deals exclusively on the motive of self-interest. Since contracts are entered into voluntarily, we can be assured that all parties to the agreement must be satisfied that their interests are being met. Thus, no protections additional to the terms of the contract are required. In fact, they are strictly forbidden since this would amount to a violation of the contract, a change to the terms of the agreement after the fact and to the benefit of one party at the expense of the other. It would be, to return to the game metaphor, a change in the rules of the game, exactly the sort of thing that would undermine the sanctity of contracts and, thereby, undermine the trust that is lubricant of market activity. We saw similar reasoning at play in Nozick's example of Wilt Chamberlain and the $2.50 surcharge willingly paid by fans to attend his games (Chapter 7). Attempts to redistribute this money back to the customers amount to a violation of the original terms of the implicit contract.

Finally, we come to what seems to be the argument of paramount concern to Friedman and which is at the heart of all the others, which I shall refer to as the Freedom Argument. In its most basic form, this is the claim that insofar as freedom of the markets is logically prior to (specifically, a necessary condition of) political freedom, we must actively promote and nurture market freedom wherever and whenever we can. Only in this way can we hope to bring about the democratic rights and freedoms we cherish in places where they do not presently exist.[26] Moreover, insofar as the existence of autocratic states constitute an existential threat to our freedoms here and now, at least in part by the very fact of the allure their propaganda holds for some of our own citizens, the promotion of market freedoms is also central to the protection of our own political rights and freedoms right here.[27]

Now, as we have seen, encouraging the entrepreneurial activities of capitalists depends on the reward of profit returned in the form of dividends that offset the risk of investment. Thus, Friedman's conclusion that defending market freedoms starts with protecting the rights of owners to maximize their profits. In fact, as he says, talk of the social responsibility of business, at least when taken seriously, is "a fundamentally subversive doctrine." Subversive because it quite literally threatens the foundations of capitalist democracies, "Few trends could so thoroughly undermine the very foundations of our free society as the acceptance by corporate officials of a social responsibility other than to make as much money for their stockholders as possible."[28] And from his op-ed,

> Business men who talk this way (i.e., about the need for business to engage in promoting desirable social ends – WIH) are unwitting puppets of the intellectual forces that have been undermining the basis of a free society these past decades.[29]

Or this, "Whether blameworthy or not, the use of the cloak of social responsibility, and the nonsense spoken in its name by influential and prestigious businessmen, does clearly harm the foundations of a free society."[30]

We will consider criticisms of these arguments in our discussion of the Stakeholder view, a theory developed, in part at least, on the basis of perceived shortcomings of Friedman's view to deal adequately with corporate governance and CSR. Here I avail myself of the opportunity to focus on criticisms of this last argument, the Freedom Argument, which is, in my view, guilty of some basic logical fallacies. In the first place, it is guilty of presenting readers with a false dilemma, as though the choice is between a full-blooded laissez-faire Capitalism or communist-style socialism. Consider the following quote,

> Fundamentally, there are only two ways of coordinating the economic activities of millions. One is central direction involving the use of coercion – the technique of the army and of the modern totalitarian state. The other is voluntary co-operation of individuals – the technique of the marketplace.[31]

This is not only a logical fallacy but also false in fact. There is, presently active, an entire range of mixed economies, from the statist economies of South East Asia (e.g., South Korea, Japan), to the social welfare states of Northern Europe (e.g., Norway, Sweden), to the relatively free market economies of the United States and the United Kingdom.[32]

Additionally, though, Friedman is guilty of *Straw Man* reasoning here, a tendency buttressed throughout the book by the language used to describe alternatives to his own vision of what a genuine capitalist economy should look like. This is evident when he says, for example, "Underlying most arguments against the free market is a lack of belief in freedom itself."[33] And when he dismisses concerns for the possibility of abuse of employees at the hands of employers by saying, "The employee is protected from coercion by the employer because of other employers for whom he can work, and so on. And the market does this impersonally and without centralized authority," it seems Pollyannish at the very least.[34] Even if we suppose, generously, that the labour markets in 1962 were significantly better than today at guaranteeing full employment, or close to full employment, it had to be factually false even then to suppose that there was a complete absence of coercion in the workplace. It is on such occasions, when he indulges in these misrepresentations, that Friedman can appear doctrinaire and which have probably fuelled the unfair comparisons to Ayn Rand.

The Stakeholder Model (R. Edward Freeman)

We turn now to the progressive alternative to Friedman's Shareholder Model, usually known as the Stakeholder Model. I will focus on the model of this view that has been developed by R. Edward Freeman in a book published in 1984 and in a series of papers published over the last couple of decades or so.[35] In what follows, I will rely for the most part on his paper titled, "Stakeholder Theory of the Modern Corporation," which conveniently and clearly summarizes his view.[36]

As with Friedman, we begin by getting as clear as we can on his thesis. On p.258 he says, "[M]anagers bear a fiduciary relationship to stakeholders." And a little further, "In short, management, especially top management, must look after the health of the corporation, and this involves balancing the multiple claims of conflicting stakeholders."[37] We must take note of two points. First, the notion of a stakeholder is a straightforward generalization of the notion of a shareholder. By that I mean to direct the reader's attention to the definition of a shareholder, as agreed to by all parties, wherein a shareholder is defined as a partial owner of the firm and, by virtue of this ownership, is an individual *having a stake in the firm*. Thus, as we've seen, the Reward for Risk Argument follows from the fact of having a stake in the firm and the risk this involves for the individual. Having a stake in any venture inevitably involves at least some risk; it is what it means to have "skin in the game," to use the colloquial expression. Stakeholders, too, have skin in the game insofar as their interests are affected, either negatively or positively, by the actions of the firm and by this fact of risk have a claim on the firm.[38] And the second point to note is that the term "stakeholder" can be used to refer to individual stakeholders (individuals with a claim on the firm) or to groups (the shareholders being one obvious group of stakeholders).[39]

Now in the expression of Freeman's thesis quoted earlier, it is probably most helpful to think in terms of stakeholders as groups. Taken this way, Freeman usually draws a distinction between two meanings of the term, which he calls the *narrow definition* and the *wide definition*. The narrow definition includes those groups of stakeholders "who are *vital* to the survival and success of the corporation."[40] These are the owners, management, employees, suppliers, customers and members of the local community. And the wide definition of stakeholder "includes any group or individual who can affect or is affected by the corporation."[41] These could include, for example, members of the global community or even non-human members of the biosphere insofar as the pollution effects of a corporation do not respect borders.

In trying to clarify this distinction between stakeholders as narrow versus wide, I am assuming that when he speaks of the groups in the narrow definition as "vital" to the interests of the firm, he is at the very least intending to capture the idea that these stakeholder groups and the firm are related to one another directly (i.e., without intermediaries) and, for this reason perhaps, the consequences of the behaviour in relation to the other are also more direct. So, for example, if the suppliers don't deliver requisite supplies, production ceases. If customers don't buy and demand evaporates, production ceases. But stakeholder groups taken widely have only indirect relationships to the interests of the firm and vice versa, and for this reason it can be difficult to determine the consequences of externalities as they impact these stakeholders or, for that matter, the consequences of the actions of these groups for the firm. It is difficult, if not impossible, for example, to determine the negative contribution made by the pollution effects of a firm to the health of citizens considered globally, as opposed to the direct health impacts of Union Carbide for the citizens of Bhopal. Similarly, it is difficult to determine the precise impacts on the bottom line of a firm that might result from the organization of a boycott and petition organized globally, but very loosely, via social media and often involving persons who were perhaps never consumers of the firm boycotted,[42] as opposed to the direct effects of a strike by employees. Of course, in practice, the distinction is less than perfectly clear, but for the moment we waive this concern and the discussion proceeds on the basis of the narrow definition.[43]

It should be clear that, even on the narrow definition, the mandate of the corporation can no longer be limited to maximization of profits for return to owners exclusively. In requiring management to manage the health of the corporation (which, as we'll see is to care for the long-term viability of the corporation) they are required to find the best possible balance between the (often conflicting) claims of all stakeholders. The presumption of the Shareholder Model is that the management of a firm have a fiduciary relationship to *all and only* the owners. Freeman is contradicting this: they have a fiduciary relationship to *all* stakeholders (narrowly defined) by virtue of their all having a vested interest in the firm.[44]

So corporate governance will perforce look much different than it does under the Shareholder Model, which Freeman calls managerial Capitalism. Even on the narrow definition, members of the local community, taken together, constitute a stakeholder group and are deserving of having their interests taken into account by corporate management. Intuitively, this would seem to imply that CSR will require management to govern the firm in ways that go beyond the minimal requirements of the law. In other words, we should expect there to be real social responsibilities the corporation is required to meet; the law sets the minimum standards that must never be contravened. Freeman is not specific about how this balancing of interests is to be implemented. The paper is intended to establish the plausibility of the Stakeholder Model as a *theory* of corporate governance and CSR.[45] No one denies that, in practice, balancing the (often) conflicting interests of the various stakeholder groups (even when) narrowly defined will be an extraordinarily difficult task that will require management to be sensitive to the relevant facts before them as much as the competing claims themselves. The foremost guiding principle that will orient management in its balancing of competing claims will be, I think, the long-term health of the corporation.

Having laid out his thesis, let us turn to Freeman's arguments. The first argument is called by Freeman the Legal Argument. It starts from the observation that the corporation is a legal construct, defined by law as a legal person and "existing in contemplation of the law,"[46] a fact we noted earlier in the introduction. The significance of this for Freeman, though, is that it implies that corporations have the entirety of their legal rights and obligations defined by the statutes that apply to them and, by virtue of this fact, "are constrained by law."[47] Significantly, though, for several decades now legislators have created new laws and the courts have been interpreting the law in ways that have increasingly circumscribed the actions of corporations to protect the interests of various stakeholders.[48]

An early illustrative example involves the case of a suit brought against Bloomfield Motors, a dealer selling Chrysler vehicles, in 1960.[49] As the result of a steering malfunction, an accident left the plaintiff's wife seriously injured. A suit was brought on two counts, one involving negligence on the part of the seller, Bloomfield Motors, and the other involving a breach of warranty. The negligence suit was dismissed, but the breach of warranty was found in favour of the plaintiff. The legal significance of this is that the warranty form signed by the plaintiff, which was then standard in the industry, had exculpated the seller (Bloomfield) and the manufacturer (Chrysler Motors) from any and all responsibility beyond repairing or replacing defective parts within 90 days or 4,000 miles after delivery to the customer, yet the courts affirmed that the manufacturer shall bear responsibility for ensuring that the product is safe, at least under intended use. The author of the majority opinion noted that there is *a valid public policy reason* for placing the burden on the manufacturer rather than the consumer because of the differences in bargaining power between the parties, and this despite the view then prevalent in law known as *caveat emptor*, implying that the burden of due diligence falls on the consumer.[50] The author goes on to note that these same principles apply to retailers in relation to consumers, again owing to the differences in bargaining power between them. And, notably, a disclaimer in the fine print of a contractual document does not exculpate the manufacturer or the retailer of this liability.

The significance of the ruling and subsequent legislation and various regulatory bodies brought into existence[51] is that lawmakers and the courts have recognized the need to protect the interests of consumers in cases where the traditional caveat emptor (buyer beware) would leave consumers to be victimized by market failures, in this case an asymmetry of information and an imbalance in bargaining power. It is not just that these consumers did not have all the relevant information, but no due diligence on their part could have closed the deficit. And this is obviously true of many consumer goods that are so technical it is unlikely that even the manufacturers can be said to have all the relevant information as it pertains to, for example, possible health risks. What might be the health risks associated with the use of Bluetooth devices, or in-ear monitors, for example? Or consider the pesticide product Roundup by Monsanto. Is it really plausible that even Monsanto could foresee all the possible negative health and environmental effects of this product at the time it was released to the market?

As Freeman documents, analogous arguments can be constructed in regard to the interests of employees (the creation of the National Labor Relations Act) and local communities (the creation of the Clean Air Act, the Clean Water Act, etc.). Governments in many jurisdictions have created agencies, such as the Canadian Food Inspection Agency, the National Labor Relations Board and Environmental Protection Agency (US), and charged them with the mandate to ensure implementation and enforcement of the law. These agencies have been granted considerable powers to inspect for compliance and develop policy. A fruitful way of framing Freeman's argument here is to see this as a response to Friedman's argument that all stakeholders other than shareholders have their interests protected by contracts (what I have referred to earlier as the Sanctity of Voluntary Contracts Argument). Essentially, Freeman is saying that legislators and the courts have recognized the need, and have progressively acted, to circumscribe this view.[52]

The conclusion follows that the law recognizes the rights of various stakeholder groups, including consumers, suppliers, employees and local communities, to not be harmed and to not be subjected to avoidable risks in the absence of fully informed consent and has acted to protect these rights by circumscribing the activities of firms in ways that ameliorate such risks for these stakeholders,

> I have argued that the result of such changes in the legal system can be viewed as giving some rights to those groups that have a claim on the firm. . . . It raises the question, at the core of a theory of the firm: In whose interest and for whose benefit should the firm be managed? The answer proposed by managerial capitalism is clearly "the stockholders," but I have argued that the law has been progressively circumscribing this answer.[53]

Consider now the Economic Argument. It is helpful in this instance to see Freeman's argument as a response to the Rising Tide Argument (RTA). Recall the gist of the RTA is that maximization of profit on behalf of the shareholders is at the same time the best way to benefit society as a whole, to maximize the aggregate good, as it were. The Economic Argument counters that this has been repeatedly falsified by the facts. Though we can point to instances of positive externalities, it is simply not true as generalization. Freeman's argument raises three challenges to RTA: monopolies that have consistently functioned to maximize benefits for wealthy owners at the expense of everyone else, externalized costs that disproportionately offload the burdens of pollution and other harms onto the general public and moral hazards, which I will focus on here insofar as we have already explored monopolies and externalities.

A moral hazard may occur in any case where the individual or group purchasing a good or service can pass the costs of that good or service to another or others. An example would be a system of government procurement wherein a government department, say defence, has a contractual arrangement with a particular producer, say a manufacturer of aerospace technology, and where the costs are picked up by the tax-paying public. In this case, neither the consumer (i.e., the military) nor the producer has any reason – at least none that arise out of self-interested motives which, according to the model, are supposed to be given free rein to determine all our market decisions – to economize because the costs are paid by public tax revenues. The moral hazard in this case is the perverse incentive for the manufacturer to inflate costs and for the purchaser to ignore this (after all, careful monitoring of costs imposes burdens of its own). Consistently with Freeman's point that RTA has been frequently falsified by the facts, we can here point to a real example. In the mid-1980s, President Reagan instituted the Packard Commission to investigate cost overruns on military procurement. In its report, it claimed to have found evidence that the military had, for example, paid \$435 for a hammer.[54]

The potential for an unregulated market economy to generate moral hazards may be the enduring lesson of the Great Recession of 2008/2009. The story begins with years of successful lobbying by the financial sector in the United States to be deregulated that led ultimately to the repeal of the Glass-Steagall Act[55] and permission for mergers and acquisitions that had the cumulative effect of resulting in what has been called the financialization of the American economy.[56] The Glass Steagall Act, signed into law by Franklin D. Roosevelt in 1933, had been introduced in part to address the risks arising out of speculative investments by financiers and which had been responsible in large part for the crash of the Wall Street Stock Exchange in October of 1929, an event which arguably played a crucial role in the Great Depression that followed. Glass-Steagall was intended to address this by partitioning high-risk investment financing and traditional banking that deals principally in small business and consumer loans. The bill also introduced government insurance to guarantee consumer savings deposits to prevent a future run on the banks, as had happened several times in the early 1930s following the crash. The repeal of Glass-Steagall did away with the partition allowing financial institutions to combine high-risk investment financing and consumer loans under one roof, while the mergers and acquisitions had the effect of creating a set of financial institutions that became "too big to fail." Yet the government backing remained in place with the government becoming, essentially, the "lender of last resort."[57]

And so, continually in search of high returns, these institutions, in their role as investment bankers, began bundling mortgages together as investments, into what are known as mortgage-backed securities. Moreover, these securities qualified as low-risk investments and were rated as such by the major ratings agencies.[58] After all, the investments were secured by the real properties. This new investment device, the mortgage-backed security, perversely incentivized these same financial institutions, now in their role as consumer banks, to indulge in the proliferation of sub-prime mortgages, often to people who would not have qualified under the criteria in place prior to the repeal of Glass-Steagall.[59] This in turn generated the real estate bubble as demand for property drove the escalation of property valuations. The increase in property valuations in turn meant individuals were getting rich from their homes and were actively encouraged to use the equity to finance home equity loans, which significantly elevated consumer debt.

To put it metaphorically, the bubble economy was a house of cards built on a mountain of consumer debt that vastly exceeded the value of the real underlying properties that secured the debt. Moreover, the mortgage-backed securities, sold as low-risk investments, had been gobbled up by investors globally, including many pension funds, and the inevitable crash spilled far beyond the radius of the Wall Street financiers to shake the world economy to its foundations, the impacts of which are still being felt to this day.

The problem was compounded by the fact that in its role as lender of last resort, and under the justification that these deregulated financial institutions had become "too big to fail," these same institutions who had brought ruin to the world economy were bailed out to the tune of trillions of dollars by governments in the United States and Europe: profits internalized and costs externalized or, as it was labelled at the time, profits privatized, costs socialized.[60]

Thus, the conclusion of Freeman's Economic Argument is that legislation, public policy and jurisprudence must be the tools to prevent such perverse incentives and the externalities that result. These tools must be used to force corporations to embrace governance models that require them to balance the interests of all stakeholders, where this balance is oriented by the principle of ensuring *the long-term viability of the corporation as a source of value creation for all those with a stake in the firm.*

Though not labelled as such, there is a third argument in Freeman that is essentially Kantian in nature. This argument, and the perspective at its heart, runs throughout Freeman's paper. In speaking of the basic motivations for why the rights of stakeholders other than shareholders must have their interests considered in the management of the firm, Freeman says, "each of these stakeholder groups has *a right to not be treated as a means to some end*, and therefore must participate in determining the future direction of the firm in which they have a stake."[61] Clearly, Freeman means to imply that these stakeholders, considered now as individuals, have a right to not be treated as the means to some end not of their own choosing. This is essentially the reasoning we saw to be at the heart of Taylor's argument in Chapter 4 for the claim that our fellow organisms in the community of life are deserving of respectful treatment because it is their right, as individuals of intrinsic worth, to not be treated as the mere means to our ends.

Now, if one looks at the issue of CSR through the lens of a reconstructed Preference Utilitarianism, this argument won't carry any force.[62] You would insist that the market as a whole is to be judged by its tendency to maximize opportunities for agents to act on their preferences. Leaving individuals to act from purely selfish motives will have the effect of distributing the goods of society in a way that is said to be Pareto optimal in the sense that it will be impossible to improve the situation of any individual without simultaneously making someone else worse off.[63] But if one is of the view that the basic rights of individuals are inviolable, then this argument is likely to be decisive. For ultimately this is the implication of Friedman's Shareholder Model: the owners have the right, by virtue of their ownership of the firm, to use the other stakeholders as the means to their own ends, the maximization of their self-interest. Looked at this way, it is unlikely that one will be appeased by the thought that being used in this way might result in a situation where one's material situation could not be improved. And this is so even if we suppose RTA to be true.

One might suppose we have an irresolvable stand-off here between competing views, but it is interesting to note that rather a lot of recent research has demonstrated that productivity depends on people feeling that they are being treated fairly, so the best way to maximize productivity is to treat all stakeholders respectfully, giving full recognition to their autonomy. That is, to treat them in ways generally thought of as Kantian. In discussing the findings of T. Peters and R. Waterman,[64] Freeman notes that their research establishes a clear connection between performance and reputation,

> I would argue that Peters and Waterman have found multiple applications of Kant's dictum, "Treat persons as ends unto themselves," and it comes as no surprise that persons respond to such respectful treatment, be they customers, suppliers, owners, employees, or members of the local community.[65]

Of particular interest, though, is that this reveals a tension at the heart of Classical Liberalism. On the one hand, the ideology presumes Preference Utilitarianism. It is at the heart of the central argument, the Rising Tide Argument, that rationalizes and justifies the model. After all, it's understood that the rising tide maximizes *aggregate welfare* while simultaneously relegating many to the functional role of being the means to this end. On the other hand, as we have seen to be affirmed by Locke and Nozick, the fundamental premise of the view is that the rights of each individual are sacrosanct and constitute the very foundations of the view. Thus, everyone has the right to not be used as the mere means of another. One is reminded of Nozick's argument for suggesting that the minimal state is the most that can be justified because the redistribution required to move beyond the minimal state would essentially give some individuals (the recipients of the state's redistributive efforts) a partial ownership of those forced to pay.

We pursue this tension at greater length in Part III, but returning to Freeman, the cumulative force of these arguments is for the conclusion that the purpose of the firm must be redefined. Under the terms of the Shareholder Model, the purpose of the firm, as we've seen, is defined as the maximization of profits for return to shareholders. Thus, management are presumed to have a fiduciary duty to run the firm in the interests of the owners exclusively, and they have only such duties to other stakeholders as are set out in contracts with these parties. But on the Stakeholder Model, the purpose of the firm is to benefit the interests of all stakeholders, that is, all individuals with a claim on the firm. Thus, management is charged with the fiduciary duty to all stakeholders to balance the often-competing interests among them.

This fiduciary duty to stakeholders in turn carries the implication that the first duty of management is to the long-term health of the corporation. The significance of this cannot be overstated. In my view, it marks the most significant change in perspective of corporate governance and management of the last 40 years (at least). The Shareholder Model, insofar as it is committed to maximizing the interests of shareholders exclusively, has had the effect of focusing management on the short term. Management look to corporate valuations as the best indicator of their performance, and valuations look no further ahead than the results of the next quarter. By contrast, insofar as the Stakeholder Model starts from the conclusion that the purpose of the firm must be redefined in relation to finding the best balance of the interests of all stakeholders, the emphasis shifts to the long-term health of the corporation, and this because it is the continuation of the entity that is the firm, in a condition of health, on which the interests of all stakeholders depend. This is a commitment to the long-term ability of the corporation to create value for all stakeholders.

Now there is some unclarity around exactly how we are to define the health of a firm, but I would suggest the following considerations. First, to be considered healthy as a firm, it must be sustainable. And this in turn requires that it be competitive. Also, to return to the point about balancing the interests of stakeholders, this will mean, in part, that the firm is flexible in a way that is responsive to market conditions and adaptable in balancing the competing interests of various stakeholders, but with an overarching ethical perspective to do no harm.[66] This is, I think, what Freeman is getting at when he says management of the firm cannot separate the "business" part of a firm's value creation activities from the "ethical" part. This returns us to the point about Friedman's choice of metaphor, business as a game, where it was noted that this suggests that business is always conducted in the context of zero-sum outcomes. Thinking of business as a game invites one to artificially partition firm management from ethical considerations about the impacts of the firm's activities on stakeholders because the focus is exclusively on winning, where winning means the maximization of profits. Conversely, the Stakeholder Model is committed to the claim that firm management is built on a *normative core*. Speaking of the normative core at the heart of the Liberal notion of equality and fairness, he says,

> The normative core for this redesigned contractual theory will capture the liberal idea of fairness if it ensures a basic equality of stakeholders in terms of their moral

rights as these are realized in the firm, and if it recognizes that inequalities among stakeholders are justified if they raise the level of the least well-off stakeholder. The liberal idea of autonomy is captured by the realization that each stakeholder must be free to enter agreements that create value for themselves, and solidarity is realized by the recognition of the mutuality of stakeholder interests.[67]

Let us now draw some tentative conclusions from our discussion thus far.

Summary and conclusions

I am of the opinion that the only approach to corporate governance and CSR that will be compatible with a sustainable form of Capitalism must be some version of the Stakeholder Model.[68] There are at least two reasons for this. First, only a stakeholder model can bring into being the alignment of corporate and public values that will be required to see the project through to successful implementation of the new business model at the heart of Natural Capitalism.[69] And second, because a sustainable form of Capitalism will require a coordinated global response, which will in turn depend upon the willing consent and participation of citizens everywhere. As the arguments of this chapter and Chapter 7 show, the willing consent of agents requires that we follow Kant in respecting the intrinsic dignity of the individual.[70] Defending this priority of the Stakeholder Model, there are at least two objections that must be addressed.

First, one might argue that Freeman's Stakeholder Model is not really a theory of corporate governance and CSR at all. It is too vague, with no indication at all of how one might reasonably implement the model to achieve this balance of competing interests. By contrast, Friedman's Shareholder Model has the virtue of perfect simplicity and clarity with respect to the principles of corporate governance and the roles of all participants in relation to the firm and its proper goals. As Friedman says, in relation to the contractual arrangement between executives and shareholders, "the persons among whom a voluntary contractual arrangement exists are clearly defined," and, as he implies, the same point would apply to all the terms of the agreement under his model.[71]

In my view, this clarity is entirely illusory. Simplicity is sometimes a very good thing and always desirable when you can have it, but sometimes it obscures the hard truth that things are not so simple and will require a subtle and nuanced approach that must be responsive to a situation that is in continual flux. Never is this more the case than when we are required to negotiate amongst competing interests of autonomous individuals in a complex organizational setting. In my view, it is this simplicity imposed on what is, in reality, a very complex situation that misleads Friedman into missing so much of importance and can make him sound doctrinaire. For example, and as noted earlier, he oversimplifies greatly when supposing that the choice between economic arrangements is the stark one between a radically laissez-faire model and an authoritarian central command economy. But we also see this in his characterization of market freedom as a check against the abuse of political power. He gives no indication of recognizing the need for a check on the accumulation of economic powers. As illustration, consider the following quotes,

> The difficulty of exercising "social responsibility" illustrates, of course, the great virtue of private enterprise – it forces people to be responsible for their own actions and makes it difficult for them to "exploit" other people for either selfish or unselfish purposes. They can do good – but only at their own expense. . . . In a free society, it is hard for "evil" people to do "evil," especially since one man's good is another's evil.[72]

One can only wonder how he could sincerely believe such a thing, but I suspect it is, at least in part, a consequence of living at a privileged time in history. I think the same point could be made in regard to Nozick, particularly when he argues that the Night Watchman State is actually inspiring, the closest thing to utopia humans are capable of realizing. They could believe this sincerely precisely because they lived at a time when the mixed economy of the Western world was doing a very good job of generating economic growth consistently, and the distribution of goods, and of equality of opportunity in particular, was much fairer than the present reality. Under these circumstances, one could look around and plausibly believe that people could, for the most part, be trusted to restrain themselves within reasonable limits. Guarantee citizens their freedom, and the principles of the laissez-faire market will provide all the discipline society can require. As discussed in Chapter 7, though, this has been falsified by the historical facts of the last four decades.[73]

In any event, Freeman's Stakeholder Model is framed quite deliberately at the level of general principles, a level of abstraction that is not the same thing as being vague. Rather, it is quite specific in specifying the necessary conditions for any particular application of the theory in a corporate setting to count as a version of the model. And Freeman approaches it in this way for at least two separate reasons. First, as already noted, he is concerned to argue for the philosophical soundness of the approach as an alternative to the Shareholder Model, and this requires that one establish the consistency and credibility of the general principles. But also, I suspect, it is because it is not possible to approach the matter otherwise. If I am granted the point about the complexity and contextual sensitivity involved in any complex organizational setting, then it follows that any particular approach to corporate governance on the part of an individual firm will have to be framed in a way that maximizes flexibility to meet continually changing circumstances. The point to emerge is that under such conditions, it is impossible in principle to delimit in advance what a particular application of the Stakeholder Theory must look like. In Chapter 3 we took note of a similar point in relation to the need for regulatory frameworks to be framed in ways that would set the boundary conditions for sustainable practices while framed sufficiently broadly to leave it to businesses to find their own solutions that will maximize competitiveness and profits while respecting the goals of sustainability.

A second objection one hears rather often, though, is that, in practice, there is no difference between the two models insofar as the shareholder advocate might adopt CSR as a strategic posture. Indeed, Friedman seems to suggest as much when he says, "Of course, in practice the doctrine of social responsibility is frequently a cloak for actions that are justified on other grounds rather than a reason for those actions."[74] Later on the same page, he refers to such practices as hypocritical window-dressing and, though he finds that he "cannot summon much indignation to denounce them," he deplores the duplicity for the negative impact it can have on the foundations of free societies, a point noted earlier.

I think, though, there is a very real difference insofar as Friedman's model permits, in fact requires, that all stakeholders other than the shareholders be treated as the means to the ends of shareholders. So we can expect the real difference between the two models to manifest itself behaviourally in almost all cases where there is no advantage to adopting a posture of social responsibility. Under these circumstances, the two strategies will likely have different outcomes. Thus, we can always ask, "Would they have behaved this way (i.e., with an apparent concern for social responsibility) even if no profit was to be realized by doing so?" Well, if the firm is genuinely committed to the long-term health of the firm, and finding the best possible balance of the interests of all stakeholders, the answer is "yes." If not so committed, then it is likely to be "no." Certainly, under the Shareholder Model they will behave contrary to the principles of the Stakeholder Model in any case where, as a direct result of doing so, it is expected that there will be a loss of profits and typically will do so even in cases where that loss may be redressed by greater gains over the long term.

For further insight into these claims, I introduce the arguments of John M. Darley from a paper titled, "How Organizations Socialize Individuals into Evildoing."[75] I caution the reader, though, that I am merely scratching the surface of this profound and insightful paper and doing so with my own purposes in mind. Having said that, Darley's thesis may be summarized as follows: corporate evil doing is largely a matter of corporate culture, the culture that elevates profits above all else. The striking insight, first drawn to our attention by Hannah Arendt, is the banality of evil. Evil is rarely done by persons who are themselves evil. Rather, it is done by ordinary persons, not so different from you and I, acting in a social context, an organizational setting as it were, where the evil doing enhances bonding among the members and helps to advance the goals of the organization. In short, persons are *socialized* into wrongdoing and the socialization within the culture serves to normalize the wrong doing. Additionally, though, organizational settings have a tendency to exacerbate the problem by incentivizing the bad behaviour by means of linking such behaviours to performance bonuses and possibilities for career advancement. In fact, this incentivizing is itself part of the enculturation and normalization of the bad behaviour: not only is this behaviour normal, but also it is good and deserves to be rewarded.

We saw this illustrated with our example of the perverse incentives that led to the crash of the world economy and the Great Recession of 2008/2009.

Focusing on corporate socialization in itself, Darley recounts a particularly engaging story, originally from Michael Lewis.[76] The book tells the story from the perspective of Lewis's experiences on becoming a bond sales agent for Salomon Brothers. In this case, the socialization was explicit, though branded as education/ training, and subjected new recruits to several weeks of intensive exposure to the wisdom of senior partners and special guests. This education was dedicated to nurturing values of corporate loyalty, cut-throat competition and the exaltation of profit above all else. We subsequently follow Lewis through the progress of his career as a bond salesperson. In a telling anecdote, he reveals the details of a programme to sell junk bonds to clients on the false premise that they were actually a great deal flying below the radar and about to go big. In fact, the senior partners knew the firm to be sitting on a huge stockpile and had inside information that they were about to fail. Young sales agents were incentivized to sell them – unload them quickly before the imminent failure – and the subsequent fallout from the crash of the stocks left many clients in financial ruin.[77]

In addition to the role played by perverse incentives and the resulting moral hazards, and the enculturation of young recruits to think of their clients as "sheep to be fleeced," another important point revealed by this story concerns the asymmetries of information built into the very organizational structure of the firm. There were details about the bond's imminent failure that even the partners should not have had access to, but it was crucial to the "success" of the scam that this information was withheld from the sales agents and their clients. This feature of information flow within the organization, namely that it does not flow freely, is often intrinsic to organizations that adhere to the Shareholder Model precisely because it enhances the organization's goals to maximize short-term profits.

As Darley points out, there are various elements of hierarchical organization that go beyond information flow and contribute in an essential way to organizational goals and wrongdoing in meeting those goals. These organizational settings are not merely complex but are often organized in ways that mimic military organizations. In particular, there are typically strict reporting lines for personnel. Labourers report to immediate supervisors, who in turn report to middle management, who in turn report to upper-level management. It is usually frowned upon at the very least, and often invites serious reprisals, to violate these strict reporting lines. This is further exacerbated by rigid job descriptions that strictly limit the duties and responsibilities of personnel.

The outcome of this hierarchical organization is twofold. First, there is the diffusion of information within the organization. Thus, the strict reporting lines often mean that information relevant to addressing corporate wrongdoing fails to make it to the right persons that might act on the information effectively, and at an early stage, to prevent worst outcomes. The information gets trapped, as it were, in specific "locales" within the firm. As Darley says, "Those who make decisions are often walled off from information that is vital to good decision making."[78] And second, there is the diffusion and fragmentation of individual responsibility. This means that, and insofar as job descriptions strictly limit responsibilities and the possibility of personal initiative on the part of anyone below upper-level management, responsibility for wrongdoing falls to no one until it is too late to take effective action. At this point, the only possibility is cover up and indemnification, where the deniability that flows from restricted information flow plays a crucial role.

Suppose, for example, that a low-level employee is directed by their immediate supervisor to dump some toxic waste into a local estuary ("and don't get caught"). And perhaps this is because the supervisor was directed, with obvious impatience, to "deal with the problem" but not given any specific directions on how to do so. Imagine further, though, that management would never condone such behaviour. The problem is, they will never know. Under this kind of organizational structure, it is highly unlikely that the information will ever make it to the persons with the responsibility for directing resources to find a better solution. The original employee directed to dispose of the waste can absolve themselves of responsibility, "Hey, I was just following orders." Similarly, the supervisor can say to himself, "It's not part of my job to worry about it. I did what I was told. I took care of the problem."

So what conclusions can we draw from Darley's discussion if we wish to avoid corporate wrongdoing? Or, more pointedly, are there any principles of corporate governance and organization that might minimize the sort of corporate wrongdoing that results from firms structured in these ways? Well, first I believe, firms must be encouraged to avoid adoption of CSR as a window dressing or as merely a strategic directive. When adopted seriously, as the basis for a reworking of corporate culture to bring it into line with the values of the Stakeholder Model, it has been found to bring innumerable long-term benefits to the firm. Although it would take me too far afield to get into a detailed discussion of the examples, the reader is encouraged to investigate.[79] From such established firms as G.E. and Lego, to such relative newcomers as Starbucks and Google, one can find examples of corporations that are reimagining what it means to speak of corporate responsibility. Moreover, the data clearly establishes that doing so can be profitable, not to mention less costly to the extent that problems are avoided.

While none of this is meant to imply that their practices are beyond reproach, that is not the point. The point is about embracing the idea that corporate culture and governance must be brought into line with the values of the Stakeholder Model: taking into consideration the needs and values of all stakeholders with the goal of balancing these competing interests to maximize the long-term health of the firm and the creation of value for all stakeholders. From this perspective, any firm will be continually re-evaluating its practices with the goal of making further refinements at every iteration to bring its business practices closer to the ideal.

Second, every effort must be made to open lines of communication, to enhance the likelihood that information will flow freely within the organization. Further, hierarchical organization is to be avoided. These two points are related insofar as transparency in communication depends upon avoiding strict and hierarchical structures. Flattening hierarchical structures as much as possible while empowering individuals to act responsibly within the domain circumscribed by their duties is the best way to engage employees. People want to be empowered, and denying them this opportunity has the effect of demotivating them. As an additional benefit, empowering individuals is known to enhance their commitment to the firm and its goals in ways that go beyond what is possible when individuals are motivated exclusively by self-interest.[80]

In closing, we return to the question with which this chapter began: "In whose interests are corporations to serve?" I have argued that it must be civilized society as a whole, or put differently, all stakeholders. And this is so because all stakeholders can be shown to have skin in the game; they are all affected, either positively or negatively, by the actions of the firm. If the corporate sector is to expect permission to operate within the jurisdiction of a civilized society, and to benefit from all the services that only civilized societies can provide, from the protection of the Rule of Law, to infrastructure essential to the conduct of commerce, not to mention an educated class of potential workers with the necessary skill sets to maximize productivity, then they can't expect to be granted permission to maximize profit for the few. Not only is this to regard the overwhelming majority of stakeholders as nothing more than the means to the ends of shareholders, but, as we have seen, it has the effect, over the long term, of eroding productivity and profits, not to mention public trust.

I note in passing that all the themes raised here, from decentralized authority and transparency of communications to respecting the intrinsic dignity of all stakeholders were at the heart of the sustainability narrative of Biosphere Consciousness sketched in Chapter 4. As now established by several lines of argument, any model of Capitalism that is to be sustainable will require the willing participation of citizens everywhere, and this will require that we respect their intrinsic dignity. I close this chapter with a quote from Mazzucato:

> The private sector can be transformed by the simple but profound expedient of replacing shareholder value with stakeholder value. . . . Stakeholder value recognizes that corporations are not really the exclusive private property of one group of providers of profit-sharing financial capital. As social entities, companies must take into account the good of employees, customers and suppliers. They benefit from the shared intellectual and cultural heritage of the societies in which they are embedded and from their governments' provision of the rule of law, not to mention the state-funded training of educated workers and valuable research. They should in return deliver benefit to all these constituencies.[81]

★★★

In Part III, we investigate Classical Liberalism and globalized Capitalism and the economic system it licenses from a normative perspective. I will argue that Classical

Liberalism, like any other ideology, is in need of the resources that can be provided only by a normative theory. The fact is that as soon as we secularize Locke's theory by severing it from all appeals to God and teleology, it is no longer obvious what motivation we have for following the prescriptions that flow from the ideology that results. Put differently, it is no longer clear what connects the theory to our values. The connection to our values and, thus, the moral justification of the actions and policies of the ideology must come from elsewhere. The goal in Part III is to enquire which of our moral traditions, if any, might be best suited to this task.

In Chapter 9, we focus on the Consequentialist family of theories known as Utilitarianism. For purposes of clarity, it is helpful to begin with the most influential version of this theory, known as Rule Utilitarianism, attributed to John Stuart Mill. We will trace its development into a contemporary version known as Preference Utilitarianism. This latter version is the normative tradition most likely to judge Classical Liberalism favourably, though even here the agreement is far from perfect. Preference Utilitarianism serves admirably to give a moral dimension to the arguments supportive of the policy prescriptions of Classical Liberalism (it's about making the world a better place). Unfortunately, in one important respect it is at odds with the conceptual foundations of Classical Liberalism. Utilitarianism denies that there are such things as inalienable rights which the theories of Locke and Nozick presume both as the very foundation of property rights and the freedoms required for genuine agency in the marketplace.

In Chapter 10, we turn to Kantian Duty Theory, a very influential theory that is probably closer to the moral intuitions of most people and which shares with Locke the view that people are endowed with rights to autonomy and should be free to choose for themselves. As it turns out, though, for Kant this implies that persons have an intrinsic dignity that must never be sacrificed for any goals we might set for ourselves. Thus, the supposed benefits of globalization come under intense scrutiny. The benefits of globalized Capitalism can never be justified in cases where there is abuse of individuals and, in particular, cases where these individuals become the *means* to benefit the wealthy few.

As Part III will show, there is no consistent normative theory that can do for this ideology what it needs. In particular, there is no normative theory I am aware of that can offer consistent normative justifications for its policy prescriptions and the actions it licenses. In my view, this underscores the fact that Classical Liberalism is a mature ideology and the internal inconsistencies in its logic are now manifest and unavoidable. It is demonstrated to be inconsistent with our (non-market) values and corrosive of democratic institutions wherever it takes root and grows unchecked. It is incompatible with genuine sustainability and, to that extent, is ill-suited to meeting our urgent need to address the issues raised by climate change. Thus, the ultimate conclusion of the book is reinforced that Classical Liberalism, as the dominant cultural narrative, is no longer suited to the needs of our present circumstances. A new narrative is needed urgently.

Notes

1 An illustrative example of this can be found in *Onward: How Starbucks Fought for Its Life*, by Howard Schultz (2011).

2 This brief history is drawn from a variety of sources. While I have cited these where I can, many of them I have forgotten.

3 In practice, though, the situation is very complex. Some companies, the East India Company being a prominent example, were granted leave and the means to protect themselves and may be said to have established something rather like a paramilitary protectorate (until this role was taken over by the British government for egregious mismanagement). See Chapter 5 of Ellen Meiksin-Woods, *Empire of Capital* (2003).

4 https://en.wikipedia.org/wiki/Joint_Stock_Companies_Act_1844 (retrieved March 27, 2023).

5 Though again it must be said that, in practice, the story of the emergence of the modern concept of limited liability is a complex one. See Ron Harris, "A New Understanding of the History of Limited Liability: An Invitation for Theoretical Reframing," available at: https://corpgov.law.harvard.edu/2019/08/29/a-new-understanding-of-the-history-of-limited-liability-an-invitation-for-theoretical-reframing/(retrieved October 31, 2022).

6 Case notation: 118 U.S. 394.

7 Case notation: 558 U.S. 844.

8 For an example of a less flattering portrait than the one I will be giving here, see Part 2 of Klein (2007) *The Shock Doctrine: The Rise of Disaster Capitalism*.

9 For a fascinating portrait of this period and its anxieties, see Steil (2019) *The Marshall Plan: Dawn of the Cold War*, and Woloch (2019) *The Postwar Moment: Progressive Forces in Britain, France, and the United States After World War II*.

10 A riveting depiction of the Roosevelt administration and the New Deal era is found in Brinkley (1995) *The End of Reform: New Deal Liberalism in Recession and War*.

11 I am thinking, in particular, of her mischaracterization of Kant's ethical doctrine, which is the most egregious example of poor scholarship of which I am aware (e.g., see, *The New Left: The Anti-Industrial Revolution*, New American Library (1971)).

12 Friedman (1962, p.133).

13 Friedman (1970, p.1).

14 In my view, the term is not used with a consistent meaning in these various uses, a point I will discuss *in situ*.

15 Friedman (1962, p.135).

16 Friedman (1970, p.1).

17 Friedman would, I think, aver that if the laws of a particular society do not grant exclusive rights to owners to the disposition of their property, restrained only by Mill's Harm Principle, then it isn't really a capitalist society at all.

18 Nozick (1974, p.238); italics in original.

19 Ibid.

20 In particular, see Friedman (1962, p.127). It is also hinted at in Nozick. See the lengthy discussion of the risk associated with starting a firm and how these risks are differently apportioned by capitalist societies, p.255 *et passim*.

21 The same justification has been used in connection with, for example, the implementation of laws that limit liability for investors, offer preferential tax rates for dividends etc.

22 An insightful article on Reagan on entrepreneurs can be retrieved at: https://www.ff.org/ronald-reagan-on-the-importance-of-entrepreneurs/(retrieved October 31, 2022).

23 As it turns out, the Soviet Union was also in economic and political decline, though this was not widely known at the time. Indeed, in retrospect it seems very likely that Gorbachev's initiatives were motivated in large part to salvage the Soviet Union.

24 The classic expression of this triumphalism is Fukuyama (1992). For an overview of the Washington Consensus, see: https://www.intelligenteconomist.com/washington-consensus/(retrieved October 31, 2022).

25 R. Edward Freeman in his article, to which we turn next, refers to something very similar which he calls the Privity of Contracts argument.

26 This line of argument was implicit in much of the rhetoric during the negotiations of trade deals with China in the years immediately following the suppression of what became known as the 1989 Democracy Movement in Tiananmen Square. In response to critics who alleged that pursuing trade relations with Beijing would reward the regime's bad behaviour, the claim was that we are better to engage them in trade. This will be the quickest and most effective way to bring about democratic reforms. As I write this (January 2022), recent events in Hong Kong incline one to scepticism on this point, at the very least.

27 Again, it pays to remember that at the time Friedman wrote these works, at the height of the Cold War, the entire Eastern bloc, consisting of the USSR, Communist China and their satellite states – including Cuba located in direct proximity to the United States – meant that at least one-third of the human population lived under direct communist rule. And by virtue of their official doctrine, they did, ideologically at least, constitute an existential threat to Western democracies.

28 Friedman (1962, p.33).

29 Friedman (1970, p.1).

30 Op. cit., p.5.

31 Friedman (1962, p.13).

32 I am using the term *mixed economy* here in the standard sense to refer to any economic system that combines public and private enterprise. Indeed, if one were to use the term free market strictly to refer to what Friedman appears to mean by the term, it might be that there are none anywhere. By this criterion, all economies would be socialist, or mixed, economies.

33 Op. cit., p.15.

34 Op. cit., pp.14–15.

35 The book is, *Strategic Management: A Stakeholder Approach*. Boston, Pitman (1984).

36 The reference I have available for this paper is from a textbook, *Ethical Challenges to Business as Usual*, edited by Shari Collins-Chobanian, Prentice-Hall (2005, pp.258–267).

37 Both from Freeman (2005), the latter quote on p.263.

38 See Freeman (2005, p.261).

39 In what follows I will usually follow standard practice of leaving it to context to indicate which of these senses are intended.

40 Ibid.; italics added.

41 Ibid.

42 For example, a boycott of a corporation involved in the production and distribution of meat products organized by an NGO or a conglomerate of NGOs that are broadly based on a membership of persons who consider themselves vegans or vegetarians of some description.

43 As Freeman says, "I shall begin with a modest aim: to articulate a stakeholder theory using the narrow definition," Op. cit., p.261.

44 Notice, I am interpreting Freeman to be making the claim that management have a fiduciary duty to *all* stakeholders narrowly defined but not to them *only* as is the case with Friedman's thesis as it pertains to the Shareholder Model (*all and only* shareholders), and I interpret him this way because it could turn out that in some cases at least, management will have a fiduciary duty to stakeholders considered widely. Freeman's thesis, in other words, is deliberately framed in a way that leaves the question of to whom exactly the management is bound by fiduciary duties open-ended and sensitive to context whereas Friedman's is not.

45 To anticipate a point to be explored at greater length in the following, the Stakeholder Model discussed here is really a meta-theory, or a theory of theories, as it were. By this is meant that it is expected there will be a variety of particular theories, adapted to the

individual circumstances and determinative of the social responsibilities of each individual firm that will count as versions of the Stakeholder Model. It is for this reason I refer to Freeman's theory as the Stakeholder Model.

46 Op. cit., p.259.
47 Ibid.
48 Freeman's discussion focuses exclusively on the United States, but evidence suggests a similar approach in European and Canadian legislation and jurisprudence.
49 Case notation: Henningsen v. Bloomfield Motors, Inc. 32 N.J. 358 (1960). See, https://law.justia.com/cases/new-jersey/supreme-court/1960/32-n-j-358-0.html   (retrieved November 1, 2022).
50 Additionally, though not mentioned in the opinion, there is a significant asymmetry of information between the parties that is a common feature of market exchanges and often leads to market failures.
51 As examples of such agencies, in Canada, there is the Bureau of Consumer Protection and in the United States, there is the Consumer Product Safety Commission.
52 Op. cit., p.260.
53 Ibid.
54 https://en.wikipedia.org/wiki/Packard_Commission (retrieved November 1, 2022).
55 It was repealed on November 12, 1999 by President Clinton when he signed into law the Gramm–Leach–Blilely Act.
56 See, for example, Niall Ferguson (2008) *The Ascent of Money: A Financial History of the World*.
57 For details, see Adam Tooze (2018) *Crashed: How a Decade of Financial Crises Changed the World*. Also, Stiglitz (2010) *Freefall: America, Free Markets, and the Sinking of the World Economy* and Mariana Mazzucato (2017) *The Value of Everything: Who Makes and Who Takes from the Real Economy*.
58 All three major rating agencies, Standard & Poors, Moody's, and Fitch, rated these as AAA investments, the highest rating possible, signalling to investors that these were among the safest investments [Tooze (2018) and Stiglitz (2010)].
59 It was widely reported in the post-crash analysis that the running joke among bankers selling these sub-prime mortgages was, "if they've got a pulse, give 'em a loan." See, http://www.nbcnews.com/id/29827248/print/1/displaymode/1098/(retrieved November 1, 2022).
60 Shockingly, in the United States, much of this money was used to pay their top executives performance incentives, a bitter twist of irony if ever there was one [Stiglitz (2010) and Mazzucato (2017)].
61 Freeman (2005, p.39); italics added.
62 We consider the details of Preference Utilitarianism in Chapter 9. For present purposes, it is sufficient to note that Preference Utilitarianism is concerned to maximize aggregate welfare at the expense, if necessary, of the rights of individuals.
63 Named after the Italian economist, Vilfredo Pareto (1848–1923), who first articulated the principle.
64 From, *In Search of Excellence*, (1982).
65 Freeman (1984, p.43). Such research is also discussed and affirmed by Mazzucato (2017) and Mark Carney (2021) *Values: Building a Better World for All*.
66 In practice, this may mean "do no harm where it is avoidable."
67 Freeman (1984, p.265); the reader will notice the allusion to Rawls in this quote in the point that such inequalities as arise will be justified if they can be seen to raise the level of the least well-off stakeholder.
68 And by "sustainable" I mean socially and politically sustainable in addition to environmentally sustainable, as per our discussion in Chapter 2.
69 For example, see Carney (2021) and Mazzucato (2017).
70 We will clarify the meaning of this claim and discuss the arguments in Chapter 10.

71 Friedman (1970, p.2).
72 Op. cit., p.4.
73 On this point, see Jacob S. Hacker and Paul Pierson (2016).
74 Friedman (1970, p.5).
75 The original citation for this work is, Darley, John M. "How Organizations Socialize Individuals into Evildoing." *Codes of Conduct: Behavioral Research into Business Ethics*, edited by David M. Messick and Ann E. Tenbrunsel, Russell Sage Foundation, 1996, pp. 13–43. *JSTOR*: www.jstor.org/stable/10.7758/9781610443913.6 (retrieved November 2, 2022). I have been unable, thus far, to get a copy of this original version, though, and have relied on a copy from Shari Collins-Chobanian (ed.), *Ethical Challenges to Business as Usual*, Pearson Prentice Hall (2005, pp.211–223).
76 Lewis (1989). The reader is strongly encouraged to read the book. It is shocking, enraging and hilarious. It's hard to imagine finishing the book without a sense of being gobsmacked.
77 Darley (1996, p.222).
78 Op. cit., p.214.
79 Some places one might start include the following (retrieved November 2, 2022):
https://www.prezly.com/academy/relationships/corporate-social-responsibility/10-examples-of-exemplary-csr-initiatives#
https://www.classy.org/blog/6-socially-responsible-companies-applaud/ https://digital marketinginstitute.com/blog/corporate-16-brands-doing-corporate-social-responsibility-successfully
https://youmatter.world/en/top-100-companies-best-csr-reputation2019-28108/
80 For example, see Goetsch (2011) *Developmental Leadership: Equipping, Enabling, and Empowering Employees for Peak Performance.*
81 Op. cit., pp.268–269.

PART III

Classical Liberalism through a normative lens

9
CONSEQUENTIALIST MORAL THEORIES

Historical context and motivations

Jeremy Bentham is usually credited with giving the world the initial explicit formulation of Utilitarianism. This theory was articulated in his book titled, *The Principles of Morals and Legislation*, published in 1780. I say "explicit" because there are certainly anticipations of Bentham's theory in the earlier work of the Stoics and David Hume, in particular, though Adam Smith may also have had an influence on Bentham.

This influence stemmed from the theory of moral sentiments as developed by Hume and Smith and not Smith's market theory which we examined in Chapter 6. The theory of moral sentiments holds that we are, in the first instance, driven to action by sentiments. The choice of the term "sentiments" is quite deliberate here. It includes what we would today call the emotions, and these play a prominent role. But it is broader than that and includes such intentional states as desires, anticipations and so forth. The theory holds that we are driven to action, *motivated* to act, because of desire in the first instance, but our emotions and affective states, how we feel about things, are direct indications of the strength of our motivational states. The status of these desires, as motivational states, is that they are grounded in instincts but susceptible to being shaped, or moderated, by learning.

Prominently for Bentham, our motivational states are as they are because they enhance our survival. As such, there is no question that our desires are largely determined by what is in our self-interest. Our self-interest is not exclusive in determining our behaviour, though. We can also be moved by feelings of empathy, and, like any other moral sentiment, our empathy can be shaped by experience. So, based on the motivational structure of our behavioural repertoire, we are inclined to pursue the things we desire and avoid the things that could harm us. We seek what gives us pleasure, and avoid what gives us pain, but within these broad parameters, we

DOI: 10.4324/9781003388555-12

are capable of self-sacrifice to various degrees to the extent that our empathy for another is engaged.[1] This is the cluster of ideas inherited by Bentham as he sets his mind to reforming morality and social custom, particularly as it concerns legislation and punishment.

In addition, though, note the date of publication: 1780. This puts Bentham's period of activity as roughly contemporaneous with Smith. Bentham, though, is a lawyer and legislator. As such, he is particularly concerned with reforming laws and jurisprudence quite generally to make it both more equitable and better reflective of such norms and values as would be likely to have the effect of improving the human condition. In particular, Bentham was concerned with jurisprudence insofar as it is directed at punishment. Bentham was of the view that the then-contemporary theories of punishment were beholden to an older theory of morality, Christian morality and punishment as retribution, which no longer served the interests of humanity. More strongly, I think he might have said that it was utterly without foundation.

This last point raises a real concern that had been brewing since the advent of modern science, and the metaphysics that accompanied it, in the early 17th century. The issue concerns the source and legitimacy of moral norms and values. By this time, science has abandoned the teleological explanations of medieval science that grounded all explanation of natural phenomena in appeals to the ends, or objectives, of God. This sacerdotal system has been replaced by non-teleological, or naturalistic, explanations in terms of nothing other than the mechanical properties of the particles, or corpuscles, of which everything is held to be constituted. According to Francis Bacon (1561–1626), for example, no appeal at all to such things as goals, wants or desires is needed in our explanations of natural events. More strongly, such appeals are frowned upon as unscientific appeals to the unobservable. And, he adds, if there is any truth to such teleological explanations at all, it will be the case that these will be reducible, without loss of any essential content, to non-teleological, causal explanations. That is, there will be an alternative story in the mechanistic terms of modern science that will be able to explain the same phenomena and allow us to capture whatever truth there might be in these teleological explanations without loss of anything important. So, as far as the explanation of natural events goes, teleological explanations are either false or otiose. The presumption is that, at the end of the day, every natural event is to be explained entirely and exclusively in terms of such mathematically quantifiable properties of bodies as their position, mass, state of motion and so forth.

The problem this introduces is that the older medieval metaphysics, which had grounded the sacerdotal system of the Church and state, and had defined the relationships between persons, between persons and state and ultimately between persons and God, had grounded values ultimately in reference to God's plan for us. It is God's intentions for us that legitimize the objective moral norms implicit in Christian Duty Ethics, for example. From this perspective, one might say that a moral vacuum of sorts has opened with the programme of modern science. Thus, it is no longer clear what does, or even could, provide a ground for objective moral norms.

Famously, Descartes had developed a dualistic metaphysical system that attempted to obviate this difficulty by essentially circumscribing the domain of science to apply to nothing other than the mechanical world of moving parts, such things as solar systems, billiard balls, humans tumbling down the stairs and such. But, said Descartes, we are dual beings. In addition to our physical bodies, which are mechanical, we are also souls (or persons), which are held to be spiritual beings that are in no way mechanical. These immaterial beings share a spiritual domain with God and the angels. Descartes held that the two domains, material and spiritual, need never come into conflict. Science can do its thing. The conclusions of science apply exclusively to the physical realm, and the Church continues to have exclusive domain over the spiritual realm.

There are, of course, serious difficulties with this view, not least of which there is no plausible explanation of how our material bodies are able to interact with our immaterial, spiritual selves. Descartes was also required to make the preposterous claim that we, as spiritual beings, are utterly unlike other non-human animals insofar as they are mere mechanical beings, to be explained in all their operations by appeal to purely mechanical causes. Together with Bacon's metaphysical ruminations, in particular the view that the purpose of science is to bring nature under human control, we see the origins of the modern view that nature is something "other than" us.[2] We are separate from nature; it exists "out there" subject to our control. We have already seen evidence of the rapid diffusion of this view in Locke's claim, noted in Chapter 5, that nature, in and of itself, is virtually worthless.

Returning, though, to the specific issue of the moral vacuum this has created, it is not possible to overstate the significance of this as a problem for nascent science and Western European culture more generally. Newton himself was greatly exercised by it, as was Locke and, as we shall see, Immanuel Kant. In fact, it continues to resonate to this day.[3] Bentham is strongly motivated to close this vacuum with a theory that will ground objectively true moral prescriptions about what we should do and which will do this in a way thoroughly compatible with the way science approaches its study of nature. As I put the point, we would capture everything about our moral discourse that is true and useful in terms consistent with modern science while expunging all reference or appeal to the supernatural. Indeed, as we shall see, Utilitarianism gives us a theory of moral decision-making that is a kind of cost–benefit analysis. Bentham himself calls it a *felicific* (or, *hedonistic*) *calculus*. On this view, the objectivity of the *norms* of Utilitarian moral theory is held to follow from certain facts about human nature, as discussed in the theory of moral sentiments of Hume and Smith (and confirmable by science) and a relatively modest assumption about the point of ethics. And, so far as the individual prescriptions of the theory are concerned, we calculate the result that is best because it gives us the best overall balance of benefit over risk (or pleasure over pain).

Rule Utilitarianism

We now turn our attention to the reconstruction of Utilitarianism. This will proceed in two stages, for reasons that will become clear as we proceed. I begin with a version of the theory that is probably closest to the classical version of Rule Utilitarianism articulated by John Stewart Mill, because it is usually held to be the most influential and was developed quite specifically by Mill to address perceived difficulties with Bentham's system.

Stage 1: the standard version of John Stuart Mill

Utilitarianism is a species of the kind of normative theory usually called Consequentialist.[4] The feature common to all Consequentialist theories is that they focus exclusively on the consequences of our actions for deciding what to do and how we should judge what people do. This shared feature reflects the unique starting point of this kind of moral theory, namely that we should always strive to make the world a better place.

I think the point is easily motivated in the following way. A Consequentialist can argue that, in contemplating a normative theory, what we are really doing is contemplating willingly placing constraints on ourselves. For remember, as Locke would put it, in the State of Nature we exist in a state of perfect freedom and equality. Normative theories place constraints on this freedom for they rule many actions as impermissible and others as mandatory, even when these prescriptions are at odds with what is in one's immediate self-interest. So why would we willingly invest the time and energy to develop such a theory and then place ourselves under its constraints? The answer is, we recognize that in doing so, we can all be better off. In short, we can make the world a better place. The restrictions and limitations on our behaviour have the overall effect of guaranteeing all of us a certain level of peace and security which, in turn, has the effect of making possible the kind of cooperation that makes us all better off.[5] Surely, the Consequentialist will say, this is all that could possibly matter when considering a normative theory: will it have the effect of making the world a better place?

And so, I present what I take to be the opening premise of Consequentialism generally, and of Utilitarianism in particular:

P1: The purpose of a normative theory is to make the world a better place.

Now, this goal of making the world a better place is always future-oriented. By that I mean to say that no matter how good things may be at this moment, it is always possible that things could be better. So we always have our eye on the future, always asking ourselves, "What can we do now that would have the effect of moving us further in the direction of making the world a better place?" In this sense, the theory is aspirational. Moreover, it is this future orientation that explains why these theories are called Consequentialist, for when we are asking what we should do, the only meaningful standard for deciding will be to evaluate the predictable consequences of our proposed action with respect to the likelihood that these consequences can be expected to make the world a better place. Nothing else could possibly matter if your starting point is P1.

However, this raises a question: when evaluating the expected consequences of our actions, what should we be looking at? More pointedly, the content of P1 implies a moral valuation; the word "better" is itself a moral term that denotes a value judgement. So the point of the present question might be reframed as: what moral features of the expected consequences of our actions should we focus on to determine whether these would count as a genuine case of making the world a better place? Or, alternatively, what would count as a better world? These questions are answered by the second premise:

P2: The only thing with intrinsic worth is pleasure and the avoidance of pain.

So the short answer to our question is that the relevant moral features of our actions are their tendency to produce pleasure and/or avoid the production of pain. The world will be a better place to the extent that there is more pleasure and less pain. But one is entitled to ask, why this should be the case?

To fully understand the significance of the answer being proposed to this question, we must note that this premise introduces a profound conceptual notion, the notion of *intrinsic worth*, sometimes called *inherent worth*, and this will require an extended discussion.[6] To understand this notion, you must consider it in connection with its opposite, *extrinsic worth*, or, as it is sometimes called, *instrumental worth*. I will define these notions as precisely as I can and then discuss them:

- *Intrinsic worth* is the value an object, or an action, has in and of itself. That is, the intrinsic worth of an object or an action is completely self-contained, it is not in any way relational or dependent.
- *Extrinsic worth* is a relational notion of worth. It is the value of an object, or an action, in relation to that which has intrinsic worth. The extrinsic worth of an action is *dependent upon* that which has intrinsic worth.

These definitions have been framed in a very general way so as to apply to intrinsic worth independently of the claims about the character of intrinsic worth as given by any particular normative theory. Let us now consider an example to illustrate these definitions, one that will be favourable to understanding the Utilitarian notion of intrinsic worth. For the Utilitarian, money has merely extrinsic worth. It has no worth in itself – no kind of self-contained worth. Its value is measured by what we can use it to accomplish, for example, to buy food when we are hungry, or to find shelter. When we are hungry, food makes us happy, and shelter makes one happier than one would be living on the street. So happiness is at the end of the chain, so to speak, for the Utilitarian. That is how you can tell that it is the real thing with intrinsic worth. Putting it metaphorically, we might say that, for the Utilitarian, all roads lead to happiness (or its opposite, pain).

Now, money is a good example to illustrate the Utilitarian assumption about intrinsic worth precisely because many people mistakenly suppose that money is valuable in itself. But this is simply wrong. The Utilitarian is aware that many people value money a great deal, and the Utilitarian is not denying this. Moreover, people value money for a lot of reasons, in addition to being able to buy food when you are hungry. For example, some people like to accumulate money because having a lot of it is a source of power and prestige. But for the Utilitarian, even power and prestige have a merely extrinsic worth. Some people are made happy by power and prestige. It is a fact that people are very diverse in the things that make them happy, or which cause them pain, for that matter. But at the end of the day, it is the pain and the happiness that focus our attention. You can see this by contemplating being shipwrecked on a deserted island with a trunk full of money. In this context, it would be worthless and useless, except perhaps the few moments of happiness it might provide if you burned it to keep yourself warm. You can readily see, though, that the money itself is worthless. Its value is exhausted by what it can contribute to your happiness or avoidance of pain for you. This is the sense in which we can say that what has extrinsic value is *instrumental*: it serves the instrumental value of leading to happiness or avoiding pain (for example, you avoid the pain of homelessness by renting accommodations). Whereas happiness itself is, for the Utilitarian, the only thing we value for itself; the only thing we value in an entirely non-relational, non-dependent way.

A further point about intrinsic worth before moving on. I mentioned earlier "the Utilitarian *assumption* about intrinsic worth," and this may cause the reader concern or puzzlement. The suggestion seems to be that Utilitarianism is merely *assuming* that it is pleasure and the avoidance of pain that have intrinsic worth, and this would be true. The oddness arises from the fact that, on the one hand, intrinsic worth is being characterized as a profound and significant notion; yet on the other hand, the suggestion is that the Utilitarian is permitted to assume that pleasure is the only thing with intrinsic worth. I will simply assert, though, that every theory must make some assumptions. These "first" assumptions constitute the starting place of our theories. To reject them is to reject the theory, and to reject assumptions entirely is to reject the very possibility of theorizing at all. Looked at this way, it might be best to call these *presuppositions*. The theory quite literally presupposes these to be true; they are the *sine qua non* of the theory. The real point of interest here, though, is to note that other normative theories may well make a very different assumption about intrinsic worth and where it is to be found.

In fact, there is a view in the literature on this point in relation to normative theories that has become, so far as I can tell, the standard narrative about the two normative traditions most influential in the Western philosophical tradition. This narrative holds that every normative theory must make some assumption about where intrinsic worth is located, and this assumption in turn determines the character of the resulting theory.[7] This standard narrative holds that the notion of intrinsic worth serves the essential function, in the context of normative ethics, of "fixing" value for the theory in a way that is analogous to the way the American dollar "fixes" the exchange rate for currencies. To clarify this metaphor, recall that normative theories are in the business of making value judgements. From within the perspective of a normative theory, we say such things as, (pointing) "This is better than that." Or "This is good, that is bad" and so forth. But this presumes we have some measure of value, a way of *ranking* the moral value of actions and judgements. It is intrinsic worth that serves this function in the context of a normative theory: it is the fixed, self-contained, non-relational and completely independent notion of moral worth against which all other moral values are measured. And, as you can see from the way I put that last point, the phrase "all other moral values" is clearly meant to refer to the things and actions that have relational, or extrinsic, worth.

To clarify further the point about intrinsic worth determining the character of a normative theory, recall that the example about money is meant to be favourable to Utilitarianism. To anticipate the discussion of Kantian Duty Theory (Chapter 10), note that this tradition locates intrinsic worth in the human individual, not in pleasure and/or the avoidance of pain. This is a truly profound point because it is what makes Duty Theory so distinctive. If you assume that intrinsic worth, which the reader must keep in mind is a completely self-contained, non–dependent kind of worth, is to be found in persons, then you will be committed to the claim that human beings are deserving of being treated in certain ways *regardless of the consequences for bringing about pleasure and/or avoiding pain.*[8] That is, this very different assumption of where intrinsic worth is located commits one, by that fact alone, to *not being a Consequentialist*; you are committed, by that fact alone, to the claim that the starting point of a normative ethics is to obey the duties we have to respect the intrinsic worth of others and demand respectful treatment for ourselves and not to focus exclusively on making the world a better place. We will take up this point again later in our assessment of Utilitarianism in relation to Classical Liberalism.

To summarize, intrinsic worth is a notion of worth that is completely self-contained and serves the function of grounding the rankings of value judgements for a theory in the sense that all else is measured relative to that which has intrinsic worth. Also, different types of normative theories make very different assumptions about where intrinsic worth is to be found, and this difference is held to determine the character of the theory that results.

Returning to the second premise, recall that it is intended to answer the question raised by the first premise. In the first premise, we are told that the purpose of normative ethics is to make the world a better place. But the question raised by this is, how do we do that? What would count as making the world a better place? And the answer, in general terms, is to increase the total amount of intrinsic worth in the world as much as possible, or better, increase the total accumulation of things with intrinsic worth as much as possible. Now since it is pleasure that has intrinsic worth, the world will be a better place precisely to the extent that there is as much pleasure as possible. Also, note that pleasure (or, as it is sometimes referred to, happiness) is held to be the opposite of pain. The two are inverses of each other. So, in fact, the second premise indicates that we will make the world a better place by maximizing pleasure and/or minimizing pain. In some cases, no matter what we do, pain will be unavoidable. In such a case, we try as much as possible to minimize pain.

A further premise is crucial, though, for the Utilitarian, and distinguishes Utilitarianism from some other kinds of Consequentialist theories, for example, egoism. Thus, we have a premise so important to Utilitarian Theory that it is sometimes given the exalted name of The Principle of Equality:

P3: No one's pleasure or pain is to count as more important than anyone else's.

As Peter Singer puts the point, pain is always bad wherever and whenever it happens. Focusing on just the pain itself and ignoring everything about who is experiencing the pain, there is no basis for thinking some pain is more significant than another pain beyond what can be accounted for in terms of features intrinsic to the pain itself, for example, the intensity or duration of the pain. To put it somewhat differently, what Singer is telling us is that when we are contemplating the expected outcomes of our actions, we are to give *equal consideration* to the pains and pleasures of all persons affected. Thus, we shouldn't weigh the pains of our friends and family more heavily than the pains of persons unknown to us who will also be affected, simply because these are the pains that will be experienced by our friends and family. We should not, in other words, act in ways that will avoid pain for our friends and family even at the cost of resulting in more overall pain than there otherwise would be. We should make our decision from the perspective of a benevolent but disinterested spectator with no vested interest in the outcome other than making it optimal in the sense of making the world a better place by producing the best overall balance of pleasure over pain it is possible to produce by that action at that moment.

In fact, Singer makes an argument to generalize the point beyond persons to include equal consideration for all *sentient beings*.[9] Here "sentient being" means any being capable of experiencing pleasure or pain. Singer's point, as noted here, is that pain is always bad wherever and whenever it happens. It rivets the attention, disrupts one's life and, in the worst cases, brings a complete arrest to all normal functioning. From the perspective of the disinterested spectator, one can ask what difference it could possibly make if the pain is experienced by a non-human animal. If we grant the Utilitarian premise that the purpose of ethics – the purpose of voluntarily accepting limits on our behaviour – is to make the world a better place, that is, maximizing pleasure and minimizing pain as much as possible, then we are required to ignore the subject-hood of pain. It doesn't matter who is experiencing the pain, it is only the pain itself that matters.

To behave otherwise, to give preference in our moral deliberations to the pains of humans, for no reason other than the fact that they are human, is *speciesism*. It is an unacceptable bias, a mere prejudice every bit as parochial and baseless as racism or sexism. In this case, it is prejudice that favours the interests of humans for no reason *other than the fact that they are human* (one of us; members of our species). For Singer, this is no different from a prejudice that favours members of one's sex, or race, for no reason other than the fact that they are of our sex, or our race. The principal question that gets begged in any prejudice is, "Why should this matter?"

What is it about the mere fact of being human that makes our pain relevant, and the pains of non-human animals irrelevant or, at the very least, count for much less? An argument is needed that does not appear to be forthcoming. Certainly, if you are a Utilitarian, it can't be relevant, for example, to say that, unlike non-human animals, we are created in God's image. In addition to the burden of appealing to religious considerations, the real issue here, if you are a Utilitarian, is that, by assumption, all that matters are the consequences of our actions – the pains and pleasures that result.

So, on the basis of this reasoning, we might rephrase P3 as follows:

P3': We must give equal consideration to the pains and pleasures of all sentient beings that will be impacted by the consequences of our contemplated action.

It is important to emphasize that equal consideration is not the same thing as equal treatment. Giving equal consideration, even to our fellow humans, does not require us to treat them in the same way, at least not always. For example, since men and women are different in some respects, there will be things done for one group that will enhance their likelihood of experiencing happiness and avoiding pain that will have no beneficial consequences for the other group. Equal consideration demands we weigh their interests equally and do what will have the most beneficial consequences from the perspective of the disinterested spectator. So, for example, consider Mill's campaign for universal suffrage.[10] Withholding the vote from women is wrong because it is a source of frustration and pain for them; it constitutes a limit to their freedom and self-realization that causes unnecessary, avoidable pain. In this case, comparing the situation of men and women, equal consideration demands equal treatment. But it would make no sense and would do nothing to improve the situation of animals to extend the vote to them. In this case, equal consideration of their pains and pleasures might demand of us merely that we leave them alone with space to live their lives according to their own natural urges.

Another way to phrase this point, and to give substance to the notion of the disinterested spectator, is that, for the Utilitarian, our ethical deliberations are to focus on states of mind, not the individuals experiencing these states of mind. In every case, our exclusive focus on the consequences of our actions requires us always to act in such a way as to maximize one particular type of state of mind, namely experiences of pleasure/happiness, and minimize another type of state of mind, namely experiences of pain. So, for example, there may be cases where the best overall result – that is, the greatest overall production of pleasure – requires us to deliberately cause pain to some individuals. This will be acceptable whenever the pain caused is offset by a greater amount of pleasure for some other individuals than could otherwise be realized. The point of P3 is to warn us that this is so even in cases where the pain might fall on our loved ones. This should make no difference to us in our deliberations, and that is what it means for the Utilitarian to speak of acting like a disinterested spectator.

To illustrate this point, consider the following example, an adaptation of the now famous mine collapse in Peru in 2010 that trapped 33 miners underground for 69 days. Imagine a similar case where we have, again, 33 miners trapped underground in two distinct groups: six in Group A and 27 in Group B. Both groups have retreated to the safe areas with food, water, medical supplies and communications. Oxygen is estimated to be adequate for a week. Let us suppose further that the two groups are separated by collapsed rubble between them that effectively isolates them from each other. Rescue of all the miners will require of us two independent operations to burrow to each group separately. There isn't enough equipment to undertake two simultaneous burrowing operations and, in any event, the geologists are cautioning that the area is too unstable. Moreover, burrowing to either group will take nearly a week. The miners will be out of oxygen and dead before the second operation can be completed.

Now a further complication emerges. Sonar scans reveal that burrowing will be very risky in any event as another collapse is imminent. The suggestion is made to use small-scale targeted, precision blasts to open an excavation hole on one side, then burrow through the rubble separating the two groups to extract the group on the other side of the hole made by targeted blasting. The geologists and engineers agree that this strategy has the best chance, by far, of saving any lives at all. The problem is the group on the side of the targeted blasting will almost certainly be sacrificed. How does one decide? For the Utilitarian, the decision is done by looking in a purely disinterested way at the actual consequences. Saving Group A saves six lives and sacrifices 27, whereas saving Group B saves 27 lives and sacrifices six. And here is the present point: it wouldn't matter if you, as the person responsible for deciding, had a family member in Group A. Make it as strong as you like. Imagine that all six members of Group A are family members and none of Group B are. This changes nothing; it is only about the numbers. This is what it means to decide as a disinterested spectator, and this is what it means to give all persons involved equal consideration.

A further point to note in the discussion of P3' is that the emphasis shifted from pleasure to pain. There is nothing duplicitous about this, and nothing is lost in the transition. Recall that pleasure and pain are seen as two sides of the same coin, to put it metaphorically. In any situation where it is not possible to produce a net increase in pleasure, the same considerations drive us to minimize pain. The reason for the focus on pain in the discussion of P3' is that pain focuses the attention in ways that, arguably, pleasure can never match. So, when thinking about the consequences of the action being considered, it is particularly relevant to stay focused on avoiding all pain that possibly can be avoided except in those cases when allowing avoidable pain to occur has the offsetting effect of producing an abundance of pleasure that could not otherwise be realized.

So, taking these premises together, we draw the Main Conclusion of the Utilitarian's argument:

MC: The Principle of Utility: always act in such a way that the consequences of your actions can be reasonably expected to maximize pleasure and/or minimize pain for as many sentient beings as possible.

The Principle of Utility

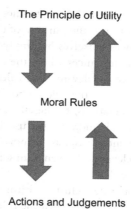

Moral Rules

Actions and Judgements

FIGURE 9.1 Rule Utilitarianism

This conclusion is given the weighty title, Principle of Utility (or, alternatively, The Greatest Happiness Principle) because it is the fundamental principle of Utilitarian moral theory when formalized as such. The resulting theory is called Rule Utilitarianism to distinguish it from other varieties of the theory.[11] Diagrammatically, the theory can be represented as follows (see Figure 9.1):

In this diagram, the downward arrows represent the direction of *prescription*: the theory prescribes certain actions and judgements. And the upward arrows represent the direction of *justification*: we can justify the actions and judgements prescribed by the theory by tracing the reasoning up through the rules and principles of the theory.

The point is that the Utilitarian argument we have reconstructed is an argument for the plausibility of this fundamental principle that is the source of the rules and prescriptions that follow from it and the ultimate justification for all actions and judgements which it licenses. Notice, first, the reference to what can be *reasonably expected* to flow from the actions one is contemplating. It is sometimes made as a criticism of Utilitarianism that we are not particularly good at predicting the consequences of our actions. More strongly, it might be said we are often very bad at this, for reasons having to do with many complications arising out of unexpected side effects, the distortions of personal bias and self-interest and so forth. Thus, it is alleged, the theory is unrealistic in its expectations of us as moral beings; we are simply not able to live up to the standards set by Utilitarianism. The Utilitarian, though, would make the claim that you discharge your moral obligations by taking your moral deliberations seriously and trying earnestly to predict, from the perspective of the disinterested spectator, the consequences that can be reasonably expected to flow from the actions contemplated. No theory can expect more than this.

Moreover, the point of the rules is that they codify the cumulative wisdom of civilized societies and in this way, discharge us of the obligation, at least in many cases, of undertaking to actually predict the outcomes reasonably expected to flow from our actions. In other words, it is the purpose of the rules to discharge us from the need to predict consequences ourselves by simply consulting the relevant rule, except in the most unusual circumstances where the event type is rare and there is not an existing rule that is directly relevant. The rules are in turn justified by the Principle of Utility, in the sense that the rule embodies a generalization (based on the experiences of previous generations) to the effect that, under relevantly similar circumstances, behaving as the rule indicates has usually had the effect of bringing about the best overall consequences.[12] So, as an individual, one's moral obligations are satisfied by following the relevant rule even if, in some particular case, it does not have the expected consequences.[13]

In relation to this point about rules, a further feature of Rule Utilitarianism is that it is resolutely forward looking in the sense that the rules are always under review for their efficacy at bringing about a better world. We might put the point by saying that Utilitarianism is reflexively revolutionary. This returns us to our point of departure: if we grant that the fundamental point of normative theories is to focus our attention on the purpose of ethics, then we find ourselves committed to the view that the purpose of ethics can only be to make the world a better place. But then, as noted at the outset, the world can always be better than it is at present, so we are always oriented to the future, striving always to make the world even better. This means that, although rules can be useful in summarizing the cumulative wisdom of previous generations, and thereby absolving us of the need to predict the outcome of our contemplated actions on each occasion, they are not infallible; they are perennially under review for their efficacy. Moreover, insofar as society is continually changing, the rules must be continually re-evaluated to ensure that they are still suited to existing circumstances.

As an example, very much at the heart of contemporary debate in Canada and other parts of the world, consider euthanasia and doctor-assisted dying. Imagine the following scenario: you are a doctor with a patient who has terminal cancer. Moreover, the patient is suffering enormously and there seems little that current pain relief can do any longer. The patient asks you, begs you, to end his life immediately and painlessly.[14] As a doctor, you can do this, and as a compassionate Utilitarian you want to. But there is a law that would count this as homicide and a professional code that prohibits doctors from ever, by their actions, doing anything that threatens the life of their patients. Moreover, all of this is sanctioned by a moral rule and a set of societal values that regard euthanasia as murder, and murder as the worst moral offense of all.

Utilitarian calculations tell us that no pleasure is possible in these circumstances. The best one can hope for is to minimize pain, and the best way to do that is to end the patient's life painlessly and to do so sooner rather than later. But to act on this initiative in these circumstances would produce much pain for possibly many individuals, flowing from the legal, professional, and social consequences, despite bringing the unnecessary suffering of the unfortunate patient to an end. So we find Utilitarian calculations at odds with what the law and conventional morality seem to require of us. What is our good doctor to do? The answer given by Rule Utilitarianism, is to follow the sanctioned rule and continue to do what can be done in terms of palliative care. But the rule, though it conveniently summarizes the expected balance of pains and pleasures in this context, ignores what is, for the Utilitarian, the real issue, which is the pain that could otherwise be avoided by simply bringing our laws and rules in line with Utilitarian values. So, while observing the law and conventional morality, our good doctor and like-minded individuals will commit themselves to leading the public discussion to review the relevant laws and social norms to bring them in line with what Utilitarian thinking says is required to minimize pain and help to make the world a better place. Indeed, this was exactly the position taken by Bentham with respect to the penal laws at the time, not to mention the treatment of animals (see the quote in Note 9). Also, John Stewart Mill in his, *The Subjection of Women*, made the Utilitarian case for universal suffrage and, more generally, the emancipation of women on the basis of arguments like this.[15]

Stage 2: Preference Utilitarianism

Rule Utilitarianism, as reconstructed here, raises two problems in regard to Classical Liberalism.

First is the idea that our moral deliberations must take at least some non-human animals into account. This is a consequence, the reader will recall, of the reading most Utilitarian theorists are inclined to put on the Principle of Equality (i.e., P3'). Classical Liberals are not concerned, at least not in their capacity as Classical Liberals, with non-human animals.[16] Moreover, animals are often a source of profit and any proscriptions of our treatment of them will be problematic.

Second, insofar as traditional Rule Utilitarianism bases its entire theory on the claim that pleasure is the only thing with intrinsic worth, a host of problems ensues from the fact that pleasure, as a state of mind, is irreducibly subjective. As noted earlier in our discussion of the example about money, it is a fact about people that we are not uniform in our response to the things we call pleasure and pain. One person likes vanilla ice cream, another hates vanilla and likes chocolate, while a third person hates both. Some people are highly motivated to acquire a lot of money and find the work required to do this exhilarating. Others get pleasure from their relationships or even donating their time to walk animals at the local pound. Pleasure and pain are subjective in another sense too. Both pleasure and pain occur as the psychological states of individual sentient beings, and these psychological states are private in the sense that they are not accessible to the inter-subjective observation of external events. This means we can't confirm the occurrence of these states except in our own case. Beyond the reports of others, we can't know what they are experiencing. Put these two features of pleasure and pain together and it is very difficult if not impossible to know with any certainty whether our actions have the effect of making the world better in the specific sense required by Rule Utilitarianism.

One approach to resolving these difficulties results in a version of Utilitarianism that goes by the name, Preference Utilitarianism, the view that we make the world a better place to the extent that we maximize opportunities for as many people as possible to act on their preferences. The underlying assumption is that if people are left free to act on their preferences, we can be sure that they are happy. Conversely, to the extent that people are limited in their freedom, prevented from acting on their preferences, this will count as an obstacle to their happiness and, more likely, will result in directly producing frustration and unhappiness. So, by simply leaving people free to act on their preferences, we are no longer stymied by the subjectivity of pleasure. Moreover, this immediately connects Utilitarianism to the laissez-faire market principles of Adam Smith because the best way to maximize opportunities for as many people as possible to act on their preferences is to leave them free to choose in the market.

Now if the market is truly free, leaving everyone to act exclusively from their self-interest, then we can be assured that the best indicator that they act on their preferences will be given by their "willingness to pay." Willingness to pay is being used as a technical term here, defined by economists as follows:

> Willingness to pay (WTP) is the maximum amount an individual is *willing* to hand over to procure a product or service. The price of the transaction will thus be at a point somewhere between a buyer's willingness to pay and a seller's willingness to accept.[17]

The point to focus on here is that, if we assume the conditions of a laissez-faire market to be in place, then whenever a deal is concluded, it must mean all participants are doing so willingly. They are all acting from self-interest, so they must be satisfying their preferences. There is a wonderful consilience of reasoning here insofar as the conditions of Smith's free market are, at the same time, the very conditions that maximize opportunities for people to satisfy their preferences and, thus, the conditions that maximize opportunities for people to act selfishly are the very conditions that make us all better off, the conditions that continually move us in the direction of making the world a better place. Moreover, note that the focus is now exclusively on the preferences of market participants. Animals are excluded in principle. They aren't market participants, they are commodities, which, from the Classical Liberal perspective, is as it should be.[18]

Why Preference Utilitarianism is such a good fit for Classical Liberalism

With this reconstruction of Preference Utilitarianism before us, it's time to understand what it does for Classical Liberalism, which is, at the same time, to understand why it is such a good fit for Classical Liberalism. Indeed, of the normative traditions we will look at, Preference Utilitarianism is clearly the best fit.

First, and foremost, it provides Classical Liberalism with the resources to justify the policy prescriptions it advances and the actions taken under its direction by various players, most prominently global multinational corporations. This is clearly no small thing. It will be useful to consider an example. I will frame the example theoretically; it is not meant to bear a resemblance to any particular real-world event, though it may bring to mind several of recent memory. So imagine a clothing manufacturer with global sales, call it Company X, that announces its intention to relocate at least some of its manufacturing facilities to a third-world country. Moreover, the new production facilities will be structured under a contractual arrangement with a manufacturer in the host country who will, in turn, be responsible for delivery of a product that meets X's quality requirements at the agreed price and schedule of delivery and will further assume all responsibility for the facility, staff, working conditions and such local requirements of law in regard to environmental protections, working conditions and so forth. Any arrangement like this always has consequences for many parties, including workers in the home country who can expect to lose their jobs, government in the country where the company is domiciled that may experience loss of tax revenue, shareholders, consumers, various interests in the destination country, in particular employees, and so forth. Any action with consequences that reach this broadly must be justified, and of course the justification may differ in the details depending on the target audience. But generally, we can expect the justification to appeal to the Utilitarian argument that by acting in this way, the corporation is improving the lives of many and thereby making the world a better place.

Consider how this argument is to be made. Certainly, there will be some pain, for example, the home country workers who lose their jobs, but the argument goes, this pain will more than offset by the greater good produced for many others. The workers in the third-world country will now have regular employment at a level of remuneration that, while well below the remuneration of the displaced home country workers, will still make them better off than they were previously. Indeed, this has been at the heart of Classical Liberal arguments in support of the liberalization of global markets for decades, that it is lifting third-world countries out of poverty. Indeed, this was noted in our discussion of the Brundtland Report in Chapter 2. These workers are now on the way to joining the capitalist economies of the developed world and, it is urged, this is far better than any programme of short-term aid could ever accomplish, not least because it also begins the process of developing an industrial capacity in the host country.

As far as the displaced home workers go, this loss is actually the start of a redistribution of jobs in the economy. In time, as the economy grows under the stimulus of global development, these jobs will be replaced by others that are part of the service sector, or even higher up the knowledge skills hierarchy, meaning better jobs for more individuals. The lower-level production jobs have been moved to where they find their natural level of remuneration (allocative efficiency at work). And, of course, as consumers, everyone benefits from being able to buy these consumer goods at cheaper prices. In fact, the argument is often extended to claim that the money saved by cheaper consumer goods leaves more money available for consuming more, and, in some cases, reinvestment in the economy. Moreover, insofar as X's profits are enhanced, the shareholders benefit as well.

All of this has the beneficial effect of growing the economy, producing more goods for consumption on a global scale than would otherwise be produced and stimulating ever more investment, thus bringing more prosperity to more people than would otherwise be expected. Generally, the rhetorical phrase that is used to present these justifications is "more choice." By its actions, Corporation X has enhanced the choices available for more participants; more opportunities for more market players to act on their preferences. This is Preference Utilitarianism at work to rationalize a system that is for our mutual benefit. On its face, this is a powerful argument in favour of the free market prescriptions of Classical Liberalism.

The attentive reader might well interject here that, insofar as the theories of both Locke and Smith make prescriptions, they too must be normative theories. Or, to put the point slightly differently: why does Classical Liberalism need Utilitarianism to justify its policy prescriptions? Aren't these already adequately justified by the theories of private property and the free market? This introduces what is really the most powerful resource provided by Utilitarianism, which is that *it gives a moral force to these policy prescriptions*. The reader will recall that, insofar as we have framed the theories of Locke and Smith for a contemporary audience, we have expunged all appeals to God and, more generally, all appeals to purpose or goals in those arguments. But it was precisely the appeal to God that gave a moral force to Locke's argument. So, for Locke, the reason we must honour the rights of all to acquire what they need from the Commons by means of their labour is that God makes it so. This is the point of the Fundamental Law of Nature. Expunge God and there is no longer any reason to follow these prescriptions other than pure self-interest of the unbridled and unrestrained sort that would result in Hobbes's "war of all against all". As for Smith, as the reader will recall, there is an appeal to what we have called the Rising Tide Argument. But on its own this does not provide a *moral justification* because we are left to wonder what makes this rising tide a *good* thing. If we simply assume it to be so, we are begging the question by assuming what requires to be justified.

Utilitarianism answers these questions. It justifies these prescriptions with a very powerful argument that has moral content: we must do these things because in doing so, we make the world a *better* place. And we are told why making the world a better place is a good thing. Or, put slightly differently, we are told exactly what it means to say the world is a better place. The world will be a better place precisely to the extent that we have maximized opportunities for as many individuals as possible to act on their preferences. We are enhancing liberty and choice for more people to give expression to their preferences which is assumed to be, simultaneously, an expression of their identities, their selves.[19] And we will know that we have achieved this goal by noting that we have set in place the conditions under which people willingly pay for these preferences.

Lurking in this reasoning is a very subtle, but important point. The assumption is that, in making the arguments of Locke and Smith friendly for a contemporary audience, we have simultaneously produced arguments that are held to be objectively true and valid. In other words, the standard narrative holds that, insofar as one accepts the premises of the reconstructed version of Locke's argument (and the same point applies to Smith), one *must* accept the conclusion, and the policy prescriptions that are its consequences. And, so far as the premises go, these are held to be statements of basic fact, utterly beyond challenge, and, thus, the conclusions too are held to have the status of objective truths. But moral claims, value claims generally, are never objective in this sense, they never follow from mere statements of fact alone. So if the arguments of Locke and Smith are objective in this sense, then whatever moral force attaches to Classical Liberalism must come from elsewhere. As we now see, this source is Preference Utilitarianism.

In addition, though, Utilitarianism is a good fit for Classical Liberalism because its method of moral deliberation aligns perfectly with standard practice in business. As we have seen, Utilitarianism conducts its moral deliberations on the basis of a kind of cost–benefit analysis. In traditional Rule Utilitarianism, we are expected to project the pleasures and pains for each alternative action being contemplated, treating the pains as negatives, the pleasures as positives and choose the alternative action available to us that realizes the best outcome, which is either the greatest overall balance of pleasure or the smallest overall balance of pain. As noted earlier, Bentham took this idea so seriously that he characterized it as a kind of "felicific calculus." He seems to have thought that human responses to stimuli, as either pleasurable or painful, could be expected to have a uniformity that would eliminate the differences between persons and allow, in principle, a numerical value, positive for pleasures, negative for pains, to be attached to these responses. He speculates on some future psychology that could develop a kind of metering device that could measure these responses in people and give us numerical values. So, in the fullness of time, it was imagined that we would approach the perfection of Utilitarianism as a kind of *moral science* that could deliver its prescriptions with infallible precision. This, as we now know, is ridiculous. Not only for the reasons noted here but also because of the well-known result of diminishing value at the margins.[20]

But this is the beauty of Preference Utilitarianism: in reimagining the theory in this way, we are no longer burdened by the implausibility of Bentham's version. We now need to pay attention to only the features of markets that enhance the opportunity for choice and leave people free to act on their preferences, and this, as we have seen, is signalled by their willingness to pay. So long as we can be assured that every deal was negotiated freely, with a complete absence of fraud, coercion or withholding of relevant information, then we can be assured that the individuals involved have acted on their preferences. This has the tendency of narrowing the focus of our moral calculations on commodities and their prices. Acting on their preferences is how people maximize their pleasure, and willingness to pay is the signal that people have done so. Pleasure itself has been commodified, our moral deliberations proceed in uniformity with all business decisions: we perform cost–benefit analysis in terms of dollars.

Lastly, the view denies that there are such things as human rights, in the sense of fundamental, inalienable entitlements that one is due simply by virtue of being human. There are only certain freedoms, liberties to act on one's preferences so long as doing so can be expected to maximize the likelihood of similar opportunities for as many people as possible or at least not interfere with the liberties of others. In terms of our example of Company X offshoring its production in pursuit of maximizing profits, the workers in the destination countries are free to accept the offer of work, or not as the case may be, but they don't have any rights to basic worker protections regarding working hours, conditions of employment, safety or exclusion of children from the labour market, beyond whatever protections they may have in law. If they accept the offer of work, willingness to pay tells us that they have acted on their self-interest and, thus, must regard themselves as having improved their condition. Appeals to such things as rights as entitlements would amount to a distortion of the price signals in the market that would, in turn, undermine market efficiency. In this sense, then, the view aligns perfectly with Friedman's Shareholder Model of CSR.

Re-evaluating the "fit" of Preference Utilitarianism for Classical Liberalism

But not so fast. It's one thing to say that the denial of rights as inalienable entitlements is a good thing from the perspective of Classical Liberalism, but this raises real difficulties for the ideology.

Let us begin by getting clearer on the underlying issue. Put bluntly, if the consequences of our actions are all that matter in the evaluation of our actions, then human rights can't matter. The reader may be confused by this claim, recalling that Mill himself was a staunch defender of woman's rights and universal suffrage in particular. And it goes without saying that defending the right of women to vote in Victorian England laid one open to considerable criticism and abuse. But here's the thing: Mill was not defending women's right to vote as a "right." That is, he was not defending this as some inalienable entitlement of their personhood. Rather, he was defending this "right" on the grounds that to deny it of women was a source of pain and frustration for them, an obstacle to their fulfilment and self-actualization. Turning this argument around, we might say that granting this "right" to women is just one way we can expect to enhance the likelihood that they will have more opportunities to act on their preferences. Moreover, there is no obvious downside, no risk of creating significant amounts of pain or frustration for others, besides perhaps the petty resentments of some men for some little while till accommodation occurs. And, of course, insofar as women constitute roughly half the population, the expectation for making the world a better place in the sense meant by the Preference Utilitarian is potentially huge. There can be no sound Utilitarian argument for not doing this.

So, while the Utilitarian refuses to recognize rights as such, they recognize the Utilitarian value of certain liberties, call them "rights" to carefully circumscribe these. Paraphrasing Bentham, "all talk of rights is nonsense on stilts."[21] By this he meant that to speak of rights is, quite literally, to speak nonsense. There can be no such things in the sense of inalienable entitlements. And the reference to stilts implies that we delude ourselves into believing that there are such things and that they are vitally important because we dress it up in the language of elevated discourse, including such terms as "inalienable." But these "rights" are never inalienable in the sense that, as circumstances change, it is always possible that we will have to reconsider these entitlements with an eye to rescinding them or perhaps limiting them. It is best not to think of them as entitlements so much as temporary freedoms suited to existing circumstances in such a way as to stand a good likelihood of making the world a better place than it otherwise would be under those circumstances.

This might seem like a good thing for those leaning in the direction of radical libertarianism, but from the perspective of the Classical Liberal there is a real tension with the other assumptions of the ideology. As remarked in Chapter 8, the ideology presumes Preference Utilitarianism to the extent that this theory is at the heart of the central argument, the Rising Tide Argument, that rationalizes and justifies the model (see the discussion of Freeman's reliance on a Kantian argument for respect of persons). After all, it's understood that the rising tide maximizes *aggregate welfare* while simultaneously relegating many to the functional role of being the means to this end. On the other hand, as we have seen to be affirmed by Locke and Nozick, the fundamental premise of the view is that the rights of each individual are sacrosanct and constitute the very foundations of the view. Thus, everyone has the right to not be used as the mere means of another. There is, so far as I know, no way of adequately resolving this tension; something must give.

A further consequence of the view raises additional difficulties for Classical Liberalism. So far as Preference Utilitarianism is concerned, there can be no commitment to *Justice* in the retributive sense of the term. That is, Utilitarian justice is never a matter of giving people what they deserve. To see this point, it will be helpful to consider punishment from the perspective of the Utilitarian. Punishment essentially involves the deliberate infliction of pain, something the Utilitarian is loath to do. When we ask ourselves, "What could possibly justify the deliberate infliction of pain?" the only answer acceptable to the Utilitarian is that this pain will be offset, compensated if you will, by the production of a greater amount of pleasure elsewhere (i.e., for individuals other than those being punished).[22] There are two ways this could be accomplished by a well-designed theory of punishment and system of justice that would implement and enforce this theory. These two ways require us to focus on *deterrence* and *rehabilitation*.

First, regarding deterrence, we must note a distinction between specific and general deterrence. The *specific deterrent* of a punishment is the deterrence of the individual being punished for the duration of the punishment, presumably by being incarcerated, or confined to home under electronic surveillance, or any form of punishment that could be expected to have the effect of making it impossible for that individual to reoffend, at least for the duration of the punishment. At the extreme end of specific deterrence is execution of the criminal; obviously, this is maximally effective as a specific deterrent. The *general deterrence* of a punishment is the effect it can be expected to have on deterring others from that crime. This is likely to be the sense that would matter most to the Utilitarian and it will require some sophistication in how we set up a system of jurisprudence to administer this. The reason being, a punishment can be effective as a deterrent only if there is a real expectation amongst potential criminals that there is a very good likelihood of being caught, and upon being caught, a very real likelihood of being convicted and punished. There must be the perception of a real threat of punishment to be effective as a deterrent. This is a well-established fact about human nature. There are also questions about how to apportion punishments to be severe enough to actually deter. And all of this requires society to devote rather a lot of resources to ensure enforcement, conviction, delivery of the punishments and so forth, which leaves fewer resources for other positive goals we might have. A very tricky set of calculations to say the least.

Second, there is rehabilitation. The idea here is that if we are going to deliberately inflict pain on some individuals, we want our punishments to have the effect of reducing recidivism. Many types of techniques are possible, depending on the individuals involved and the nature of the crimes. These can include anger management therapy, chemical castration or even more intrusive interventions.[23] Utilitarian reasoning could, in principle, justify any interventions that might be expected to "rehabilitate" the criminal, regardless of how brutal they might be, and regardless of the violation of the autonomy of the individual being rehabilitated, so long as it would be offset by enhanced security for the rest of society, not to mention the very real possibility that the rehabilitated individual might now be fitted to a normalized life as a productive member of society.

But nowhere in this is there any acknowledgement that retribution plays a role in punishment. It is never, and can never, be a matter of giving the criminal what they *deserve*. As a matter of Utilitarian calculations, effective deterrence and rehabilitation can result in cases that under-punish and cases that over-punish. For an example of the first sort, consider a case involving significant and systemic corporate crime, such as investment vehicles made available to the public that involve significant risk and deception about those risks.[24] We could even imagine widespread use of Ponzi schemes defrauding the investing public. In a case like this, we can easily imagine prosecutors and the courts taking a relatively lax attitude to prosecution and punishment. They might prosecute a few high-profile cases to "make an example" and give the illusion that more is being done (for example, Bernie Madoff), but there would be good Utilitarian reasons for not prosecuting to the full extent of the law and to not systematically investigate all possible offenders. First, there is the systematic integration of the financial sector in the broader economy and the unknown impacts this might have on the economy as a whole. But also, the expense of the investigations and prosecutions themselves would gobble up a huge reservoir of resources better used elsewhere to maximize the scope and extent of markets, not to mention the difficulties involved in successfully securing a prosecution.

For an example of over-punishment, consider the case of privatized prisons.[25] Here is a sketch of the justification of private prisons. First, as we have seen, the assumption is that the private sector always does everything more efficiently, leaving more resources available for growing the economy. Running prisons is expensive, but especially in the public sector where wages and benefits are higher, and collective bargaining requires employers to maintain certain working conditions. Typically, a private sector security firm will be given the contract on the basis of a bidding process that awards the contract to the lowest bid. They achieve this result by slashing costs: contractual employees without pensions or benefits and lower wages without job security. But also, there are savings to be extracted in the operating costs: low-quality food for prisoners, cutting back on laundry services and so forth. Moreover, some 2.2 million prisoners in these facilities in the United States are employed to provide cheap labour. They do a variety of jobs, from production to agriculture, at remuneration in the range of 12–40 cents per hour.[26]

All of this may seem, at first blush, like a good result looked through the lens of Classical Liberalism. First, the government saves a lot of money that would otherwise be required to run these facilities as state-run prisons. This money can be used to provide other services or, better yet, to provide tax cuts that leave more money available in the hands of citizens to spend on goods. So more preferences actualized than would otherwise be the case and a stimulus to production that produces jobs, not to mention the additional money in the hands of investors to reinvest in growing the economy. And these spin-off benefits are additional to the growth of the private security industry itself, particularly as the industry diversifies and expands globally. It is now a $350 billion market in the United States alone.[27] Lastly, it should be mentioned, the prisoners benefit from being given the opportunity to develop job skills and the discipline necessary to re-engage as a productive citizen upon release. This is rehabilitation at work in the service of public interest.

As it turns out, though, this is actually a mixed blessing. For one thing, it puts the theory at odds with the views of most people that retribution matters. Punishment should be, at least in part, a matter of giving people what they deserve. One might reply that they are simply wrong to insist on this, and that was certainly Bentham's view. We'll take up this point again in Chapter 10, but, in any event, there is a bigger problem that cannot be so easily sidestepped. For notice, there is now an implicit pressure to maintain, in fact grow, the total number of prisoners in these privately run institutions to maintain this low-cost labour source, what is essentially slave labour, a new twist on the old Workhouse Prisons and chain-gang arrangements.[28] Pressure on governments from lobbyists, representing not only the private security firms but also the corporations for whom they provide the cheap labour, often drives legislation and policy. In other words, the privatization of prisons, at least under existing models, is another example of a perverse incentive of the sort discussed in Chapter 6, where the example focused on healthcare delivery. Parenthetically, it is worth remarking that in both cases the evidence does not clearly support the claim that these ventures in privatization actually lower costs.[29]

This perverse incentive and the policies that flow from the lobbyist pressure manifest themselves in the increasingly harsh treatment of low-level petty criminals in the United States, the United Kingdom, Australia and elsewhere, reflecting disproportionate numbers of people of colour and poor whites. The treatments typically imposed under stringent budgets – less food and food of poorer quality, less yard time, more confinement to cells, less recreational activities, lesser quality and frequency of laundry services – arguably result in inhumane conditions and have resulted in riots, sometimes with loss of life. It also imposes unsafe conditions on staff. Lower than adequate staffing and inadequate training have been known to result in more frequent assaults on staff and inmates. And negligible, or even absent, investments in real rehabilitation mean increased risk to the public from demonstrable increases in recidivism associated with privatized prisons.[30] All of this is at odds with the rights of everyone to never have their rights sacrificed for the benefit of others, in this case their rights to fair treatment by the legal system.[31]

Worst of all, perhaps, is that the perverse incentives that entrench this systemic abuse of rights undermines public trust in the system of jurisprudence and, indeed, in government generally. It is to take the administration of justice and put it into the hands of private profiteers who are not themselves directly accountable to the citizens. As businesses run for the purpose of generating profits, this is to set the stage for the corruption of justice whenever profits clash with the principles of justice, which is the predictable result when such perverse incentives are operative. Justice, as represented by a blindfolded Madame Justice holding scales, is required to be unbiased and fair at every level and where this is threatened, the predictable result is the erosion of public trust. Also, this criticism applies with equal force to the case of under-punishment insofar as the message that is sent is that money can allow one to escape justice. This is a very damaging result. After all, it is trust that is essential for the system to work as it should.

Finally, consider that Utilitarianism takes no consideration of our relations to one another except insofar as these have direct implications for the consequences expected to unfold. More accurately, one might say that Preference Utilitarianism instrumentalizes our relationships with one another. As noted in the mining example, the prescriptions of the theory flow from a bare consideration of the numbers involved and it doesn't matter who the people are or one's relations to the people involved. In fact, the requirement that we view the situation from the perspective of the benevolent but disinterested spectator is simultaneously a requirement to deliberately ignore our relations to others.

In the mining case, it might seem that this gets it exactly right: surely it would make no sense to blast the hole on the other side killing 27 miners to save six. But consider a different case. You live in a three-storey tenement building, a walk-up with no elevator. You are a single parent living at the back of the building, on the top floor with your five-year-old daughter.[32] You are also a full-time student with a part-time job struggling to make ends meet while you finish your education. One night finds you up very late studying for a test the next morning. Your daughter is asleep in her bedroom at the back of the apartment. There is a fire escape that runs down the back of the building just outside her window. Finally, at 2 a.m., you are ready to get a few hours of sleep when you realize there is no milk for breakfast. There is a small corner store at the end of the block and it will make life a lot easier to go get it now rather than try to do it in the morning while getting yourself and your daughter ready for the day. You decide to run to the store and to leave your daughter asleep. No need to wake her and dress her at this hour; you'll be back in less than 15 minutes.

But you return home to find the building ablaze. It is an old building and it is burning fast. You hear no smoke alarms, no sirens in the distance, and you left your cell phone on the table. You estimate that, at the rate the fire is progressing, you will have just enough time to run around to the back of the building, mount the fire escape, and run up the stairs to extract your daughter from the building. You will worry about raising the alarm as soon as she is safe. But, upon mounting the fire escape, you see four young children asleep in the ground-floor bedroom. What do you do? Taking the time to save the four children first makes it very likely that you won't have time to rescue your own child. From the Utilitarian perspective, though, the fact that the one child on the top floor is your child is irrelevant (the Principle of Equality). All that matters are the numbers, and the numbers are clear: it's a matter of four deaths versus one.

The problem this raises for Classical Liberalism is that the economic system at the heart of Consumer Capitalism is built on relationships. We have obligations, duties to one another, that arise out of our relationships and are dependent on the specific circumstances of our engagements. For example, deals are negotiated that bind the parties to honour the terms of these agreements. Negotiation of the deals themselves is reliant on trust between the parties and such trust would be an elusive thing indeed under any attempt to instrumentalize the relationships involved. Of course, no self-respecting Classical Liberal would deny this. In practice, though, this feature of Preference Utilitarianism is typically used to rationalize the instrumentalization of only some relationships. For example, employees can often find themselves treated in these ways, especially if they are foreign workers; or the prisoners treated as slave labour in privatized prisons; the clients injured by deliberate corporate malfeasance and subsequent attempts at denial and cover-up. All this places a strain on the credibility of the system, undermines public trust and erodes fair and free competition resulting in a system that is arguably less efficient and less productive than it otherwise might be, despite pronouncements to the contrary.

★★★

And so, Classical Liberalism needs the support of some normative theory to give its prescriptions moral force and to provide an adequate rationale for these prescriptions that connects to our values. Initially, Preference Utilitarianism might seem like a perfect fit. It conveniently accepts the commodification of non-human animals without blushing. It rationalizes the Rising Tide Argument by connecting it to the opportunity for consumers to maximize their preferences as demonstrated by their market behaviour, in particular their willingness to pay. Moreover, it denies the existence of rights as inalienable entitlements while pushing liberties to the forefront of its moral deliberations. And the focus on moral deliberation as a kind of cost–benefit analysis aligns nicely with the way business conducts its decision-making and does so in a way that instrumentalizes punishment and our relationships to others. On closer examination, though, this places the ideology under an enormous burden in at least two respects. First, it creates tensions within the ideology itself insofar as it is built on a foundation of inalienable rights and relationships with one another built on trust. Second, these judgements frequently fly in the face of popular opinion, especially in the case of justice as retribution.

Let us see if we can't find a normative theory that might better fit the needs of Classical Liberalism. In Chapter 10 we turn our attention to the Duty Theory of Immanuel Kant.

Notes

1 We can also be moved to self-sacrifice by feelings of group loyalty. For recent research on the psychology of group loyalties as they pertain to self-sacrifice, see Haidt (2012).

2 As Bacon said, "ipsa scientia potestas est" (knowledge itself is power) in his *Meditationes Sacrae* published in 1597.

3 For a recent example, see Alvin Plantinga (2011) *Where the Conflict Really Lies: Science, Religion and Naturalism.*

4 An older terminology calls these theories *Teleological*, from the Greek root word, *Telos*, usually translated as "end," "goal" or "objective." The overlap in meaning between teleological, on the one hand, and Consequentialist, on the other, is that we are urged to focus exclusively on our normative goals as the realizable consequences of our actions. Details to follow.

5 This is obviously highly reminiscent of Hobbes and Social Contract Theory generally. There is a significant difference, though, insofar as Consequentialism does not share Hobbes's bleak and pessimistic view of human nature. This is the influence of the theory of moral sentiments: human beings have a capacity to respond empathetically, and this response can be nurtured and cultivated and it is the job of a properly constructed normative theory to do.

6 The reader will recall we have encountered this notion of intrinsic worth in Chapter 4, when discussing Taylor's Biocentrism, where we agreed to keep the discussion informal. It is now time to come to grips with the details.

7 When I speak of two *normative traditions* I am referring to types or families of normative theories. One of these, the Consequentialist (or teleological) family of theories contains the various Utilitarian theories within it. The other family of theories, Duty (or Deontological) theories, of which Christian Duty Ethics is likely to be the example familiar to most readers will be discussed at length in Chapter 10.

8 There are other kinds of Duty Theory that raise the issue of whether intrinsic worth ought to be located in such a way as to apply to some non-human animals, or as in the case of Biocentrism considered in Chapter 2, all of them.

9 Singer (1990) – the first edition was published in 1975. It must be noted that Singer follows Bentham on this. Here is a quote from Bentham,

The day may come when the rest of animal creation *may* acquire those rights which never could have been withholden from them but by the hand of tyranny. The French have already discovered that the blackness of the skin is no reason why a human being should be abandoned without redress to the caprice of a tormentor. It may one day come to be recognized that the number of legs, the villosity of the skin, or the termination of the *os sacrum* are reasons equally insufficient for abandoning a sensitive being to the same fate. What else is it that should trace the insuperable line? Is it the faculty of reason, or perhaps the faculty of discourse? But a full-grown horse or dog is beyond comparison a more rational, as well as a more conversable animal, than an infant of a day or a week or even a month old. But suppose they were otherwise, what would it avail? The question is not, Can they *reason*? nor Can they *talk*? but, Can they *suffer*?

[Bentham (2007, p.309)]

10 Mill (2006).

11 One quick point to note is that there is a simpler version of Utilitarian moral theory that dispenses with the moral rules. This version of Utilitarianism is called Act Utilitarianism, whereas the version we are investigating is called Rule Utilitarianism. Act Utilitarianism, the version initially given by Bentham, is so called because it is held that we can dispense with moral rules and simply calculate the actions prescribed directly from the Principle of Utility. There is a debate underlying this distinction that is important to the development of the most plausible version of Utilitarianism, but which need not concern us here insofar as the version that has influenced Classical Liberalism is Rule Utilitarianism.

12 A significant problem here is to clarify what is to count as "relevantly similar" but we pass over this difficulty here.

13 This is exactly what the Act Utilitarian denies. But this dispute, internal to Utilitarian theorizing, is not of concern to us here. Again, we are focusing on Rule Utilitarianism because it is the version that has more directly influenced Classical Liberalism (see Note 11).

14 This example is adapted from James Rachels (2003) *The Elements of Moral Philosophy (fourth edition)*.

15 Mill (2006).

16 I put the point this way quite deliberately to note that there are probably many individuals that would consider themselves Classical Liberals and would at the same time affirm a love of animals. Or at least some animals (e.g., domestic pets). This is merely to mark out, though, the sense in which their personal feelings and values in one context can be at odds with their Classical Liberal values operative in other contexts. For some insight into the role context, or frames of reference, can play in our moral deliberations, and the resulting contradictions it can produce, see George Lakoff (2008) *The Political Mind: A Cognitive Scientist's Guide to Your Brain and its Politics*.

17 https://blog.blackcurve.com/9-factors-that-affect-a-customers-willingness-to-pay (retrieved November 3, 2022).

18 It must be pointed out that this is a particular interpretation of Preference Utilitarianism that a Utilitarian theorist like Singer would not agree with.

19 I discuss these issues in "Consumer Capitalism and Self-Identity," *Analytic Teaching and Philosophical Praxis*, Volume 36, (2015–2016), pp.55–61.

20 On diminishing value at the margins, imagine you get lost hiking off-trail and spend a couple of days without food and water until being rescued. You finally get to a restaurant and order your favourite meal. The experience is so overwhelmingly pleasurable that you immediately order another. We all know that, even if you are hungry enough to finish the second order, the pleasure will be far below what it was for the first.

21 I have read this quote many times as attributed to Bentham but couldn't actually locate the quote. The original source, so far as I know, is the following: J. Bowring (ed.) (1843) "Anarchical Fallacies" in *Works* vol. 2. It is also available in *The Collected Works of Jeremy Bentham*, (11 volumes) Oxford University Press.

22 For purposes of this example, it is easier and more natural to speak in the terms of Rule Utilitarianism (i.e., in terms of pains and pleasures), but the point applies as readily to Preference Utilitarianism and could be translated into the terms of that version of the theory.

23 On this point, and as an example of the extremes this idea can be taken to, one is put in mind of Alex, the protagonist in *A Clockwork Orange*. See Burgess (1962).

24 As per the example considered in Chapter 8 when discussing Lewis (1989) and his account of his experiences at Saloman Brothers.

25 The following discussion draws on an essay from a former student of mine: Derren Roberts (2019) *Capitalism and Prison Privatization* (unpublished).

26 Article here: https://corpaccountabilitylab.org/calblog/2020/8/5/private-companies-producing-with-us-prison-labor-in-2020-prison-labor-in-the-us-part-ii (retrieved November 3, 2022).

27 This is reminiscent of our discussion of the growth of the waste management industry in Chapter 2. Article here: https://ripusa.com/blog/the-increase-in-private-security/ (retrieved November 3, 2022).

28 Article here: https://www.theatlantic.com/business/archive/2015/09/prison-labor-in-america/406177/(retrieved November 3, 2022).

29 For a discussion of this evidence in relation to prisons, see Friedman, A. (2014).

30 Logan (1990).

31 See: https://www.sentencingproject.org/publications/capitalizing-on-mass-incarceration-u-s-growth-in-private-prisons/(retrieved November 3, 2022).

32 An adaptation of a case from Rachels (2003).

10

KANTIAN DUTY THEORY AND THE DEONTOLOGICAL NORMATIVE TRADITION

Kant's historical context and motivations

Immanuel Kant (1724–1804) was born in Konigsberg, Prussia. He was a lifelong academic and is widely regarded as the most important philosopher of the modern era. Although a contemporary of Adam Smith, Kant's career as a writer matured rather late in life. As such, he is a late Enlightenment figure who stands at a unique juncture in the history of the Western intellectual tradition. The buoyant optimism of the early Enlightenment period is giving way to the scepticism of Hume and the romanticism of Rousseau. Kant is influenced by both but sees himself as a staunch defender of Enlightenment values and is strongly motivated to challenge scepticism in all its forms. Now, when I say of Kant that he was a staunch defender of Enlightenment values, I mean specifically to attribute to him a rather strong view about the value of our capacity for reasoning. In his view, it is this capacity in us that marks us out as unique beings possessed of dignity and moral worth. Additionally, though, he believes this capacity can be employed for discovering important truths about the world, including moral truths about the realm of human action. Moreover, it is held that these truths are objective, that is true for everyone universally and not in any way dependent for their truth on what we happen to want to believe.

The Critical Philosophy he developed is the project intended to respond to this scepticism emergent in the late 18th century. Beginning with *The Critique of Pure Reason* (*CPR*, published in 1781), this first salvo is Kant's attempt to overcome scepticism by finding a way to reconcile the dominant traditions of the early modern period of European philosophy. These traditions typically go by the names of Rationalism and Empiricism.[1]

Rationalism, as an approach to knowledge, attempts to ground knowledge claims on pure reasoning that is independent of any appeal to particular experiences. To

DOI: 10.4324/9781003388555-13

the extent that experience plays a role at all in the acquisition of knowledge, it is a merely formative role. Empiricism, on the other hand, emphasizes the claim that all knowledge must ultimately be justified by tracing it back to its origins in experience. Rationalism lands itself in scepticism to the extent that it regards the evidence of the senses as prone to illusion and error and, thus, finds itself forced to claim that any knowledge of the natural world must be certified by some authority independent of the senses themselves. Thus, for example, one strategy is to escape scepticism at the cost of having to rely on God to act as the guarantor of human knowledge.[2]

On the other hand, Empiricism, as it is articulated and defended by John Locke, lands itself in scepticism because it claims that our experiences of the world, which are held to be the origins of all knowledge, are admitted to be mere representations of the objects of the world, representations that stand at the end of long causal chains between the objects themselves and our experiences of them. On this view, it becomes difficult to understand how we can ever claim to know that the objects resemble our experiences of them. The assumptions underlying Empiricism ultimately results in the scepticism of David Hume, who claims that our pretensions to knowledge can never go beyond claims about these representations or ideas in the mind.

Kant is of the opinion that Rationalism is tantamount to the defense of a kind of dogmatism and Empiricism is, by its very assumptions, unable to transcend Hume's sceptical conclusions. Thus, philosophy finds itself mired in scandal. As he says in the preface to *CPR*,

> [I]t always remains a scandal of philosophy and universal human reason that the existence of things outside of us . . . should have to be assumed merely on faith, and that if it occurs to anyone to doubt it, we should be unable to answer him with a satisfactory proof.[3]

Notice Kant's casual assumption of "universal human reason" in this quote. Ironically, though, Kant's reconciliation of the two traditions then dominant in Western philosophy leaves him forced to defend a position that sets significant limits to human knowledge, and he has been accused of scepticism himself. As he says, "Thus I had to deny knowledge in order to make room for faith; and the dogmatism of metaphysics . . . is the true source of all unbelief conflicting with morality, which unbelief is always very dogmatic."[4]

Notice that Kant specifically connects scepticism to lack of faith and conflicts with morality in that last quote. As noted in the previous chapter, in relation to the void that has opened in the realm of moral thought with the overthrow of teleological explanations of natural phenomena, this scandal is about the naturalistic explanations of science on the one hand and scepticism about the possibility of moral value on the other. The question for the modern era becomes, "How is value possible in a world that consists of nothing beyond particles in motion and determined in all their behaviour by causal forces?" This issue has exercised the best minds of the previous two centuries, notably Newton, whose work in physics has exacerbated the problem. The depth of the problem is indicated by the emergence of deism as an apparently credible alternative to theism and atheism.

Deism is the view that God has made the world and all there is and set the great clockwork mechanism in motion. But then, there being nothing left for God to do, s/he has departed the scene. For many, this position is regarded as more destructive of moral norms than atheism insofar as it purchases a role for God at the expense of admitting that the world itself really is nothing more than a mechanism which, once set in motion, no longer has any need for interventions on the part of God. Thus, moral relativism has become a live option, the idea being that moral norms are merely artefacts of human creation, norms adopted for reasons of social cohesion, but essentially, they have no foundation in anything other than human needs. As such, they vary from culture to culture, where they function as the local adaptations to the needs of social cohesion and cooperation.

This is the view that Kant sets himself to refute in his celebrated work, *The Groundwork of the Metaphysic of Morals*. His goal is to refute the scepticism implicit in moral relativism and to argue that moral norms are objective and absolute. *Objective* insofar as they are the real grounds of morality that in no way depend on what we want to believe to be true. Without these objective moral norms, there can be no such thing as morality, and such norms cannot in any way depend on what we want because "wants" are always subjective. And they are *absolute* in the sense that there will be one set of moral norms, or better, one universal principle of morality that will determine the norms for everyone, everywhere and at all times. He is also setting himself in opposition to Utilitarianism because, in his view, it presumes to justify (in principle) any action that brings us closer to our objective, identified by them as the goal of making the world a better place. For Kant, this appeal to our goals, even so noble a goal as the desire to make the world a better place, is just another kind of want. Thus, it too is tantamount to a kind of subjectivity that is at odds with the requirement that morality concerns itself exclusively with objective and absolute norms.

The theory reconstructed

I will now present a simplified version of Kant's normative ethical theory. Kant is notoriously difficult to interpret and simplification presents its own difficulties as there is always the possibility of distortion because of what gets left out. With that caveat, though, we proceed keeping front of mind our purposes here. These are, first, to give the reader some sense of the other major normative tradition of the early modern era, the Deontological tradition (or Duty Theory as it is otherwise known), as an alternative to the Consequentialist tradition, of which Utilitarianism is the best-known example. And second is to understand the nature of the differences between these two normative traditions as it applies to their respective fitness for supporting the framework of Classical Liberalism.

As was the case for Utilitarianism, the argument we will be reconstructing is intended to justify the central principle of the resulting theory. This principle occupies the determinative position in the schematized version of the theory (see Figure 10.1), such that all lower-level moral rules and prescriptions for action and judgement are presumed to follow from it as implications (in some sense of the word "implication"). Recall that in the Utilitarian tradition the ultimate principle goes by the name, The Principle of Utility, or alternatively, The Greatest Happiness Principle. In Kant's case, the ultimate principle is called The Categorical Imperative, or alternatively, The Practical Imperative, though these are not merely alternative names for the same principle. Rather they are distinct formulations.

In fact, Kant presents three formulations of his central principle. In addition to the two noted here, there is The Formula of the Kingdom of Ends, sometimes referred to as the Formula of Autonomy. Kant says, "The aforesaid three ways of representing the principle of morality are at bottom merely so many formulations of precisely the same law."[5] He goes on to imply that these different formulations serve the subjective purpose of bringing the ideas implicit in the principle "nearer to intuition . . . and so nearer to feeling."[6] By this he means to indicate, at least in part, that these alternative formulations help the reader to better understand how they are relevant to their lived moral experiences. And when he says that these distinct formulations are "merely so many formulations of precisely the same law" he seems to be suggesting that they would license the same normative prescriptions in all cases, a claim which has been widely disputed. Counterexamples have been presented that seem to show they do not license identical normative prescriptions in all circumstances. While an interesting debate, we ignore it in what follows and assume, with Kant, that they are functionally equivalent in this sense.

In any event, as noted by Alan Wood, the strategy of generating counterexamples to Kant's fundamental principle presupposes that this principle is intended by him as a kind of test for actually determining what our duties are, and a compelling argument is made by Wood that this is to seriously misconstrue Kant's purpose.[7] At the very least, I think it's fair to say that Kant thinks these distinct formulations give him an opportunity to present the central ideas of his theory in a variety of different ways that will enhance the reader's understanding. More importantly, they seem to constitute a progressive unpacking of the ideas implicit in the central principle of morality. Though I will comment briefly on all three formulations, my strategy here will be to focus on the second formulation, the Practical Imperative, because, as Kant says, this formulation concerns the *content* of the central principle of morality. I begin, though, with a brief discussion of the first, the Categorical Imperative, to draw out a few distinctive features of Kant's theory.

The Categorical Imperative: the fundamental law in its first formulation

Normative theories are concerned, in large measure, with assertions about "oughts" or "shoulds," assertions about what we should do or what we ought not to do and so on. We shall agree to call such normative assertions *imperatives*. Kant is quick to emphasize, though, that not all imperatives are ethical in nature. In other words, Kant wishes to draw an important distinction within the realm of imperatives between those that are ethical and those that are not. And his way of capturing this distinction is to frame it in terms of another distinction between hypothetical imperatives on the one hand and categorical imperatives on the other. It turns out that the ethical imperatives consist of all and only the categorical imperatives. Hypothetical Imperatives (or, as I prefer – for reasons that will become clear shortly – Conditional Imperatives) are so called because they depend for their imperative force on satisfying a certain condition.

Consider the following example:

If you wish to be a good guitarist, then you should practise a lot.

Now here is an imperative that directs the subject to invest a lot of time and energy in practising guitar. We ask, where does the imperative force come from in this case? That is, why would the subject feel any compulsion to obey this imperative? And the answer is given by the *antecedent clause*, that is, the "if" clause, which reads: *if* you wish to be a good guitarist. This clause specifies the *antecedent condition* that must be satisfied for the recipient to feel the imperative force of the assertion, which is conveyed in the *consequent clause*, the part following the comma: *then* you should practice a lot. This clause expresses the consequence of accepting, or satisfying, the antecedent condition; the action required of one should you satisfy the antecedent condition. The point is, you will feel no compulsion at all and will have no reason to obey the imperative expressed in the consequent clause, if you fail to satisfy the condition specified by the antecedent clause. In the event that you simply have no desire to be a good guitarist, you have no reason to invest your time and energy in practising a lot. The same point would apply to someone who loses the desire to be a good guitarist. The imperative loses all motive force as soon as one ceases to have the relevant desire.

Kant's point is that such hypothetical imperatives are not ethical imperatives precisely because they depend for their force on having the relevant subjective desire or want; one must satisfy the condition specified by the antecedent clause. Ethical imperatives are, according to Kant, unconditional or categorical. As he says,

> [A]n action necessary merely in order to achieve an arbitrary purpose can be considered as in itself contingent, and we can always escape from the precept if we abandon the purpose; whereas an unconditioned command does not leave it open to the will to do the opposite at its discretion and therefore alone carries with it that necessity which we demand from a law.[8]

Now when Kant speaks of the "necessity which we demand from a law" in this quote, he means a *moral law* and the necessity is the categorical, or unconditional, nature of the demands of morality that set them apart from the merely subjective imperatives which address our wants and desires. You can always escape the force of these latter imperatives by abandoning the relevant want or desire, but no rational being can escape the demands of morality precisely because they are, as noted in our introductory remarks, objective and absolute.

Thus, we have Kant's formulation of the Categorical Imperative (CI) in its first presentation:[9]

CI: I ought never to act except in such a way *that I can also will that my maxim should become a universal law.*[10]

Notice here the term "maxim." When an agent performs a moral duty, they act on a *maxim*. The term "maxim" refers to the rule that *could be* formulated explicitly in any particular instance to capture the proposed action, the intended goal of the action, and the underlying reasons for the agent's acting in accordance with that rule. Kant is not suggesting that we always do this – perhaps, it is only rarely that we pause to reflect and formulate explicitly the maxim we are following in acting as we do – but that we *could* do so. Equally important, we always can do so after the fact, and when we do, the person's motives are revealed. Thus, we judge the actions of a person on the basis of the maxim that summarizes the principle underlying their action, which is to judge them by their motives, albeit indirectly.

Further, according to Kant, our motives can be broadly grouped into three categories, which he calls "self-interest," "immediate inclination" and "done for the sake of duty." In this context, Kant hints that *self-interest* is to be understood as those cases where one's motive is focused principally on personal advantage. By *immediate inclination* Kant seems to mean something like your unreflective, knee-jerk response. It is, as I understand it, how you are *inclined* to respond, all else being equal, and without pausing to give thought to what your reasons might be. Thus, it is how you are predisposed to respond, where this predisposition can be shaped by a variety of factors including, prominently, your character but also your upbringing, cultural context and specific situational features of the case before you. If your character predisposes you to be helpful and kindly, you may be expected, all else being equal, to give help to others in a variety of circumstances and perhaps even at some disadvantage to yourself, even when others might not do so. The important point for Kant is that immediate inclination is *not the result of rational deliberation.* The last category of motives, indicated by the awkward phrase "done for the sake of duty" is intended to capture those cases where a person's motive arises out of their recognition of what moral duty requires of them in the circumstance *and they do it for this reason.* One might say, putting it into Kantian terms, in such cases you choose to do the right thing simply because you wish to exemplify a good will.

When we turn to the evaluation of individual actions, Kant insists that we partition the moral worth of actions into three categories. First are cases where one knowingly violates their duty – their maxim is in violation of what duty requires in this instance – and these are cases where we properly judge that what was done was bad or morally wrong. Such actions are judged to have *negative moral worth.* But even in cases where the person's maxim aligns with the requirements of duty, this fact in itself is not sufficient to judge that the action is morally good or of positive moral worth. Kant insists there are two possibilities in such a case. First, one might do the right thing but for the wrong reasons/motives.[11] That is, your maxim is in accordance with what duty requires, but your motive is grounded in either self-interest or immediate inclination, and in such a case we say that the moral worth of the resulting action is *neutral* at best. As Kant says, we may judge in such cases that, insofar as you have done the right thing, your action deserves praise and encouragement, but it is not worthy of moral esteem.[12] It is only in cases where one's action aligns with the requirements of duty *and* one does so for the right reasons that we judge the resulting action to be morally good, or, alternatively, of *positive moral worth*, as Kant sometimes puts it. And, as noted, the right reason is to do your duty for the sake of duty itself, that is, simply because you recognize it to be what duty requires in this case and your motive is to do the right thing.

A frequent misunderstanding arises here to the effect that Kant is insisting that one must always act for the sake of duty *exclusively*. It would without doubt be an unacceptably harsh moral theory that insisted that one cannot act morally unless one's motives are pure and exemplify duty for its own sake exclusively. For clarity, though, let us distinguish between two possible situations involving multiple motives. The first concerns those cases where multiple motives are operative, but they are all of them aligned with what duty requires. In principle, it is possible that, in any particular case, all three categories of motive could be operative, and your self-interest and your immediate inclination align with the requirements of duty. The point for Kant is that you have exemplified good will in such a case, and for this reason your action has positive moral worth, *if you would have acted in accordance with the demands of duty even if no motive other than duty's sake had been operative*. Nowhere does he claim that your action would lack moral worth simply because it was also in alignment with your self-interest or immediate inclination.

Second is the more troubling case where these other motives are in conflict with duty. In such a case, your action has positive moral worth and is particularly worthy of moral esteem by virtue of the very fact that you have set aside the motives of self-interest and, perhaps, immediate inclination. Misunderstanding on this point is probably the result of the examples Kant uses to illustrate his theory. These examples typically portray some individual in a moral dilemma that requires of them that they act in accordance with duty at enormous personal expense and/or in ways that are at odds with their immediate inclinations. They are forced to set aside all considerations of personal interest or immediate inclination and to choose duty despite this. This feature of Kant's presentation has had the effect of leaving readers with the impression that his theory is dour and puritanical, that in moral deliberations our feelings must not be allowed to play any role at all, that it is all a matter of some rarefied and pure rational deliberation. It is my view that Kant's purpose, though, is merely to emphasize the point that our moral duty will sometimes require this of us, and these are the moments when we shall truly be tested. These are the moments that will determine whether we are genuinely of good will, and that this will require that the sole motive that determines our action will be the will to do one's duty for the sake of duty itself.

So now we see how all these ideas come together in Kant's conception of the Categorical Imperative. The Categorical Imperative demands that we do our moral duty for the sake of duty itself, and this clearly implies that we are not permitted to make exceptions for personal reasons. To say that you must always act in such a way that you could will that your maxim become a universal law is to say that you could consistently will that all persons who found themselves in morally similar circumstances should act on your maxim. You could will this *consistently* and *truthfully*. That is, your maxim is consistent with the objective moral principles that apply and is truthful in the sense of being absent of any self-deception that might give bias to your own favour. Your maxim would be the moral law in all such cases and in this is revealed how it is the expression of a moral law that is both objective and absolute. A maxim that corresponds to the motive of duty, insofar as it appeals to the categorical requirements of duty alone, is *objective* because it does not in any way depend on our subjective wants or desires, as we noted to be the case for all hypothetical imperatives, and is *absolute* by virtue of applying to all persons in morally similar circumstances, regardless of time or place.

Truthfully, though, stated in this way Kant's principle retains an air of austerity that can make it alienating insofar as it makes little connection to our moral motivations, our moral sentiments as Hume and Smith would call them. It's one thing to insist that one must do their moral duty for the sake of duty itself, but why should one *care* about the demands of duty? The answer, I think, is hinted at when Kant tells us that this formulation of the supreme moral principle concerns the *form* of our moral maxims, and the form is such that they must be unconditional; as we've seen this means the moral imperatives have the form of being stated categorically. We now turn our attention to the second formulation of the supreme principle of morality, the Practical Imperative, which concerns the *content* of the moral maxims, what they are *about*, and it is in the content of moral maxims that it is revealed why we should care.

The Practical Imperative: the fundamental law in its second formulation

It is here, in the discussion of the Practical Imperative, that I will reconstruct Kant's supporting argument to better reveal the direct connection to our moral motivations. Let's begin, though, by stating this version of what Kant refers to as the Supreme Moral Principle, our destination as it were, before turning our attention to the supporting argument. Here is the Practical Imperative (PI) in its first presentation, which, as a formula(tion) of the supreme principle of morality, is called the Formula of the End in Itself:

PI: Act in such a way that you always treat humanity, whether in your own person or in the person of any other, never simply as a means, but always at the same time as an end.[13]

It is implicit here that we should *care* to do our duty because it concerns directly our treatment of humanity, ourselves and others. That is, we should be moved to do our moral duty by the recognition that humans *deserve* to be treated in ways that reflect our *caring*.

The supporting argument is stated very succinctly in Chapter 2, but it begins much earlier with a claim made at the very outset of the main part of the book in Chapter 1 and immediately following the *preface*. In my view, this claim is *the* foundational claim for Kant's entire argument in the *Groundwork*. Here is that premise as Kant states it:

P1: It is impossible to conceive anything at all in the world, or even out of it, which can be taken as good without qualification except a *good will*.[14]

When Kant speaks of something that is good "without qualification" he means to speak of that which is good in itself or, equivalently, that which is possessed of inherent worth. So Kant is affirming that, in his theory, intrinsic worth is located in a good will. Kant is a pluralist regarding intrinsic worth, though, and is prepared to accept the Utilitarian premise that pleasure, or happiness, is amongst the things with intrinsic worth. It is his recognition of this that grounds his claim that we have a duty to advance the happiness of others. Other things with intrinsic worth include the aesthetic value of art. All of these things have a worth in themselves and not a merely instrumental value. But Kant's view is subtle and he is insisting on a distinction within the realm of intrinsic worth such that a good will is the only thing with intrinsic worth whose value is absolutely *unconditional* or good without qualification. His meaning here is that we value happiness for itself, and not because it is the instrumental means to something else we value, but the worth of happiness is not unconditional because, as he says, one must be possessed of a good will to be deserving of happiness, "consequently a good will seems to constitute the indispensable condition of our very worthiness to be happy."[15]

We know already, from our discussion of the Categorical Imperative, that a good will is one that does its duty for the right reasons, namely for the sake of duty itself. But this doesn't answer the question of where in the world a good will is to be found. For this, we turn to the discussion of the Practical Imperative itself from Chapter 2. There Kant says:

P2: The will is conceived as a power of determining oneself to action *in accordance with the idea of certain laws.*[16]

Notice the phrase "determining oneself to action." It is implicit here that this power, called the will, is a power to set oneself to action in the world with the intention to accomplish some goal. That is, we set out to do something that will have consequences, and we do so with the deliberate intent to bring about those very consequences. That is, an *action* is the behaviour of a being for which that being is to be held accountable, and this because actions embody *intent*. A being that performs actions does these things with deliberation and intent; you calculate the means to bring about your desired goal. This is different from being caught up in *events* determined by natural law and where the resulting behaviour does not involve deliberate intent. Thus, if I trip on the stairs and bump into you, knocking you down the stairs, this is regrettable, but I am not held culpable in the same way I would be if I were to have deliberately pushed you down the stairs. So the will is a power had by beings capable of determining themselves to action in ways that are independent of the determining forces of natural law.

But now focus on that final phrase, "in accordance with the idea of certain laws." The laws Kant refers to here are the categorical laws of morality, and such beings as are possessed of a will are capable of determining themselves to action on the basis of the *idea* of such laws. This means that such beings can determine themselves to action on the basis of *contemplation* of the situation before them and what morality requires of them in such a case, and they can decide *on this basis* to act in accordance with such laws and in ways that emancipate them from the impulsions of their subjective desires and wants.

Now what kind of beings are these, possessed of such a power? The answer is given by the following:

P3: And such a power can be found only in rational beings.[17]

Immediately following this Kant elaborates, pointing out that all non-rational beings are mere things because they are without will.[18] They have not got a power to determine themselves to action in contemplation of the idea of a moral law. Thus, in all their behaviours, they act exclusively from the impulsions of their instinctual drives. As such, they are without inherent worth. Their worth is exhausted by the instrumental worth they have for rational beings in relation to the goals rational beings set for themselves. So Kant is locating unconditional inherent worth exclusively in rational beings, whom he also calls "persons." And this is because it is persons, to be thought of as individuals possessed of a capacity for rational thought and autonomous judgement, that are also possessed, in principle at least, of a good will.

Now, in this world, this is to locate inherent worth exclusively in human beings. There may be rational beings elsewhere in the universe, God, angels, rational aliens perhaps, but on Earth it is we humans, and only we humans, that meet the relevant standards. Kant might have been open to the possibility that some non-human animals here on Earth are possessed of these capacities, so far not recognized by us as the persons they are in fact. Whales might be a candidate for such status, if it should be discovered, for example, that they speak to one another using a language.[19]

So, to reaffirm, it is exclusively persons who are capable of determining themselves to action on the basis of a (rational) capacity for contemplating moral laws. And from this there follows an implicit sub-conclusion, which I state as follows:

P4/SC: Thus, unconditional inherent worth is found exclusively in rational beings (persons).

The significance of this point is emphasized when we compare Kant with Utilitarianism on this point. The reader will recall that Utilitarianism locates intrinsic worth in felt pleasures, states of mind as it were (and pains as their inverse). As such, for the Utilitarian the world is made a better place by maximizing the one kind of state of mind and minimizing the other, but it's not about who gets to enjoy these pleasures. In a very real sense, individuals are irrelevant because the focus is on aggregate pleasure. So in any situation where this aggregate pleasure is to be increased, but only at the expense of creating pain for some, this can be acceptable so long as the pain suffered by some is offset by significantly larger amount of pleasure for others, and we take care to not violate the Principle of Equality by ensuring those who experience the pain are not deliberately selected by any criterion other than the Utilitarian cost–benefit calculation of aggregate pleasure.[20]

Finally, we get the following:

P5: Now I say that man, and in general every rational being, *exists* as an end in himself, *not merely as means* for arbitrary use by this or that will: he must in all his actions, whether they are directed to himself or to other rational beings, always be viewed *at the same time as an end*.[21]

Here it is implied that as beings of unconditional inherent worth, persons must never be subject to arbitrary use by any will. In all our dealings with persons, it must always be front of mind that they are beings of unconditional intrinsic worth and, as such, must never be subjected to use as mere means. Note, too, that this applies to oneself as well as to others. That is, one must never treat another person as the means to satisfy one's goals, as would be acceptable in the case of non-rational beings (according to Kant), but nor is one permitted to allow oneself to be used as means by another.[22]

This is the sense, noted earlier, that our recognition of the humanity in ourselves and others demands of us that we recognize that all persons *deserve* to be treated in ways that reflect our *caring*. And it is in this recognition that a direct connection is made to our moral feelings, the sentiments that motivate us to act. I think this is hinted at in the following quote,

Suppose, however, there were something whose existence has in itself an absolute value, something which as an end in itself could be a ground of determinate laws; then in it, and in it alone, would there be the ground of a possible categorical imperative – that is of a practical law.[23]

As I read Kant here, if it can be established that there is something that has absolute, unconditional value this would be enough to establish that this "something" would at the same time be the ground of categorical, therefore moral, imperatives. And by "ground" Kant means, at least in part, that it would be the sufficient reason in itself to demand of us moral treatment. And so, speaking of persons as beings who must always be considered to be ends in themselves, and never regarded as mere means, he goes on to say, "[u]nless this is so, nothing at all of absolute value would be found anywhere. But if all value were conditioned – that is, contingent – then no supreme principle could be found for reason at all."[24] His meaning here is that, if there were nothing in the world possessed of unconditional inherent worth, there would be no such thing as morality at all. The only imperatives would be the hypothetical imperatives that are contingent on our subjective wants and desires.

And so, it is from these premises that we get the Practical Imperative, stated as the conclusion that follows. We have, in short, reconstructed an argument that establishes the Practical Imperative (PI) as the supreme principle of morality, and we have done so from the perspective of the content of our moral maxims (what they are about), and, thus, from the perspective that directly engages our moral motivations. For ease of reference, I restate the argument here, altered slightly for purposes of clarity:

P1': Unconditional inherent worth is found exclusively in a good will.

P2': The will is power of determining oneself to action on the basis of contemplation of what morality requires (i.e., on the basis of the idea of moral laws).

P3': Such a power is found only in rational beings (persons).

P4/Sub-Conclusion: Thus, unconditional inherent worth is found only in rational beings (persons).

P5': Rational beings, as beings of unconditional worth, must never be used as mere means nor allow themselves to be so used.

Main Conclusion: (PI) Always act in such a way that in your treatment of rational beings, whether in your own person <u>or</u> in the person of any other, they are never treated simply as a means but always at the same time as an end.

And, when presented schematically, the theory looks like this (Figure 10.1):

Focusing for the moment on the first premise, one might ask whether Kant is implying that only those persons who manifest a good will have intrinsic worth. Assuming, for sake of argument, that evil persons do not have a good will, are they possessed of intrinsic worth? Put slightly differently, is Kant drawing a distinction within the family of persons between those who have intrinsic worth (i.e., those with good will) and those without? And, if so, what might be the status of these persons without a good will? Insofar as they are possessed of a capacity for rational thought and autonomous judgement, they qualify as persons, so surely they can't be lumped together with non-human animals as mere things.

It is my view that Kant is to be understood here to mean that such persons are to be ranked among those with intrinsic worth; the family of persons is a unified one for Kant. Thus, the above formulation of P1 in terms of "a good will" is unfortunate and potentially misleading. For Kant, the crucial point about being a person is that one is to be held responsible for one's actions. It is because persons are *capable* of determining themselves to action in accordance with the idea of certain

The Practical Imperative

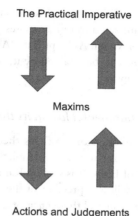

Maxims

Actions and Judgements

FIGURE 10.1 Kantian Duty Theory

laws (i.e., moral laws) that they are possessed of unconditional worth. So, on my interpretation, Kant's phrasing here in terms of a good will is not meant to deny that persons who act immorally have intrinsic worth. Rather it is deliberately chosen to emphasize at least two points. First, when we evaluate the actions of others, the focus should be on their motives and not on the consequences of their actions. As Kant says, if a person sets out to do a good action but fails to accomplish their goal, we judge them by their intentions and motives and not by what they failed to accomplish.[25] Second, he wishes to emphasize that there can be nothing more significant than aspiring to be a good person, a person who does morally good deeds for the right reasons. And from the aspirational perspective, as it were, everyone is capable of a good will (in principle).

This underscores one more respect in which Kant is distancing himself from Utilitarianism or, more generally, any kind of Consequentialist approach to normative ethics. The Consequentialist asserts that we must judge others exclusively on the basis of the consequences of their actions. They would maintain that a person's motives are necessarily private. We can never know with any assurance what a person's motives are, perhaps not even in our own case given the possibility of self-deception, so let us ignore them in our ethical deliberations and evaluations. I would assert that Kant's emphasis on motives plays an important strategic role that is additional to the constitutive role it plays in his theory. That is, in addition to making a point essential to his supporting argument, it is intended to signal from the very outset that his theory will not look to consequences to determine either what we should do or how we should judge the actions of others.[26] More generally, we can note that Kant's argument does not start, as the argument for the Principle of Utility does, with a premise to the effect that, "the purpose of ethics is . . ." This is quite deliberate insofar as Kant wishes to emphasize that the starting point of normative ethics is to properly locate unconditional moral worth and to insist that this worth is a feature of the world that is unrelated to our purposes or goals; this is precisely the sense in which it is *unconditional* worth. As he puts it, "An action done from duty has its moral worth, *not in the purpose* to be attained by it, but in the maxim in accordance with which it is decided upon."[27]

The kingdom of ends: the fundamental law in its third formulation

Before turning our attention to the fitness of this theory for Classical Liberalism, I want to consider briefly the third and final formulation, which Kant refers to as the Formula of the Kingdom of Ends. It is said by Kant that this formula combines the first two. That is, it is the fullest expression of the central principle of morality insofar as it expresses both the *form* and the *content* of the moral law. Insofar as Kant does not state this formula explicitly, I rely on the wording given by the translator of the edition I am using, H.J. Paton:[28]

> *The Formula of the Kingdom of* Ends: So act as if you were through your maxims a law-making member of a kingdom of ends.[29]

This formulation is said by Kant to follow from the presupposition that underlies the possibility of morality, namely that we must be autonomous, that is free to choose the maxim that will guide our actions. This must be presupposed if we are to regard ourselves as the makers of the very laws that we are to follow. This is what it means to *choose* the maxim of our action and to choose in a way that could be universal, that is categorical in form. So for Kant this is a presupposition of the *possibility of morality itself*. Kant is clear that he does not presume to prove that we are free but merely that it is not impossible and, in any event, if we are not autonomous, morality would itself be impossible.[30] There would be only behaviour under natural law, the same natural law that determines the outcome of events involving inanimate objects.

The reader may well ask, "How is this a combination of the Categorical and Practical Imperatives? And what can it mean to say that?" Well, first, there is this: it is this autonomy that grounds our unconditional worth and this because we are subject to the very laws that we make for ourselves. That is, insofar as we choose to follow the maxim of categorical duty, we set aside the impulsions of wants and desires, and it is this that sets us free from that determinism that is characteristic of natural law. In so doing we are subject to a law of our own making, and it is this which confers on us an unconditional worth and dignity. Moreover, it is only when we choose this path for ourselves that we become members of a possible community of such law-givers, a community of beings possessed of intrinsic dignity, beings that must be regarded as ends and never merely as means.[31]

Now, as the Categorical Imperative informs us, this means that the form of our maxims must always be categorical because we must always choose to act in such a way that we would permit our maxim to become a universal law for this community of law-givers, this kingdom of ends as he puts it. And, as the Practical Imperative tells us, we are moved to choose this by virtue of our recognition of this dignity in ourselves and the other members of this community. Form and content come together to emphasize that it is morality, the possibility of intentional action for which we must be held responsible, that is the fullest realization of our shared dignity. Quoting Kant,

> Now morality is the only condition under which a rational being can be an end in himself; for only through this is it possible to be a law-making member in a kingdom of ends. Therefore morality, and humanity so far as it is capable of morality, is the only thing which has dignity.[32]

And so, it is this possibility of the fullest realization of our intrinsic dignity that ought to move us to embrace our membership in a possible community that is a kingdom of ends. When Kant speaks of the kingdom of ends as a *possible* community, the point, I think, is not to suggest that this is aspirational but rather that it functions as an ideal. To suggest it is aspirational would be to suggest that it is possible in the sense of being a concretely realizable goal, and I'm not convinced Kant thinks this. But to suggest that this community is possible in the sense of existing as an ideal is to suggest that it functions in a *regulative* sense. It functions as a regulative principle of reason that can guide us should we choose to accept the challenge. As a regulative ideal, the moral law has a validity that transcends our taking an interest in being moral. Choosing to act in accordance with the moral law is the only way to realize our dignity as ends, the only way to manifest our intrinsic worth. Quoting Kant,

> This much only is certain: the law is not valid for us *because it interests us*. . . the law interests us because it is valid for us as men in virtue of having sprung from our will as intelligence and so from our proper self.[33]

Thus, to think that the law will be valid for us only if we take an interest in being moral is to get the reasoning exactly the wrong way around. To choose to ignore the moral law is to choose a life that is less than human. It is the recognition of our dignity as law-givers in a community of shared value, the validity that the law has for us if we are to be truly human, that is the reason it interests us.

The nature of our duties: perfect versus imperfect duties and duties to others versus duties to self

But what of our duty? What does duty actually require of us? And how is our duty to be determined? It is a subtle matter, and it seems clear that Kant never claimed that determining the maxim of one's action in some particular case would be a simple matter of deducing it from the central principle alone, in whatever formulation one considers. The central principle, in other words, does not specify the sufficient conditions for a maxim to be moral. It is worth emphasizing that it was not Kant's goal in the *Groundwork* to answer the question of how our duties are to be determined. His singular goal, as he tells us, is to discover and establish the supreme principle of morality, "The sole aim of the present Groundwork is to seek out and establish *the supreme principle of morality*."[34] Ignoring this point has occasioned much misunderstanding.

Having said that, though, Kant does present several examples intending to illustrate how the determination of duty is to work in practice. We consider these examples briefly in part because of the illumination they cast on his theory. Kant partitions our duties into four categories using two distinctions. First is the distinction between *duties to others* and *duties to self*. As we saw earlier, Kant is insistent that we must always demand respectful treatment for ourselves. If you willingly and knowingly permit another to use you as the means to their ends, you are as deserving of moral blame as the perpetrator. This implies that we have duties to self in addition to our duties to others. And second, there is a distinction between *perfect* and *imperfect* duties. Perfect duties are exception-less and specific with respect to both the intended benefactor(s) and the duty owed to that benefactor(s). Imperfect duties introduce an element of discretion. In regard to our imperfect duties to others, for example, there may be discretion around who shall be the benefactor(s) of one's action and also some discretion around exactly what is owed to the benefactor(s). It has been suggested that the imperfect duties are meritorious, whereas the perfect duties are strictly required.[35] Let's illustrate these points by briefly considering Kant's own examples.

The first concerns a *perfect duty to one's self*. In this case, we are to consider a man who, although healthy in body and still possessed of an undiminished capacity for rational deliberation, has reached a point of despair in his life that drives him to the brink of suicide. The reader is free to indulge their imagination on what misfortunes could have this effect with someone otherwise healthy and still possessed of their ability to engage in rational deliberation. We are to imagine, though, that he is sufficiently reflective to ask himself whether his self-destruction might be a violation of his perfect duty to himself. Specifically, would suicide in his case be immoral? Well, it matters a lot how one formulates the maxim one is contemplating universalizing as moral law, but suppose it is this: "From self-love I make it my principle to shorten my life if its continuance threatens more evil than it promises pleasure."[36] Framed this way, the objective assessment of reason is that this could never be universalized as a genuinely moral maxim because,

> It is then seen at once that a system of nature by whose law the very same feeling whose function is to stimulate the furtherance of life should actually destroy life would contradict itself and consequently could not subsist as a system of nature.[37]

One is struck immediately by the appeal to teleological purposes here. Kant is supposing that self-love has the *proper function* of stimulating the preservation of life. Thus, for him, the appeal to self-love to rationalize suicide is a perversion of self-love, a misdirecting from its proper function. This appeal to teleology is, in my view, not fatal to the basic ideas at the heart of Kant's theory: the idea that morality is a matter of duty and is not to be determined by an exclusive appeal to the consequences of our actions, *and* that intrinsic worth is to be found in individuals, and not in mere psychological states of mind such as pleasures or the preferences of agents.

There is room, though, for being critical of Kant's conclusion if one assumes him to be implying that suicide could never be acceptable, even in cases involving individuals who are not suffering from some incurable physical ailment. We are much more inclined today to think that suicide can be acceptable in some circumstances and that it can be a vitally important manifestation of one's autonomy, perhaps the most important.[38] Kant has been accused of importing the values of his Pietist Lutheran upbringing into his normative theory and merely giving them the veneer of a secular rationalization. I think this is too harsh and, in any event, it misses the main point at issue here. Kant is not intending these examples to be illustrations of how the Categorical Imperative is used to universalize a maxim as a general duty, in this case a general and sweeping prohibition against suicide. Rather, these are meant to be illustrations of *moral reasoning in action*. Could you, really and truly, universalize this maxim you are contemplating right now into a universal law of nature? Could you, really and truly, be willing to say that your maxim should become a law of the kingdom of ends? How one answers these questions will depend a lot on how one formulates the maxim one is contemplating, which in turn involves consideration of the goal one has set for oneself, one's motives and also the specific circumstances. I see no reason to think that Kant would deny this. We can't state with certainty, on the basis of what Kant says here, that he wouldn't permit suicide under contemplation of a different maxim. To think otherwise is to succumb to the mistake noted earlier of assuming that Kant is presenting the Categorical Imperative as a more-or-less straightforward test for determining our moral duties. Quoting Wood,

> [T]he universalizability test as Kant uses it is never more than a permissibility test for maxims. As such, it cannot be used to ground any positive moral injunctions or any classes of moral duties. The most the test could show, for instance, is that it is wrong to commit suicide or make false promises *on the specific maxim under consideration*. . . . In order to show that suicide and false promising are wrong in general using these tests, one would have to show that there is no possible maxim involving these kinds of acts that could be willed as universal laws (or laws of nature). Kant never attempts to do this, nor is there any clear way in which anything of the kind might be done.[39]

The second example concerns a *perfect duty to others*. This case involves a person driven to borrowing money on the basis of a false promise to pay it back, presumably under conditions of some agreed time frame and rate of interest. Again, the reader can indulge their imagination here, but the point is that, though Kant does not specify the circumstances, the *need* to borrow money is not trivial. Kant himself calls it a "need" and not merely a "want." We ask, though, could such a lie be a violation of moral duty, even in a case such as this? Kant formulates the maxim as follows: whenever I believe myself short of money, I will borrow money and promise to pay it back, though I know that this will never be done.[40] And the response of reason is, "I then see straight away that this maxim can never rank as a universal law of nature and be self-consistent, but must necessarily contradict itself."[41] This inconsistency follows from the fact that lying gains an advantage for the liar only against the backdrop where the norm is truth-telling. And this because the trust that makes communication, reciprocity and social cooperation possible depends in an essential way on the assurance that truth-telling is the social norm. So any attempt to universalize lying as a law of nature would undermine promise-making generally and, thus, cease to be an advantage even to the liar. Perhaps, it would have the effect of nullifying reciprocity and social cooperation altogether. A dire result indeed. We take up this point again later.

Again, we can be sceptical of Kant's reasoning on this specific example, and we can easily imagine cases where we would probably find lying preferable to any available alternative. For example, imagine that you need the money to buy life-saving medications for a loved one. If we suppose that the need for money in this case is real, as Kant himself stresses, then the phrasing of the maxim seems cavalier ("Whenever I believe myself short of money . . .") failing utterly to convey the need implicit in these imagined circumstances. Put this way, though, this situation could result in a phrasing of the maxim quite different from the one Kant uses for his example. This situation is further complicated by things Kant has said in other contexts but pursuing this would take us too far afield.[42]

In these examples, Kant repeatedly says things that imply that on careful consideration of the maxim, one sees "straight away" that it must be a violation of duty because it involves a kind of contradiction or inconsistency. It would not be unreasonable for the reader to conclude that by this phrase Kant means that, on careful consideration, it is just obvious that there is some single correct answer regarding what duty requires of us. And so, the objection is often made that Kant has likened moral deliberation to solving a logic problem, but I think that's a mistake. A more subtle reading suggests that Kant's meaning here is that one sees *straight away* only when one has used rational deliberation to distance oneself from the impulsions of self-interest and immediate inclination. In other words, the scales of self-deception have fallen from one's eyes and you are evaluating the situation objectively. In these circumstances, you see clearly (straight away) what could never have been seen at all as long self-interest or other impulses of one's character intervened. Remember, too, that the object of contemplation is the *specific maxim* under contemplation. Presumably, what one sees clearly when the scales have fallen from one's eyes is that the maxim as formulated is a self-serving rationalization that could never be consistently universalized.

If Kant is guilty of an error here, it is overconfidence in our capacity for rational deliberation in the sense meant by him, the sense in which objectivity is, to use a phrase popularized by Thomas Nagel, "the view from nowhere." This tendency, as I hinted in my introduction to this chapter, is a legacy of Kant's Enlightenment values and the optimism that assumed that reason could be used to achieve objectivity in this sense. We are today much less sanguine that such a form of reasoning is even possible and much more inclined to think that in matters of normative ethics, the best we can hope for is what can be achieved collectively as a result of reasoned public debate.

In any event, let us return to Kant's examples. The next two concern our imperfect duties. Consider, then, an *imperfect duty to self*. For this, we are to contemplate a situation where you are possessed of a talent, or possibly multiple talents, which you fail to cultivate, giving yourself instead to the pursuit of hedonistic pleasures. The relevant maxim is framed by Kant as follows: "Does my maxim of neglecting my natural gifts, besides agreeing in itself with my tendency to indulgence, agree also with what is called duty?"[43] While such a maxim

> could subsist as a law of nature, he cannot possibly *will* that this should become a universal law of nature or should be implanted in us as such a law by a natural instinct. For as a rational being he necessarily wills that all his powers should be developed, since they serve him, and are given him for all sorts of ends.[44]

Here again, we see the appeal to teleological reasoning playing an important role. Our natural talents have been "given" to us, and for the purpose of "serving us for all sorts of ends," presumably whatever ends we might, as persons who are ends-in-themselves, set for ourselves. Thinking of this case differently, though, let us suppose that our talents are merely the result of the natural lottery. As a matter of one's genetic heritage, say, you find yourself possessed of some talents that reflect your strengths, those capacities that you could develop, with the disciplined application of effort, striving, practice and find yourself to be a top performer relative to the average performance. Moreover, as we know, the distribution of talents on the natural lottery is uneven. Some are born with multiple talents, the quintessential polymath says, while others are less fortunate. Now, on this understanding, absent of any appeal to teleological purposes, could we honestly say that it would be acceptable for someone to ignore their talents completely and resign themselves to a life of hedonism?

Anecdotally at least, the answer seems to be "No." Most of us find such cases to be lamentable, at the very least, and highly regrettable in extreme cases. We might fall short of claiming that it is a *strict* moral duty to develop one's talents, but then Kant is not insisting that it is strictly required either. That's the point of this being an *imperfect* duty to self and is perhaps the sense in which imperfect duties are meritorious. As he says, the neglect of this duty could subsist as a law of nature, in the sense that if it were universalized, it's not as though this would result in a system of moral laws that was inconsistent, a system of moral laws that would threaten social cohesion and the possibility of morality. So it's not like the earlier example, where universalizing lying would be inconsistent with the norm of truth-telling, the very thing that confers an advantage on lying. But, if Kant is right, we couldn't consistently *will* that this should be the case or that it should be implanted in us as a natural instinct in the sense that you couldn't consistently *want* it to be the case that we were, all of us, lacking in any discipline or will to accomplish.

Here an important insight comes out of a recent paper by John Russell where he notes that our character and sense of self-worth is intimately connected to what he calls "striving." As he defines it, "striving involves an agent's significant conscious effort expressed in action to master or overcome challenges or obstacles or difficulties in the way of achieving, fostering, or preserving sought-after ends that are importantly valued."[45] He argues persuasively that striving is an activity we value for itself because it is intimately connected with our conceptions of happiness and flourishing. "[H]umans are fundamentally strivers and . . . their sense of a meaningful, happy or flourishing life and a sense of dignity is bound up with a conception of themselves as strivers."[46] And this is true of us independently of any appeal to teleological purposes.

One final point to consider in relation to this example, as I have framed it, is that it also reveals how our imperfect duties introduce an element of discretion. If we are to suppose an individual who is rich in the natural lottery, with an abundance of talents that *could* be developed, one is free on Kant's conception of imperfect duties to choose those gifts to be developed. The shortness of life makes such choices necessary. Einstein, for example, might have focused his striving in the first instance on his musical talents and left his ability to excel at math and physics as a mere hobby, an indulgence. The world might have been just as well served, perhaps, and, in any event, Einstein would have met his imperfect duty to himself to have not left his talents dormant. The moral wrong would be to live a life without striving at all: a life devoid of any disciplined focus on the development of one's talents; no attempt made to reach one's potential. But there is no strict duty to develop some specific talent, not even if we suppose it might be of greater Utilitarian benefit to society at large. I am confident that Kant would agree with this.

Last, consider an example of an *imperfect duty to others*. This involves an individual who, although they do nothing to interfere directly with others, also do nothing to help them, even when doing so would be within their means and involve no significant hardship. Kant refrains from formulating a specific maxim here, speaking instead of the attitude underlying the behaviour. Admitting a universal law could subsist consistently with such an attitude, presumably because, at the very least, the attitude does not advocate direct harm to or interference in the interests of others, "yet it is impossible to *will* that such a principle should hold everywhere as a law of nature."[47] And this because,

> For a will which decided in this way would be in conflict with itself, since many a situation might arise in which the man needed love and sympathy from others, and in which, by such a law of nature sprung from his own will, he would rob himself of all hope he wants for himself.[48]

If I understand Kant correctly here, universalizing the maxim to, say, "never either interfere, or offer help to others" would not result in a system of moral laws that was inconsistent in itself. This would not have any impact on exchanges based on reciprocity. The implicit promise to at least not interfere or deliberately harm others would leave it to each to be self-sufficient, and there is nothing contradictory about a system of moral laws arranged in that way. So this underscores at least one sense in which this is an imperfect duty. But he seems to be saying, universalizing such a maxim would put us into conflict with ourselves insofar as it would make impossible caring in all cases that did not involve benefits for all parties involved or, at the very least, the promise of future repayment. In other words, universalizing such a maxim would make impossible the kind of caring that depends on empathy and not reciprocity. In such a world, help from others would be something one could not even hope for. We would, all of us, be *required* to be self-sufficient in all things, all of the time, though we might often find ourselves wishing for help. And more than help, you could not expect even the slightest hint of sympathy on the part of anyone else, ever. That is what it means to be self-sufficient in this sense, after all. I think Kant would insist that this is inconsistent with our social nature as human beings, our need for forming bonds of attachment to others. Therefore, it is inconsistent with what we would want for ourselves.

There is another sense, though, in which this is an imperfect duty and involves an element of discretion. If we admit to a duty to help others within the limits of what we are able to do, there is discretion as to who shall be the beneficiary of one's help, and in regard to what one might do in discharging such a duty. For example, consider the person who, because they work such hours as make hands–on help impossible – they have no leisure time above and beyond their commitments to family duties and career, say – but they have money. Thus, they choose to discharge their imperfect duty to help others by donating to worthwhile charities. Moreover, there is discretion with respect to those charities to which they choose to contribute and, within limits, how much.

Having done my best to present a simplified version of Kant's Duty Theory, and to make it as plausible and compelling as possible, we must now turn our attention to how it stands in relation to Classical Liberalism.

Kantian Duty Theory and Classical Liberalism

With this basic understanding of Kant's theory in hand, let us ask, "How does it stand in relation to Classical Liberalism?" Well, first there is the fact that, in con-tradistinction with Preference Utilitarianism, Kant's view is in agreement with the foundations of Locke's theory insofar as both are premised on the basic right of all individuals to autonomy. If anything, Kant's theory goes further than Locke's in establishing this right as the very precondition, the *sine qua non,* as it were, of being an agent and, thus, of being fully human.

Further, as we have seen, both Locke and Kant emphasize our relations to one another, as implied by the fact that for both thinkers the foundation of civil life is a social contract that depends for its success on trust. For Kant, this also implies that the nature of our duties depends in part on our relations to one another. There is no embrace of the Utilitarian interpretation of equal consideration that looks to only aggregate pleasures and pains. To see this, let us return to the example of the burning building from the previous chapter. Recall that you have returned from a quick run to the corner store to pick up milk for breakfast to find your building ablaze with your child asleep on the top floor and you have forgotten your cell phone on the dining room table. Having mounted the fire escape to rescue your daughter, you have spotted four other children asleep in a room on the ground floor. As we saw in Chapter 9, the Utilitarian insists that duty requires you to attend only to the aggregate pleasures and pains involved, and this requires you to rescue the four children unknown to you before turning your attention to your child. And this is because, as we saw, in the calculation of aggregate pleasures and pains, relationships are irrelevant.

But a Kantian would insist that you must save your child first. That is your duty *because it is your child.* As a parent, the relationship between you and your child is a relation of dependence, which implies duties to that child take precedence over your duties to others. This has nothing to do with parental love or any kind of special regard that arises out of one's feelings. It is determined by the nature of the relationship itself. The same duty would apply to you even if you had no feelings for your child at all. In recognizing that this is your duty, you must act in accordance with the requirements of duty *because* it is your duty. That is, your motive should be to do your duty because that is what duty requires.

As for non-human animals, one might infer that the Kantian Duty theorist will not be troubled by the use of animals in commercial agriculture and in other ways as commerce demands, assuming there are no negative impacts for persons. Of course, though indirect in nature, negative impacts for persons are easily imagined under circumstances of industrialized agriculture. In addition to the negative impacts on human health resulting from the degradation of soil, water and air quality, and the significant contribution to GHG emissions, there are regular outbreaks of zoonotic diseases whose probability and potential severity are significantly increased by the density of animal populations in proximity to humans. A defender of the use of animals for human purposes could claim, though, that these problems are technological in nature and can be solved by refinements of our practices. It is, at least, not a moral problem with respect to the treatment of the animals themselves.

This casual assurance may be premature, though. To make the point explicit in regard to indirect harms, Kant held that we have *indirect duties to animals* that arise out of our direct duties to persons, and by this he meant that in all my actions I must be vigilant to never treat an animal in any way that will harm a person, even indirectly. How might this happen? Well, an obvious example involves ownership. If I harm your dog, I do you a harm. In this respect, it is no different from any case where I inflict damage on your property. But there are subtler ways of harming others by our abusive treatment of animals. Thus, Kant was concerned that in treating animals cruelly, or even with casual disregard, we run the risk of desensitizing ourselves to cruelty and disregard generally. This insight lies at the heart of prohibitions in many jurisdictions that restrict individuals who work in abattoirs from serving jury duty in a murder trial where a guilty verdict could result in the death penalty. The fear is that individuals working in such conditions become inured to death and, more generally, inured to cruel treatment. It is not clear to me what Kant would say were he to witness the scope and the nature of the abusive treatment of animals in the context of industrialized agriculture, and on a scale involving tens of billions of animals a year, but it's not obvious that he would blithely accept it as morally irrelevant.

In any event, it is beyond speculation that Kant will not accept any cases where persons are used as mere means in advancing the interests of others, and I think the implication is that he would find globalization as normally practised to be highly problematic. And no argument to the effect that it is merely a technological problem to be fixed by refinements of our practices can be made to work here. The practices are systemic and routine and are in place precisely because they enhance profits. It is easily shown that the majority of persons on Earth are in servitude to the profit motives of transnational commerce and finance and in ways that often put them in harm's way. Nor can there be any bargaining for such measures on the grounds that we are slowly lifting everyone out of poverty. In addition to the claim being dubious on its face, it is, from the Kantian viewpoint, to admit the very point at issue. It is to admit that we put our developmental goals ahead of our duty to ensure the respectful treatment of all stakeholders.

Moreover, there are the manifest harms that are the result of these routine business practices, often resulting in the death of hundreds, even thousands. We have already discussed several examples in the earlier pages, from the collapse of Rana Plaza in Bangladesh to the explosion at the Union Carbide plant in Bhopal, India. These cases, significant as they are, belie the fact that persons are harmed every day as a matter of routine business practice. The International Labour Organization estimates that some 2.3 million persons a year succumb to work-related accidents or illnesses traceable directly to their conditions of employment. This amounts to 6,000 persons every day.[49]

There are really two problems here. One concerns the point that a Kantian will insist that we must always ensure that the conditions for any transaction will guarantee the fully informed consent of all stakeholders and that the transaction will be to the mutual benefit of all parties. As we have seen, Kant locates intrinsic worth in the individual moral agent, and this worth confers on us an intrinsic dignity. As a result, Kant insists that we are all deserving of a zone of freedom that guarantees we *can* act with moral agency and we are owed this by virtue of our humanity; our intrinsic dignity as moral agents requires it. Thus, any economic system that puts persons in positions where it is impossible for them to act from genuine autonomy is guilty of systematic usury and moral wrongdoing and of the worst possible sort. It is, quite literally, to treat the overwhelming majority of humans in ways that deny them their human dignity.

The second problem concerns our motives. All of this usury and the attendant harms are the inevitable outcome of a system that is organized to maximize profits for the few that find themselves in the enviable position of wealth and influence.

Moreover, the operation of globalized commerce, organized in ways intended to maximize profits, is carried on in full recognition of the harms being done and with a complete disregard for the interests of others beyond the minimum required by law. A more blatant example of self-interest as the exclusive motive would be hard to find.

Summary and conclusions

So where does this leave us with respect to our normative traditions in relation to Classical Liberalism and Consumer Capitalism? Like every ideology, Classical Liberalism is in need of the resources only a normative theory can provide to give substance to the justifications of the economic activity and policy prescriptions that flow from the model by connecting them to our values. Unfortunately, neither Utilitarianism nor Duty Theory is a particularly good fit for this purpose.

Preference Utilitarianism suggests that making the world a better place requires us to maximize opportunities for as many agents as possible to act on their preferences in the marketplace. Superficially at least, this appears to align moral justification with Smith's theory of the free market. Conveniently, this excludes animals from consideration since they aren't market participants, and we avoid problems associated with the subjectivity of pleasures and pains. We now need to pay attention to only the features of markets that enhance the opportunity for choice and leave people free to act on their preferences, and this, as we have seen, is signalled by their willingness to pay. Pleasure itself has been commodified, our moral deliberations proceed in uniformity with all business decisions insofar as we conduct our moral deliberations in terms of cost–benefit analysis of market transactions. This provides a powerful moral justification for the Rising Tide Argument that is at the core of Classical Liberalism.

Additionally, the view denies that there are such things as rights understood as inalienable entitlements. There are only certain freedoms, liberties to act on one's preferences so long as doing so can be expected to maximize the likelihood of similar opportunities for as many people as possible or at least not interfere with the liberties of others. This is convenient for globalization under the terms of Classical Liberalism insofar as the maximization of profits under this model requires the systematic suppression, if not the outright violation, of the rights of the majority of stakeholders. But, on the downside, this very denial of rights puts Preference Utilitarianism at odds with the most basic assumptions of Classical Liberalism, the assumptions at the heart of Locke's theory of private property and Smith's theory of agency.

Preference Utilitarianism is also at odds with the idea that justice must serve a retributive role. The Preference Utilitarian can simply disregard this (as we would expect Bentham to do), but this puts them at odds with the prevailing opinion of most people. Thus, private prisons serve as a cautionary tale of the tensions in relying on Preference Utilitarianism to justify the prescriptions of Classical Liberalism. We have here an example of a system of jurisprudence that ignores the retributive role of punishment while systematically exploiting poor, underprivileged and racialized communities and denying the most basic rights of personhood. Worst of all, perhaps, is that the perverse incentives that entrench such systemic abuse of rights undermine public trust in the fairness of jurisprudence and, indeed, undermines trust in government generally. This is the sense in which Classical Liberalism and unconstrained globalization have become corrosive of democracy itself.

Finally, Preference Utilitarianism takes no consideration of our relations to one another except insofar as these have direct implications for the consequences expected to unfold. More accurately, one might say that Preference Utilitarianism instrumentalizes our relationships with one another, just one more feature that, superficially at least, seems to make it well suited to the needs of Classical Liberalism. As noted in several examples, the prescriptions of the theory flow from a bare consideration of the numbers involved and it doesn't matter who the people are or one's relations to the people involved. The problem this raises for Classical Liberalism is that the economic system at the heart of Consumer Capitalism is built on relationships. We have obligations, duties to one another, that arise out of our relationships and are dependent on the specific circumstances of our engagements. For example, deals are negotiated that bind the parties to honour the terms of these agreements. While the Preference Utilitarian can easily justify a rule to honour contracts, the negotiation of the deals themselves is reliant on trust between the parties and such trust would be an elusive thing indeed under any attempt to instrumentalize the relationships involved. As seen in our discussion of social capital (Chapter 7) erosion of trust significantly raises the transaction costs.

In the case of Kantian Duty Theory, we have a theory that, despite difficulties in its formulation, arguably aligns more closely with the values and moral intuitions of most people. It accepts the retributive role of punishment, and it agrees with the foundations of Locke's theory insofar as both are premised on the basic right of all individuals to autonomy. I have argued that, in fact, Kant's theory goes further than Locke's in establishing this right as the very precondition of being fully human. Further, as we have seen, both Locke and Kant emphasize our relations to one another, as implied by the fact that for both thinkers the foundation of civil life is a social contract that depends for its success on trust, and this implies that the nature of our duties depends in part on our relations to one another. Thus, there is a complete rejection of the Utilitarian view that equal consideration requires us to look only to aggregate pleasures and pains.

The issue of non-human animals is complicated by the fact that, for Kant, our duties to animals are merely indirect duties that arise out of the direct duties we have to persons. But even if one supposes that indirect harms to people as a result of our treatment of animals in industrialized processes can be dealt with by refinements of our practices, there is the further problem that, in treating animals cruelly, or even with casual disregard, we run the risk of becoming desensitized to cruelty. As noted, this insight is at the heart of prohibitions in many jurisdictions that restrict individuals who work in abattoirs from serving jury duty in a murder trial where a guilty verdict could result in the death penalty. Psychological research indicates that many individuals working in such conditions become inured to death and even cruelty more generally as a kind of coping mechanism.[50] In any event, it is not clear to me that Kant would accept such treatment as morally irrelevant, particularly given the scope of these activities and the nature of the harms that have become routinized in the context of industrialized agriculture, in particular.

There can be little doubt that Kant would object strenuously to globalization as normally practised under the terms of Classical Liberalism. The routine practices sanctioned by business under globalization involve the direct infliction of harms on a scale that is staggering. Moreover, there can be no argument to the effect that refinements of our practices could adequately address the issue. Nor is there room for arguing that such measures are justified because they are necessary for the task of gradually lifting everyone out of poverty. Both responses are implausible on their face and, in any event, they are essentially Utilitarian responses that are irrelevant from Kant's perspective. It is to admit that we put our developmental goals ahead of our duty to ensure the respectful treatment of all stakeholders as our first priority.

The real problem is that the very practices that result in the denial of basic rights and visit these staggering levels of harm are systemic and routine precisely because they enhance profits. Thus, globalization as practised is an economic system that puts persons in positions where it is impossible for them to act from genuine autonomy and is thereby guilty of systematic usury. From the Kantian perspective, this is moral wrongdoing of the worst possible sort because it treats the majority of humans in ways that deny them their human dignity and denies them their very humanity. And to the extent that this is carried on in full recognition of the harms being done and with a complete disregard for the interests of others beyond the minimum required by law, it is a blatant example of self-interest as the exclusive motive. And all of this is additional to the environmental harms already covered extensively in Part I.

And so, what our investigations in Part III have revealed is that, as an ideology, Classical Liberalism harbours inconsistencies. There is no consistent normative theory I am aware of that can do for this ideology what it needs, no normative theory that can offer consistent normative justifications for its policy prescriptions. In my view, this underscores the fact that, as a mature ideology, Classical Liberalism is no longer suited to the needs of our present circumstances. It is demonstrated to be inconsistent with our values, and it is incompatible with genuine sustainability and, to that extent, ill-suited to meeting our urgent need to address the issues raised by climate change.

★★★

The reader may wonder why we have passed over Human Rights Theory in our discussion of normative traditions. There are a couple of points worth noting. First, rights theory, as such, is a diverse range of theories that differ in their views on many issues. For example, there is the issue of what rights there are (e.g., the distinction between positive and negative rights) and, also the issue of who has these rights (e.g., are any non-human animals the bearer of inalienable entitlements? and, if so, which ones?). Any adequate discussion of these complexities is beyond the scope of this book.

Second, our goal has been to survey the dominant normative traditions of Western capitalist nations with an eye to understanding what such traditions might offer in terms of support for the policy prescriptions and justifications of the practices under the terms of Classical Liberalism. Now, rights theories, all of them, take their inspiration from Kant's view that intrinsic worth is to be found in individuals and not states of mind or preferences. Because of this orientation, it is no surprise that rights theories are consistently the harshest in their criticism of Classical Liberalism and, to the best of my knowledge, there is no support to be found there. And, so, it is my view that in focusing on Kant we have covered the main points that need to be covered in relation to Classical Liberalism.

Notes

1 It is worth noting that these labels and the particular way of drawing the distinction between them are owing largely to Kant himself. Thus, it is a kind of *post facto* reconstruction of these two traditions. As such, it is important to point out that some distortions inevitably result from the application of these labels to some of the principal figures concerned.
2 Strictly, this formulation would apply (so far as I know) only to Descartes. But the point is that Rationalism, to the extent it can be given a general characterization, escapes scepticism by asserting that some claims can be seen to be self-evidently true. That is, these claims carry their own certitude "on their face" as it were and do not in any way depend on experience for their justification. This is a strategy that is clearly not available to Empiricists insofar as they are committed to the claim that all knowledge must ultimately be justified in relation to sensory experience.
3 Kant (1998, p.121).
4 Op. cit., p.117.
5 Kant (1964, p.103).
6 Ibid.
7 Wood (2010).
8 Op. cit., p.87.
9 When I speak of the Categorical Imperative in its *first presentation*, I refer to the fact that, in addition to the three distinct *formulations* of his central principle already noted, Kant presents each formulation under several distinct wordings, perhaps as a stylistic variant but more likely to, once again, give him the opportunity to enhance the reader's understanding. This is the progressive unpacking of the content of the central principle of morality alluded to earlier.
10 Op. cit., p.70; emphasis in original.
11 Clearly, in this context motives function as reasons. I will use the two terms interchangeably unless the context requires me to distinguish them.
12 Op. cit., p.66.
13 Op. cit., p.96.
14 Op. cit., p.61; emphasis in original.
15 Op. cit., p.61.
16 Op. cit., p.95; emphasis in original.
17 Ibid.
18 For Kant, the category of non-rational beings seems to include all non-human animals without exception. I disagree with this assertion, but pass over it in what follows. We take up the topic again later in the chapter.
19 That is, a language like ours in having a recursive syntax, which is a fancy way of saying "a complex grammatical structure," and, thus, not merely a kind of proto-language, or a mere imitation of human language by, say, a parrot.

20 I am assuming here that the amount of pleasure must be significantly greater than the amount of pain it offsets, though I decline to specify what is meant by "significant" in this context. I want to avoid the accusation of attempting to trivialize the Utilitarian position by presenting a Straw Man argument that could be taken to suggest the Utilitarian would blithely accept avoidable pain being inflicted on some for a merely equivalent offset or even a small surplus of pleasure.

21 Op. cit., p.95; emphasis in original.

22 Of course, you are excused from any culpability if you are the victim of fraud or deception that no reasonable person could uncover in the circumstances or subject to a force you cannot in any way resist. Put the other way around, you share in the culpability of your use as means whenever you can be said to consent to this abuse. Deciding individual cases can be difficult, but this is an epistemic issue, not in itself a challenge to the moral claim being made here.

23 Op. cit., p.95.

24 Op. cit., p.96.

25 Op. cit., p.62.

26 It is actually an oversimplification to suggest that Kant ignores all consideration of consequences, but I shall pass over this here.

27 Op. cit., pp.68–69; emphasis in original.

28 This is actually a specific version of the *Formula of the Kingdom of Ends*, which Kant refers to as the *Formula of Autonomy* (for reasons to be made clear shortly) and is implied by Kant in a discussion that occurs on pp.100–102 of the *Groundwork*.

29 Op. cit., p.35.

30 In connection with these remarks, see Reath (2010).

31 This is the point that was anticipated in Chapter 4 when discussing Taylor's Biocentrism.

32 Op. cit., p.102.

33 Op. cit., pp.128–129.

34 Op. cit., p.60; emphasis in original.

35 Wood (2010, p.305).

36 Kant (1964, p.89).

37 Ibid.

38 As Camus (1991) says, "There is only one really serious philosophical problem, and that is suicide" in *The Myth of Sisyphus*.

39 Wood (2010, p.295).

40 Kant (1964, p.90).

41 Ibid.

42 I am thinking, in particular, of the notorious Case of the Inquiring Murderer in which Kant seems to imply that it is a categorical necessity that one must tell the truth at all times and in all circumstances.

43 Op. cit., p.90.

44 Ibid; emphasis in original.

45 Russell (2020, p.421).

46 Op. cit., pp.425–426.

47 Op. cit., p.91; emphasis in original.

48 Ibid.

49 https://www.ilo.org/moscow/areas-of-work/occupational-safety-and-health/WCMS_249278/lang-en/index.htm (retrieved November 7, 2022).

50 Such workers are actually subject to a whole range of often very debilitating psychological impacts. It's certainly not the case that all of them simply become desensitized or inured to the cruelty. Here is a link to recent literature review: https://journals.sagepub.com/doi/full/10.1177/15248380211030243 (retrieved November 7, 2022).

11

CONCLUDING REMARKS

Why the Green New Deal is not the old New Deal

It has been said that the new narrative is poised to usher in a Green New Deal, and in some sense this is true. It is true to the extent that, like the original New Deal, it is intended to extend the franchise, expand the scope of democratic rights and freedoms and secure equality of economic opportunity.[1] Also, it shares with the original New Deal the idea that we cannot rely on markets alone to achieve these results. It will require the active involvement of governments everywhere to coordinate activity and set the policy and fiscal frameworks in place to facilitate the emergence of the new economic model and the deployment of the necessary infrastructure. And so, as the results of Parts II and III have demonstrated, the Progressive Liberal conception of democracy is much closer in spirit to the narrative of sustainability than the Classical Liberal conception.

Despite these similarities, current circumstances are very different. First, there is the nature of the renewable energies on which a sustainable economy will be built. These renewable energies are available everywhere, entry costs are low and marginal costs almost non-existent. If the infrastructure of the Third Industrial Revolution is widely implemented to exploit the advantages of these energies (as opposed to massive projects organized on the old model of centralized production and distribution of energy – Chapter 4), they will be laterally distributed and become the basis for a new organization of the economy. This new economy will be organized as an interconnected network of localized nodes blockchained together to the edges of continental land masses. In its fullest implementation, it is an almost perfectly democratic distribution of energy. Under these circumstances, economic opportunity is increasingly founded on the localized activity within nodes, effectively guaranteeing access to opportunity that is as close to equality as we could hope for.

DOI: 10.4324/9781003388555-14

The distribution of energy and economic activity on such a model find their natural ally in distributed governance. This returns us to the point promised in Chapter 4, Rifkin's notion of "peer assemblies." A peer assembly is constituted as a group of peers representing the divergent interests and special, localized knowledge of the citizens of a region who come together as needed to frame problems, articulate solutions and construct a road map for local deployment. Like a corporation, a peer assembly exists in perpetuity. It is not constituted *of* its members at any moment in time but *by* its members, coming together as needed in an emergent body to deal with issues that fall within the purview of local determination and incentivized by the fact that they are deciding on matters of direct relevance to their own welfare.

They are motivated, too, by the sense of empowerment that comes with the transparency and responsibility that make this model work. As noted in our discussion of Darley (Chapter 8), people want to be responsible for their own lives; they want to be empowered. When denied any opportunity to be responsible, they become demotivated. When sequestered from the relevant information, they become disconnected. Conversely, empowering individuals is known to enhance their commitment to the goals of the firm, the community or the nation in ways that go beyond what is possible when individuals are motivated exclusively by self-interest. As Rifkin says, "Peer assemblies are a way to channel a community's sense of powerlessness in the face of climate change into a sense of shared responsibility for the biosphere that we will need in the years ahead and centuries to come."[2]

At another level of inclusion and transparency, all the plans produced by the peer assemblies could be made available as open-source documents accessible to everyone in the region or even beyond, sharing broadly the ideas, expertise and best practices on the model of open-source software or the Wikipedia model of knowledge sharing and construction.

Peer assemblies do not replace nation state governments, provincial/state governments or even civic governments. There will still be good reasons for jurisdictional differentiation of responsibility into various levels of government. At the very least nation states will still have a prominent role in guaranteeing the safe zone for peer assemblies to undertake their work by the creation of regulatory and fiscal frameworks, enforcement of the statutes and the Rule of Law, ensuring national security and creating and sustaining a sense of national identity and purpose. Provincial/state levels of government will be more directly involved in building the infrastructure while civic authorities will be responsible for convening peer assemblies and coordinating their activities, among other things.

Under these circumstances, though, it is clear that politics and political institutions will undergo radical transformation. The model of a sustainable society imagined here is optimized under conditions that facilitate transparency among the overlapping nodes of localized economic activity while leaving it to each node, or region, to adapt the technologies and the regulatory and fiscal frameworks to best suit its circumstances and remaining open to ongoing adaptation. As noted (Chapter 3) regulatory frameworks will perforce be low burden, with an emphasis on transparency, clarity and ease of implementation. They will be framed broadly so as to be non-prescriptive, leaving it to peer assemblies and businesses that deploy at the local level to find their own solutions that will maximize competitiveness and profits while respecting the goals of sustainability.

Geopolitics too will change with an emphasis on ensuring smooth collaboration across interconnected nodes. Negotiations between countries sharing a border will focus on finding win–win solutions by developing systems of universal codes and standards for collaboration across interconnected networks of nodes, together with agreement on regulatory frameworks that have the effect of providing a level playing field for all participants across contiguous land masses.[3]

From this we can conclude that one very important respect in which the Green New Deal will differ from the old New Deal is that it will not be the role of government to undertake massive infrastructure projects to stimulate economic activity. Stimulating economic activity by such means will not be necessary. In fact, it would be counterproductive. Rather, the most important roles for government will be regulatory and fiscal while maintaining public consensus and a sense of shared purpose for our sustainability goals. The work of government will be guided by asking, "What are the desired outcomes?" and "Where are the incentives?" Fair Rules are required to make markets work as they are intended.

The most important point to come out of the comparison to the old New Deal, though, is to be encouraged by how quickly things can change. In the United States, it took less than a decade for political support to swing from deep commitment to the model of fiscal restraint to embrace Roosevelt's New Deal. And the spirit of the ideas spread quickly and sustained a new approach to governing that endured for decades. The implementation of the basic ideas differed according to local circumstances, as it always must, but embracing the idea that governments can, and should, spend money to stimulate economic growth and secure economic opportunity for all citizens marked a significant change in world view.[4]

Sustainability will not look the same everywhere

The model of a sustainable society considered here is intended to apply to advanced industrial economies with democratic governance models, those nations that have pushed for the liberalization of markets and the model of globalization that sustains Consumer Capitalism. Sustainability will not look the same everywhere.

First, it must be recognized that some parts of the world outside the Western capitalist nations have never developed the infrastructure of previous industrial revolutions, at least not on a comprehensive scale. In much of sub-Saharan Africa, it is not uncommon to find people who burn wood to heat their homes and cook food, and many own cell phones who never had a land line connection. Many regions in this part of the world can skip right over the infrastructure of the Second Industrial Revolution and move directly to deployment of the Third Industrial Revolution. They will need the help of developed economies to do so, though. We can do this by making reparations for decades of exploitation and voracious consumption of resources.[5] In addition, rich nations must take a leadership role in forging international consensus and undertaking to do the most they can in their own jurisdictions while lending assistance to developing economies to ensure development there does not resemble what it has been for us, in part, at least, by sharing technologies and best practices.

Another way we could help would be to facilitate safe migration from zones that become unliveable and in doing so we would also be helping ourselves. Much of sub-Saharan Africa and South Asia will be seriously impacted by flooding, drought and extreme heat that will elevate food insecurity and make some of these regions uninhabitable. The ravages of climate change will force tens of millions of people to migrate elsewhere and many of them will be young, skilled, employable and eager to forge a new future for themselves. Simultaneously, much of the developed world faces a demographic challenge: declining populations with an ageing workforce and an unsustainable ratio of elderly to working-age people. Helping climate refugees to move to places where their skills can be employed is a perfect example of an adaptive strategy to the challenges of climate change; a win-win solution when seen from the perspective of the new world view.[6]

More generally, though, we can say that the form sustainable societies will take will vary depending on historical context and the position of the state in the geopolitical sphere. As pointed out by Chandran Nair, the richer countries can achieve sustainability by focusing on resource efficiency of the sort at the heart of Natural Capitalism. The poorer countries and developing nations of the world will face a very different challenge. Their challenge will be to protect their environments and their resources from the rich nations that seek to meet their "green" targets by the ongoing exploitation of these poorer countries while simultaneously struggling to find a model of sustainable development that can offer their citizens prosperity and opportunity.[7] In a word, they will be required to pursue development and sustainability simultaneously but not in ways imagined in the Brundtland Report.

As we noted in Chapter 2, the model of globalization at the heart of the Brundtland Report rightly emphasizes the need for progress in sustainability that simultaneously addresses the issue of global poverty. We cannot achieve genuine sustainability without lifting the world out of poverty because, in the absence of this, we will never have their willing participation. The great mistake was to think that this could best be achieved by facilitating the liberalization of markets as much as possible. We now see clearly that, despite the good intentions of the report's authors, this was short-sighted. Sustainability and global development/poverty reduction must be decoupled from the model of globalization at the heart of Classical Liberalism and from the hydrocarbon economy.

What is needed is a different model of development, one that does not leave it to liberalized markets to do their thing. Nair makes a persuasive case that this will require strong states that can create and sustain political and economic systems that can deliver prosperity for their people while emphasizing collective welfare over individual rights. To pull this off, they will need to secure at least moderate improvements in the living standards of their people while preserving the natural endowment of resources for future generations. Protecting the endowment of resources from exploitation will require them to be strong enough to resist the interference of dominant nations in the geopolitical sphere. And to convince their people to accept a bargain that trades off individual rights in favour of collective welfare will require transparency and fairness of procedure that can secure legitimacy with their people.[8] A delicate balancing act, for sure, but as Nair says,

> Each country faces its own set of economic, environmental and political conditions, so the specifics of each country's transition to a sustainable state will be different. Every country needs to create an overarching political philosophy, but this philosophy, and the pillars beneath it, will be different as circumstances change.[9]

Sizing up the alternatives

The thesis of this book is that we can achieve genuine sustainability without abandoning Capitalism, but the burden of this book has been to argue that the first requisite step to achieving this goal is to replace the dominant cultural narrative with a bold, new narrative that can contextualize our sustainability initiatives and provide them with the rationalization and support needed to coalesce public opinion around a set of clearly articulated goals. In the introduction, I set out three central themes of the book that constitute the framework for this argument.

First, we must accept that the way things are is not how they need to be. As an artefact of human invention, Capitalism is a stunning accomplishment. Intended to meet our needs and enhance our lives, it is a changeable institution that has proven itself adaptable to a wide variety of circumstances. Globalized Consumer Capitalism is not the end of history; it is not the realization of the highest state of human civilization. In fact, globalized Capitalism is putting us at risk of existential peril. Thus, the *first theme* is an invitation to the reader to engage in an exercise of critical reflection that challenges some of the most basic assumptions of our world view.

Central to this is the conceptual archaeology of Part II to uncover the foundations of Classical Liberalism, presently the dominant world view that rationalizes and supports the maximal liberalization of markets at the heart of globalized Consumer Capitalism. I have several goals in mind on this point. One is to situate these conceptual foundations in their historical setting to understand the motivations of the principal sources examined here, Locke and Smith. A further goal is to trace the development of the view in recent decades with a critical examination of Nozick on the ideal liberal state (Chapter 7) and Friedman on the ideal of corporate governance and CSR (Chapter 8). To provide a context for these discussions we have contrasted Nozick and Friedman with the Progressive Liberalism of Rawls and Freeman, respectively. The outcome is a better understanding of the strengths and durability of Classical Liberalism, particularly in its fitness for addressing the problems that motivated Locke and Smith but also an understanding of exactly how and why it is no longer suited to our present circumstances. It is a world view that has truly run its course and this is, in my view, manifest in the tensions, if not outright contradictions, that accompany any attempt to reconcile the view with our moral values as discussed in Part III.

Having opened the possibility of an alternative to globalized Capitalism, the *second theme* is to argue for and articulate such an alternative: a bold new world view to inspire and coalesce public opinion. Before serious and coordinated action can be taken, we need a plan, a road map, and that plan must come out of a new narrative. This new narrative, which we are calling Biosphere Consciousness, must frame our challenges in ways that open up possibilities for discovering and implementing genuinely sustainable solutions and which can provide a context for rationalizing and supporting our initiatives by giving them a focus and coherence they would otherwise lack. More generally, the *second theme* is an extended argument for the claim that this is the crucial first step that must happen if we are to have any chance of achieving our goals: dislodging the dominant cultural narrative and replacing it with a new narrative of sustainability.

The *third theme* of this book is that reworking Capitalism to make it sustainable under the terms of this new narrative is also what is required to strengthen and reinvigorate our democracies and rebuild trust and social capital, a point that has been affirmed repeatedly throughout the book.

These themes have weaved their way through the book, the connective tissue linking the individual bits together in meeting the burden of establishing the need for, and the outlines of, a new narrative. I conclude by summarizing the alternatives.

Classical Liberalism: a world view past its shelf life

The narrative underlying Classical Liberalism is about centralization of authority and power to facilitate economic activity at a huge scale and it is committed to continual growth of the economy. The institutions of this paradigm are hierarchical, authoritarian and committed to control of messaging. It is completely antithetical to transparency.

Under these terms, the emphasis is on maximizing the productivity of labour. To achieve these goals the economy is organized around linear throughput. The process begins with resource extraction and refining, proceeds to delivery of refined inputs to production sites and then delivery of the end products to markets for consumption via an extensive network of intermediaries. Ultimately, the end goods are sent to landfill, but the process is extremely wasteful at every step with little regard for the environmental consequences. Also, for that matter, there has not been any regard for the social and political consequences. In the end, Classical Liberalism, and the economic model it supports, is revealed to be an ideology refined over the last three centuries to maximize the extraction of profits before all else.

This disregard of the environmental consequences is a natural outcome of a world view that regards nature as worthless in itself. Nature is to be brought under control, bent to our will to achieve our purposes. Value is created by the application of human ingenuity and labour to the basic resources found in the natural world. There is no recognition of the services provided by nature that are irreplaceable and essential to life. The built environment is thought to be independent of nature. It is falsely assumed that we can cocoon ourselves in our built environments and keep ourselves safe from the ravages we have brought on the natural world.

As for the social consequences, in the last 40 years or so we have witnessed the accumulation of wealth unlike anything seen since the gilded age, and it has been accompanied by the steady erosion of public trust and social capital, the very glue that makes markets work. In the worst cases, there has also been a steady erosion of economic opportunity that threatens social cohesion and unity of purpose at a time when we will need this to meet our greatest challenge.

As such, Classical Liberalism is conceptually limiting. The assumptions that we can always find substitutes for the resources we use up and that there will always be a technological solution for the problems created by these very technologies are demonstrably false in the real world but comforting because they mislead us into thinking that the way forward will resemble the way we have always done things. Reluctant to abandon our comforting illusions, they persist to determine how we frame our problems and limit what will be considered as reasonable answers. To this extent, they are determinative of outcomes. Like Edwin H. Land (Chapter 3) who realized only in retrospect that conceiving of what became known as the Polaroid camera was not even possible until he had abandoned old ways of thinking and old assumptions, we will realize the sterility of Classical Liberalism for addressing our contemporary problems only from new ground, a perspective that has abandoned its limiting assumptions.

Classical Liberalism is also at odds with our values, the things that really matter to us. As emphasized in Part III, there is no consistent normative theory that can do for Classical Liberalism what is required. It is variously at odds with even basic human rights when it insists on the aggregate benefits to come out of liberalized markets or at odds with aggregate welfare when it staunchly defends the property rights of shareholders to maximize profits by resisting all government interventions to protect even the basic welfare of citizens.

Thus, our investigation of Classical Liberalism has revealed it to be detrimental to our lives in multifarious ways far beyond the scope of the environment. It is not just that it rationalizes and supports a model of globalization that is ruinous of the environment. It is also corrosive of public trust and the very foundations of our democratic institutions. It is a pleasing and reassuring result of the analysis here that the very means needed to put sustainability on a sure footing will be the very same things that need to be done to restore this public trust and reinvigorate our democracies.

Biosphere Consciousness: a world view for a sustainable future

But, one wonders, what does it take to change a world view? The answer is that there must be a new narrative poised to take its place. This new narrative must inspire hope and offer realistic pathways to successful implementation. It must engage our imaginations. I am confident the narrative presented here does all these things.

This new narrative underlying and supporting Natural Capitalism is our new compass. The details will evolve over time, but the foundational principles of the economic model at the heart of Natural Capitalism and the infrastructural pillars of the Third Industrial Revolution required for widespread deployment of this model will usher in a new period of social and political reorganization that will change the way we live.

The emphasis shifts from maximizing labour productivity to maximizing resource productivity, and there is a corresponding shift from a linear economy to a circular economy, a closed-loop system, premised on modelling natural systems and designed to eliminate waste. More strongly, waste becomes unthinkable, and our social and economic systems are integrated into the natural environment.

This shift begins with the recognition that nature must be valued for its own sake and not merely for the instrumental reasons we might have to value it. We must finally come to terms with the realization that the biosphere is the life-sustaining envelope containing all living systems and thereby determines the limits of all activity within it. We must accept our place in this network of living systems, as individuals and as a species. We too are a naturally evolved species with membership in the community of life on the same terms as all other species. And as individuals we engage in the struggle to realize our good just as all individuals must, whether human or not.

And so, in place of the old narrative of humanity as something separate from nature, with the prerogative to bring the natural world under our control, we get Biosphere Consciousness: awareness of the biosphere as the envelope that contains and supports all life on Earth and which is itself to be regarded as a living organism and, thus, to be deserving of respect. This Biosphere Consciousness brings in its wake an enhanced empathy and humility, an enlargement of the Charmed Circle.

As foundational to social capital, it is empathy and humility that make possible the kind of collaboration that will characterize life in the post-Third Industrial Revolution world. This change in attitude is about cooperation before competition. The politics of Biosphere Consciousness will be about cooperation premised on universal and democratic access to renewable energy and the essentials of life with a corresponding diminution of the importance of private property in favour of an emphasis on the right to access. The interests and incentives of consumers and providers align, and the interests of both align with sustainability as the primary goal. Under these terms, collaboration is win–win, as opposed to zero-sum.

Built communities are designed to be integrated into the natural environment, and the restoration of nature and wild spaces proceeds as the extraction of virgin resources dwindles to insignificance. The perspective looks to the long-term, not the next fiscal quarter, with an emphasis on resilience and the healthy functioning of these integrated communities as a whole. Societies organized on these terms are adaptable and versatile, at least in part because they embrace diversity. These societies are resilient, at the level of social communities and at the level of infrastructure. As individuals, we have the luxury to be citizens first, consumers second.

As mentioned in the preface, looked at one way, our situation resembles that described by Kuhn in his theory of how scientific revolutions take place. Scientific revolutions begin with challenges that the existing scientific world view finds itself unable to resolve.[10] At some point, the problems and contradictions inherent in the older view overwhelm the irrational reasons for allegiance to the model, and transition to the new model takes place. Crucially, though, this can happen only if a new model is poised to take its place, a new model that can reframe the problems in ways that point to alternative and successful solutions. It has been my goal here to sketch the outlines of what such a new model will look like for our economic, social and political transitions to a sustainable society.

I close on a personal note to the reader. In the end the transition to a new world view will be accomplished by people, as individuals, *choosing* to adopt a new perspective and to see it as the best and most reasonable response to our situation. This means *choosing* to see ourselves as members of the community of life on the same terms as others. It means having empathy for the struggles of individuals in their pursuit of the essentials of life, whether it be climate refugees on an inflatable raft in the Mediterranean, a racoon nesting in the attic or spawning salmon struggling through streams damaged by mudslides from logging operations. We must pause long enough to ask ourselves, what will be our first response? Will it be to support initiatives to keep climate refugees out of our country? Will it be to phone Pest Control or an exterminator? Or will we embrace win–win alternatives? From the perspective of Biosphere Consciousness, we are inspired to find workable solutions that minimize the negative impacts of our choices. Perhaps we organize a drive in our neighbourhood to sponsor refugees, illustrating best practices for others to follow.[11] We might find a local rescue organization and work with them to safely relocate the racoon. We might sponsor (or even participate) in a volunteer-led initiative to clear debris and build fish ladders.

It also means coming to think of waste as unacceptable while strategizing to minimize consumption and planning for the long term. We can support local entrepreneurs who will repair items such as smartphones, CD players and other goods to keep them in service longer (such repair services are increasingly easy to find in Vancouver and Ottawa), or charitable organizations that refurbish and distribute computers and other electronic technology to schools or low-income families. We can develop a strategy to rotate clothing through a sequence that moves those no longer suitable for work to weekend attire and, at the next iteration, to the clothes one wears when working around the house and, finally, to the rags one uses when working around the house. Turn it into a challenge to find ways to use less and keep the stuff you have in service as long as possible and it becomes a game that is its own reward. Most importantly, though is to see ourselves as citizens first and foremost – citizens of the biosphere – because, this is the wellspring of the empathy needed to meet the challenge.

Notes

1 Of course, even as applied to Roosevelt's New Deal, this claim is highly idealized and needs to be qualified in many ways. For example, see Alan Brinkley (1995) and Isser Woloch (2019).
2 Rifkin (2019, p.243).
3 Op. cit., p.217.
4 Particularly informative on these points are Woloch (2019) and Steil (2019).
5 As I write this (December 2022), the COP 27 conference has just concluded in Egypt. The focus was on finding an agreement to have richer nations establish a disaster fund to assist vulnerable nations facing the ravages of climate change. It will be a long and arduous journey to an effective and satisfactory agreement, but the process has begun.
6 See Vince (2022), especially Chapter 5.
7 Nair (2018, pp.22–23).
8 Op. cit., p.34.
9 Op. cit., p195.
10 Kuhn used the term "paradigm" to refer to what I am variously calling a world view or a narrative. Notoriously Kuhn was often criticized for ambiguity, using the term "paradigm" with multiple meanings. The present point is sufficiently general that it is not threatened by this dispute.
11 For one such inspiring and heart-warming story, see Tattrie (2020).

WORKS CITED

Bentham, Jeremy (2007) *An Introduction to the Principles of Morals and Legislation*. Mineola, New York: Dover.

Bentham, Jeremy (2009) *The Collected Works of Jeremy Bentham*. J. Bowering (ed.). (in 11 volumes) Oxford, UK: Oxford University Press.

Benyus, Janine (1997) *Biomimicry: Innovation Inspired by Nature*. New York: William Morrow.

Berton, Pierre (1971) *The Last Spike: The Great Railway, 1881–1885*. Toronto: Penguin, Random House.

Brinkley, Alan (1995) *The End of Reform: New Deal Liberalism in Recession and War*. New York: Random House.

Brookes, Tim (2005) *Guitar: An American Life*. New York: Tim Grove Press.

Brundtland, Gro, et al. (1987) *Report of the World Commission on the Environment and Development: Our Common Future*. New York: United Nations Press.

Burgess, Anthony (1962) *A Clockwork Orange*. Harmondsworth, Middlesex, England: Penguin.

Camus, Albert (1991) *The Myth of Sisyphus*. Justin O'Brien (Trans.). New York: Vintage.

Caradonna, Jeremy (2014) *Sustainability: A History*. New York: Oxford University Press.

Carney, Mark (2021) *Values: Building a Better World for All*. Toronto: Random House.

Carson, Rachel (1962) *Silent Spring*. Boston: Houghton Mifflin.

Chapman, Jordan, Ahmed E. Ismail, & Cerasela Zoica Dinu (2018) "Industrial Applications of Enzymes: Recent Advances, Techniques, and Outlooks," in *Catalysts (08–00238)*. Basel, Switzerland: MDPI, pp.1–26.

Collins-Chobanian, Shari (ed.) (2005) *Ethical Challenges to Business as Usual*. New York: Pearson Prentice-Hall.

Cranston, Maurice (1985) *John Locke: A Biography*. Oxford: Oxford University Press.

Crawford, Matthew (2015) *The World Beyond Your Head: On Becoming an Individual in an Age of Distraction*. Toronto: Allen Lane.

Darley, John M. (1996) "How Organizations Socialize Individuals into Evildoing," in Shari Collins-Chobanian (ed.), *Ethical Challenges to Business as Usual*. New York: Pearson Prentice-Hall, pp.211–23.

Dennett, Daniel (1991) *Consciousness Explained*. Boston: Little, Brown and Co.

Diamond, Jared (1997) *Guns, Germs, and Steel: The Fates of Human Societies*. New York: W.W. Norton & Company.

Durning, Alan Thein (1992) "The Myth of Consume or Decline," in *How Much is Enough? The Consumer Society and the Future of the Earth*. New York: W.W. Norton & Company.

Ehrenfeld, John R. (2008) *Sustainability by Design: A Subversive Strategy for Transforming our Culture*. New Haven, CT: Yale University Press.

Ferguson, Niall (2008) *The Ascent of Money: A Financial History of the World*. New York: Penguin.

Freeman, R. Edward (1984) *Strategic Management: A Stakeholder Approach*. Boston: Pitman.

Freeman, R. Edward (2005) "Stakeholder Theory of the Modern Corporation," in Shari Collins-Chobanian (ed.), *Ethical Challenges to Business as Usual*. New York: Pearson Prentice-Hall, pp.258–67.

Frey, Carl Benedikt (2019) *The Technology Trap: Capital, Labor, and Power in the Age of Automation*. Princeton, NJ: Princeton University Press.

Friedman, A. (2014) "Apples–To-Fish: Public and Private Prison Cost Comparisons," *Fordham Urban Law Journal*, 42 (2).

Friedman, Milton (1962) *Capitalism and Freedom*. Chicago: The University of Chicago Press.

Friedman, Milton (1970) "The Social Responsibility of Business is to Increase its Profits," *The New York Times Magazine*.

Fukuyama, Francis (1992) *The End of History and the Last Man*. New York: Free Press.

Galbraith, John Kenneth (1952) *American Capitalism*. Boston: Houghton Mifflin.

Galbraith, John Kenneth (1958) *The Affluent Society*. Boston: Houghton Mifflin.

Gardner, Susan T., Anastasia Anderson, & Wayne I. Henry (2019) "Reasoning (or not) with the Unreasonable," *Analytic Teaching and Philosophical Praxis*, 19(2), pp.1–10.

Goetsch, David L. (2011) *Developmental Leadership: Equipping, Enabling, and Empowering Employees for Peak Performance*. Bloomington, Indiana: Trafford Publishing.

Greer, John Michael (2016) *Dark Age America: Climate Change, Cultural Collapse, and the Hard Future Ahead*. Gabriola Island, BC: New Society Publishers.

Gutstein, Donald (2018) *The Big Stall: How Big Oil and Think Tanks are Blocking Action on Climate Change in Canada*. Toronto: James Lorimer & Company, Ltd.

Hacker, Jacob S., & Paul Pierson (2016) *American Amnesia: How the War on Government Led Us to Forget What Made America Prosper*. New York: Simon & Schuster.

Haidt, Jonathan (2012) *The Righteous Mind: Why Good People are Divided by Politics and Religion*. New York: Vintage.

Harris, Ron (2022) "A New Understanding of the History of Limited Liability: An Invitation for Theoretical Reframing," available at: https://corpgov.law.harvard.edu/2019/08/29/a-new-understanding-of-the-history-of-limited-liability-an-invitation-for-theoretical-reframing/ (retrieved October 29, 2022).

Haskel, Jonathan, & Stian Westlake (2018) *Capitalism Without Capital: The Rise of the Intangible Economy*. Princeton, NJ: Princeton University Press.

Hawken, Paul (1997) "Natural Capitalism," *Mother Jones*, March/April, available at: https://www.motherjones.com/politics/1997/03/natural-capitalism/ (retrieved October 29, 2022).

Hawken, Paul, Amory Lovins, & L. Hunter Lovins (1999) *Natural Capitalism: Creating the Next Industrial Revolution*. Boston: Little, Brown, and Company.

Henry, Wayne I. (2015/2016) "Consumer Capitalism and Self-Identity," *Analytic Teaching and Philosophical Praxis*, 36, pp.55–61.

Hird, Myra J. (2021) *Canada's Waste Flows*. Montreal and Kingston: McGill-Queen's University Press.

Hobbes, Thomas (1968) *Leviathan*. C.B. Macpherson (ed.). Harmondsworth, Middlesex, England: Penguin.

Homer-Dixon, Thomas (2020) *Commanding Hope: The Power We Have to Renew a World in Peril*. Toronto: Alfred A Knopf.

Kant, Immanuel (1964) *The Groundwork of the Metaphysic of Morals*. H.J. Paton (Trans.). New York: Harper & Row.

Kant, Immanuel (1998) *The Critique of Pure Reason*. Paul Guyer & Allen W. Wood (Trans.). Cambridge, UK: Cambridge University Press.

Klein, Naomi (2007) *The Shock Doctrine: The Rise of Disaster Capitalism*. Toronto: Knopf, Canada.

Kolbert, Elizabeth (2014) *The Sixth Extinction: An Unnatural History*. New York: Henry Holt and Co.

Kuhn, Thomas (1962/1970) *The Structure of Scientific Revolutions* (2nd ed.). Chicago: University of Chicago Press.

Kuttner, Robert (1996) *Everything for Sale: The Virtues and Limits of Markets*. Chicago: University of Chicago Press.

Kymlicka, Will, & Sue Donaldson (2013) *Zoopolis: A Political Theory of Animal Rights*. Oxford: Oxford University Press.

Lakoff, George (2009) *The Political Mind: A Cognitive Scientist's Guide to Your Brain and its Politics*. New York: Penguin Books.

Lewis, Michael (1989) *Liar's Poker: Rising Through the Wreckage on Wall Street*. New York: W. W. Norton & Company.

Leopold, Aldo (1966) *A Sand County Almanac*. New York: Ballantine Books.

Lloyd-Thomas, D.A. (1995) *Locke on Government*. London and New York: Routledge.

Locke, John (1988) *Two Treatises of Government*. Peter Laslett (ed.). Cambridge, UK: Cambridge University Press.

Locke, John (2009) *An Essay Concerning the Human Understanding*. London: WLC Books.

Logan, C.H. (1990) *Private Prisons: Cons and Pros*. Oxford: Oxford University Press.

Lovins, Amory B., L. Hunter Lovins, & Paul Hawken (2007) "A Road Map for Natural Capitalism," *Harvard Business Review*, July–August, pp.172–183.

Mann, Charles C. (2005) *1491: New Revelations of the Americas Before Columbus*. New York: Alfred A. Knopf.

Maser, Chris et al. "Chapter 2. What We Know About Large Trees That Fall to the Forest Floor," from a publication available at the US Forest Service web site: https://www.fs.fed.us/pnw/pubs/164part2.pdf (retrieved August 31, 2021).

Mayer, Jane (2016) *Dark Money: The Hidden Story of the Billionaires Behind the Rise of the Radical Right*. New York: Anchor Books.

Mazzucato, Mariana (2017) *The Value of Everything: Who Makes and Who Takes from the Real Economy*. New York: Public Affairs.

Meadows, Donella, Jorgen Randers, & Dennis Meadows (2004) *Limits to Growth: The 30-Year Update*. White River Junction, Vermont: Chelsea Green Publishing Company.

Meiksins-Wood, Ellen (2003) *Empire of Capital*. London and New York: Verso.

Mill, John Stuart (1962) *Utilitarianism*. Mary Warnock (ed.). Glasgow: William Collins Sons & Co. Ltd.

Mill, John Stuart (2006) *On Liberty and the Subjection of Women*. Alan Ryan (ed.). New York: Penguin.

Montgomery, David R. (2017) *Growing a Revolution: Bringing Our Soil Back to Life*. New York and London: W.W. Norton & Company.

Nagel, Thomas (1986) *The View from Nowhere*. Oxford: Oxford University Press.

Nair, Chandran (2018) *The Sustainable State: The Future of Government, Economy and Society.* Oakland, CA: Berrett-Koehler Publishers, Inc.

Nozick, Robert (1974) *Anarchy, State, Utopia.* New York: Basic Books.

Peters, T., & R. Waterman (1982) *In Search of Excellence.* New York: Harper and Row.

Picketty, Thomas (2014) *Capital in the Twenty-First Century.* Arthur Goldhammer (Trans.). Cambridge, MA: The Belknap Press.

Plantinga, Alvin (2011) *Where the Conflict Really Lies: Science, Religion and Naturalism.* Oxford: Oxford University Press.

Putnam, Robert (2000) *Bowling Alone: The Collapse and Revival of American Community.* New York: Simon & Schuster.

Rachels, James (2003) *The Elements of Moral Philosophy* (4th ed.). New York: McGraw Hill.

Rand, Ayn (1971) *The New Left: The Anti-Industrial Revolution.* New York: New American Library.

Rand, Tom (2020) *The Case for Climate Capitalism: Economic Solutions for a Planet in Crisis.* Toronto: ECW Press.

Rawls, John (1971/1999) *A Theory of Justice* (Revised ed.). Cambridge, MA: The Belknap Press.

Reath, Andrews (2010) "Kant's Critical Account of Freedom," in Graham Bird (ed.), *A Companion to Kant.* West Sussex, UK: Wiley-Blackwell.

Rifkin, Jeremy (1980) *Entropy: A New World View.* New York: Viking Press.

Rifkin, Jeremy (2011) *The Third Industrial Revolution: How Lateral Power is Transforming Energy, the Economy, and the World.* New York: St. Martin's Griffin.

Rifkin, Jeremy (2019) *The Green New Deal: Why the Fossil Fuel Civilization Will Collapse by 2028, and the Bold Economic Plan to Save Life on Earth.* New York: St. Martin's Griffin.

Roberts, Derren (2019) *Capitalism and Prison Privatization.* (Unpublished manuscript).

Rosenblat, Alex (2014) *Uberland: How Algorithms Are Rewriting the Rules of Work.* Oakland: University of California Press.

Russell, J.S. (2020) "Striving, Entropy, and Meaning," *Journal of the Philosophy of Sport,* 47(3), pp.419–437. DOI: 10.1080/00948705.2020.1789987.

Schendler, Auden (2021) "The Complicity of Corporate Sustainability," *Stanford Social Innovation Review,* April 7. https://ssir.org/articles/entry/the_complicity_of_corporate_sustainability (retrieved October 30, 2022).

Schultz, Howard (2011) *Onward: How Starbucks Fought for Its Life.* New York: Rodale.

Shapiro, Ian (1986) *The Evolution of Rights in Liberal Theory.* Cambridge: Cambridge University Press.

Singer, Peter (1990) *Animal Liberation* (Revised ed.). New York: Avon Books.

Smith, Adam (2003) *An Inquiry into the Nature and Causes of the Wealth of Nations.* Edwin Cannan (ed.). New York: Bantam.

Steil, Benn (2019) *The Marshall Plan: Dawn of the Cold War.* New York: Simon & Schuster.

Stiglitz, Joseph E. (2010) *Freefall: America, Free Markets, and the Sinking of the World Economy.* New York: W.W. Norton & Company.

Stiglitz, Joseph E. (2012) *The Price of Inequality: How Today's Divided Society Endangers Our Future.* New York: W.W. Norton & Company.

Stiglitz, Joseph E. (2015) *The Great Divide: Unequal Societies and What We Can Do About Them.* New York: W.W. Norton & Company.

Stone, Oliver (1987) *Wall Street.* Hollywood, CA: 20th Century Fox.

Streeck, Wolfgang (2016) *How Will Capitalism End? Essays on a Failing System.* London and New York: Verso.

Suzuki, David, & Holly Dressel (2002) *Good News for a Change: How Everyday People Are Helping the Planet*. Toronto: Stoddart.

Tattrie, Jon (2020) *Peace by Chocolate: The Hadhad Family's Remarkable Journey from Syria to Canada*. Fredericton, NB: Goose Lane Editions.

Taylor, Paul W. (1981) "The Ethics of Respect for Nature," *Environmental Ethics* (Fall), pp.197–218.

Tooze, Adam (2018) *Crashed: How a Decade of Financial Crises Changed the World*. New York: Penguin.

Turner, Chris (2008) *The Geography of Hope: A Tour of the World We Need*. Toronto: Vintage Canada.

Uncredited, [Maser?] "The Unseen World of the Fallen Tree," US Forest Service web site: https://www.fs.fed.us/pnw/pubs/229chpt2.pdf (retrieved August 31, 2021).

Vince, Gaia (2022) *Nomad Century: How Climate Migration Will Reshape Our World*. New York: Flatiron Books.

Weber, Max (1920) *The Protestant Ethic and the Spirit of Capitalism* (Revised ed.). Stephen Kalberg (Trans.). Oxford: Oxford University Press.

Whitehead, Judy (2010) "John Locke and the Governance of India's Landscape: The Category of Wasteland in Colonial Revenue and Forest Legislation," *Economic and Political Weekly*, 45(50), pp.83–93. http://www.jstor.org/stable/25764218.

Woloch, Isser (2019) *The Postwar Moment: Progressive Forces in Britain, France, and the United States After World War II*. New Haven, CN: Yale University Press.

Wood, Allen (2010) "Kant's Formulations of the Moral Law," in Graham Bird (ed.), *A Companion to Kant*. West Sussex, UK: Wiley-Blackwell.

INDEX

Printed in the United States
by Baker & Taylor Publisher Services